Who is Who in Thermal Analysis and Calorimetry

Hot Topics in Thermal Analysis and Calorimetry

Volume 10

Series editor

Judit Simon, Budapest, Hungary

More information about this series at http://www.springer.com/series/6056

Imre Miklós Szilágyi · György Liptay
Editors

Who is Who in Thermal Analysis and Calorimetry

Springer

Editors
Imre Miklós Szilágyi
György Liptay
Department of Inorganic and Analytical Chemistry
Budapest University of Technology and Economics
Budapest
Hungary

ISSN 1571-3105
ISBN 978-3-319-37816-9 ISBN 978-3-319-09486-1 (eBook)
DOI 10.1007/978-3-319-09486-1

Springer Cham Heidelberg New York Dordrecht London

Printed on acid-free paper

Springer is part of Springer Science+Business Media (www.springer.com)

Foreword

If I easily agreed to preface this second edition of the "Who is Who in Thermal Analysis and Calorimetry", it was essentially because I consider it as a valuable tool at the service of our community. Ten years ago indeed, when Dr. Judit Simon and Prof. György Liptay's had the idea of the first edition, I must confess that I had no precise idea of its possible usefulness; but this is not any more the case since I experienced a number of situations when one could draw a real benefit from such a book:

- If you are a conference organizer (which necessarily happens for a few of us and is always a more demanding job than expected!) you find the book helpful for the selection of invited lecturers.
- Also, when you are chairing a session and want to introduce a lecturer warmly and properly, the corresponding page of the Who is Who provides you with all the desired material; and, as we know, introducing a lecturer nicely makes him feel better and allows him to deliver his lecture with more pleasure … not only for him but also for the audience!
- When you receive a request or proposal from a member of our community whom you still do not know much, having a look at his picture and page makes the contact more friendly and adequate.
- If you happen to set up a Committee or Working Group and wish it to be truly international (like in ICTAC), this book makes it easier to involve people from other countries and continents; the same if you wish to start any other type of international collaboration.
- In the more common case when you simply have a glance at the whole book, it favors your acquaintance of scientists whom you may have heard about or even already met: this allows to know each other better and develop tighter links.
- All this tends to build up a family spirit in our community and make the work in our area more enjoyable.

All reasons of interest above apply of course to the new edition, whose justification is the normal renewal of our vivid community, since about two-thirds of the scientists listed here were not in the previous edition!

I also wish to stress that the two Editors of this second edition are themselves full members of our Thermal Analysis and Calorimetry community: Prof. Gyorgy Liptay as the Editor of an original and well-known Atlas of Thermoanalytical Curves in five volumes, as the organizer of the seventh ESTAC Symposium (in Balatonfüred 1998) and as the recipient of the ICTAC Distinguished Service Award in 2008. Dr. Imre Miklós Szilágyi, as the recipient of the ICTAC—PerkinElmer Young Scientist Award in 2008 and, more recently, as the Editor of one of the two main journals devoted to our field of science.

Finally, the scientists appearing in this Who is Who are entitled to have two prides: that, well-deserved, of having been selected for publication of their picture and CV, but also that of being members of a vivid community for whom this book will be a helpful tool.

I therefore wish to thank the Editors for their initiative and for the nice work done, and to encourage the readers to keep the book in a good place for easy access in their office!

Prof. Dr. Jean Rouquerol
(ICTAC President 2000–2006)

Preface

The success of the first "Who is Who in Thermal Analysis and Calorimetry" book (2004) encouraged us that after a decade we publish a broadened, second edition. After the pioneers of thermal analysis, now the second generation of thermal analysts is about to retire. Sadly, some of the previous outstanding scientists have passed away. Parallelly, in the recent decade a great deal of young researchers have joined thermal analysis. In addition, there has been a large change in the publication and scientometric trends, and the research infrastructure and data evaluation have also developed continuously.

The broadening of thermal analysis is clearly marked by that while in the first edition ca. 240 researchers were presented, now ca. 340 of them are featured in the present, second edition. 60–70 % of the people presented in the book are new names, compared to the first edition, which clearly shows the generation change in thermal analysis. The distribution of scientists from the various countries has also changed.

The selection of the candidates was based on the outstanding publication activity in thermal analysis, professional activity, awards, etc. To find suitable candidates for the book, the authors have consulted the regional and associate editors of the Journal of Thermal Analysis and Calorimetry and several other well-known scientists in thermal analysis. Their contribution is highly appreciated and acknowledged: S.R. Bharadwaj, J. Blumm, G. Bruni, P. Budrugeac, É.T.G. Cavalheiro, E.L. Charsley, V.A. Drebushchak, J. Dweck, M. Feist, E. Füglein, C. Giancola, K. Győriová, J. Hanss, W.B. Hu, M. Jemal, L. Judovits, T. Kaljuvee, M.F. Kotkata, M.V. Kök, M. Lalia-Kantouri, V.P. Lehto, M. Liška, V. Logvinenko, D.M. Lőrinczy, B.V. L'vov, T.M.R. Maria, A. Małecki, J.D. Menczel, A. Michnik, S.C. Mojumdar, C.G. Mothé, R. Ozao, B. Pacewska, L.A. Pérez-Maqueda, A. Rotaru, J. Rouquerol, P. Šimon, P. Šulcová, J.J. Sunol, P.J. van Ekeren, S. Vecchio Ciprioti, S. Yariv, Z.W. Yu, M.I. Zaki. The help of D. Hunyadi in arranging parts of the manuscript is acknowledged.

We would like to commemorate those who passed away in the last decade, and who were key players in the development of the theory, instrumentation, and application of thermal analysis in various fields: Z. Adonyi (Hungary), T. Atake

(Japan), D. Fatu (Romania), B. Małecka (Poland), M. Ollivon (France), T. Ozawa (Japan), F. Paulik (Hungary), M.J. Richardson (UK), E. Segal (Romania), O.T. Sörensen (Denmark). E.T. Stepkowska (Poland), H.G. Wiedemann (Switzerland), B. Wunderlich (USA), W. Zielenkiewicz (Poland).

Dr. Imre Miklós Szilágyi
Deputy editor-in-chief
Journal of Thermal Analysis and Calorimetry
Department of Inorganic and Analytical Chemistry
MTA-BME Technical Analytical Chemistry Research Group
Budapest University of Technology and Economics
Budapest, Hungary

Prof. Dr. György Liptay
Consulting editor
Journal of Thermal Analysis and Calorimetry
Department of Inorganic and Analytical Chemistry
Budapest University of Technology and Economics
Budapest, Hungary

Contents

Researcher Profiles

Imre Miklós Szilágyi and György Liptay

I.M. Szilágyi (✉) · G. Liptay
Department of Inorganic and Analytical Chemistry, Budapest University of Technology and Economics, Szent Gellért tér 4., H-1111, Budapest, Hungary
e-mail: imre.szilagyi@mail.bme.hu

G. Liptay
e-mail: liptay.g@mail.bme.hu

I.M. Szilágyi and G. Liptay (eds.), *Who is Who in Thermal Analysis and Calorimetry*, Hot Topics in Thermal Analysis and Calorimetry 10,
DOI 10.1007/978-3-319-09486-1_1

Name: Magdi Fouad Abadir
Country: Egypt
Date and place of birth: 1948
Present position and address: Professor of Inorganic Industries—The Chemical Engineering Department—Faculty of Engineering—University of Cairo—Giza—Egypt
Email: magdi_abadir@yahoo.com
Website: www.scholar.cu.edu.eg/magdi
Education and scientific degrees: Ph.D. in Chemical Engineering—Faculty of Engineering—University of Cairo—1976
Workplaces: The Chemical Engineering Department—Faculty of Engineering—University of Cairo—Giza—Egypt (1976-to date); The American University in Cairo (1981–1991)
Main fields of interest: kinetics of solid state reactions, phase equilibrium diagrams of ceramic systems
Relevant categories in thermal analyses: fields (ceramics, nano); methods (DTA, TGA, DTG)
Awards and acknowledgments: Cairo University award for the highest I.F. publication (5.1)—Faculty of Engineering (2012)
Professional activities: Head of the Chemical engineering Department, University of Cairo 2005–2008; Member of the advisory board of the 12th International Ceramics Congress—Grosseto—Italy—June 2010
Publication record: papers (84), books (2), citations (270), h-index (9), sum of impact factors (56)
Equipments: Thermal analyzer (Shimadzu)
5 most important publications: [1] Kinetics of oxidation of Cr_3B_4 cemented by different metallic binders (part 1), ThermoChimica Acta, 1991, 180, 269; [2] The effect of magnesium chloride on the fire retardation of cellulosic fibers, Journal of Thermal analysis and Calorimetry, 2001, 63, 831; [3] The use of thermal analysis in the approximate determination of the cement content in concrete, Journal of Thermal Analysis and Calorimetry, 2004, 76(3), 713; [4] Catalytic oxidation of CO gas over nanocrystallite $Cu_xMn_{1-x}Fe_2O_4$, Topics in Catalysis, 2008, 47, 66; [5] Novel CdPdS/PVAc core–shell nanofibers as an effective photocatalyst for organic pollutants degradation, Journal of Molecular Catalysis, A 2012, 363–364, 186.

Name: Lorenzo Abate
Country: Italy
Date and place of birth: 1942, Catania, Italy
Present position and address: Retired from the University of Catania in the position of full Professor. Address: DII Department, University of Catania, Building 10, Viale A. Doria, 6, 95125 Catania (Italy)
Email: labate@dmfci.unict.it
Education and scientific degrees: Laurea degree in Industrial Chemistry (1966); Associate professor of Chemistry (1985); Full professor (2001) all at the University of Catania
Workplaces: Department of Chemical Sciences, University of Catania; Department of Physical and Chemical Methodologies for Engineering, University of Catania
Main fields of interest: synthesis by solid–solid interactions and thermal decomposition of complexes between inorganic compounds and organic ligands; thermodynamics of model molecules; kinetics of the thermal degradation of polymers and nanocomposites
Relevant categories in thermal analyses: fields (complexes, organic, polymers, nano); methods (DSC, TG, DTA, calorimetry, molar heat capacity, kinetics of degradation)
Professional activities: Referee of J. Therm. Anal. Calorim., Polym. Degrad. Stab., Thermochim. Acta and other Journals; member of organizing committees of several Meetings and Conferences; Honorary Board Member of Journal of Thermal Analysis and Calorimetry
Publication record: papers (83)
Equipments: Mettler DSC 20, DSC 30 and TG; Shimadzu DSC 60 and DTG 60
5 most important publications: [1] L. Abate, G. Della Gatta, G. Somsen: Heat capacities of sixteen small peptides (N-acetyl-N′-methylamino acid amides) measured by differential scanning calorimetry, Thermochim Acta, 239 (1994) 7; [2] L. Abate, S. Calanna, A. Pollicino, A. Recca: Thermal behaviour of some polyaryleneethers: a comparative study of the kinetics of degradation, Macromol. Chem. Phys. 198 (1997) 1437; [3] L. Abate, I. Blanco, O. Motta, A. Pollicino, A. Recca: The isothermal degradation of some polyetherketones: a comparative kinetic study between long-term and short-term experiments, Polym. Degrad. Stab. 75 (2002) 465; [4] I. Blanco, L. Abate, F. A. Bottino, P. Bottino, M. A. Chiacchio: Thermal degradation of differently substituted Cyclopentyl Polyhedral Oligomeric Silsesquioxane (CP-POSS) nanoparticles, J. Therm. Anal. Calorim. 107(3) (2012) 1083; [5] I. Blanco, L. Abate, M. L. Antonelli, F. A. Bottino: The regression of isothermal thermogravimetric data to evaluate degradation E_a values of polymers: a comparison with literature methods and an evaluation of lifetime predictions reliability. Part II, Polym. Degrad. Stab. 98 (11) (2013) 2291.

Name: Azzedine Abbaci
Country: Algeria
Date and place of birth: 1960, Merdès, El-Tarf, Algeria
Present position and address: Faculté des Sciences, Département de Chimie, Université Badji Mokhtar, Sidi-Amar, Annaba (23200), Algeria
Email: azzedine.abbaci@univ-annaba.dz
Website: www.univ-annaba.dz
Education and scientific degrees: B.S.: Université d'Annaba, Algeria, 1983; Master of Science: University of Maryland at College Park, USA, 1986; Ph.D.: University of Maryland at College Park, USA, 1991; Master Professional: Université Louis Pasteur, Strasbourg, France 2007
Workplaces: Université d'Annaba, Algeria; Science Expert
Main fields of interest: modeling thermodynamic properties of fluids, ionic liquids; specific heat, calorimetry
Professional activities: Advisor, Reviewer, Expert (FP7-MArie Curie Action program), editorial memberships for several journals
Publication record: papers (30), books (1), h-index (04), sum of impact factors (12.60)
5 most important publications: [1] Z.Y. Chen, A. Abbaci, S. Tang and J.V. Sengers, Phys. Rev. A 42, 4470, 1990; [2] A. Rizi and A. Abbaci, J. Mol. Liq., 64–70, 2012; [3] A. Abbaci and A. Rizi, I. M. Abdulagatov, Thermochimica Acta, vol. 567, pp. 65–72, 2013; [4] A. Acidi, M. Hasib-ur-Rahman, F. Larachi, A. Abbaci, Korean J. Chem. Eng., 31(6), 1043–1048, 2014; [5] Utilisation des technologies de l'information et de la communication dans l'enseignement et la formation [texte imprimé]/Alger: OPU, 2010. 111p; 24*17, ISBN 9961013519.

Name: Rufina G. Alamo
Country: USA
Date and place of birth: 1954, Segovia, Spain
Present position and address: Distinguished Research Professor, Florida State University College of Engineering, 2525 Pottsdamer St., Tallahassee, Florida 32310, USA
Email: alamo@eng.fsu.edu
Website: http://www.eng.fsu.edu/cbe/people/alamo.html
Education and scientific degrees: B.S., Chemistry, University of Valladolid, 1977; M.S., Chemistry, University of Valladolid, 1978; Ph.D., Chemistry, University of Madrid, 1981
Workplaces: Research scientist, DOW Chemical Europe, Spain and at the Institute of Molecular Biophysics, Florida State University, USA.; Professor at the FAMU-FSU College of Engineering, Tallahassee, Florida, USA
Main fields of interest: polymer characterization, polymer crystallization, physical properties of macromolecules, structure-properties relations of polymers, morphology of crystalline polymers
Relevant categories in thermal analyses: fields (phase transitions of polymers, kinetics, crystallization of polymers); methods (DSC, DMA, TGA, dilatometry)
Awards and acknowledgments: NATAS Award for Outstanding Achievement (Mettler-Toledo) Award (2009), American Physical Society Fellow (2012), Distinguished Research Professor, FSU (2013), Professional Development Award, FAMU-FSU College of Engineering (2000, 2005, 2008)
Publication record: papers (125), books/book chapters (15), patents (5), h-index (35)
5 most important publications: [1] Alamo, RG et al. "Thermodynamic and Structural Properties of Copolymers of Ethylene" J. Phys. Chem., 88, 6587–6595, 1984; [2] Alamo, R. G., et al. "Morphological partitioning of ethylene defects in random propylene-ethylene copolymers" Macromolecules, 33, 6094–6105, 2000; [3] Hosier, I. L., Alamo, R. G., et al. "Formation of the alpha and gamma polymorphs in random metallocene-propylene copolymers. Effect of concentration and type of comonomer" Macromolecules, 36, 5623–5636, 2003; [4] Jeon, K. Alamo, R. G., et al. "Low electrical conductivity threshold and crystalline morphology of single-walled carbon nanotubes—high density polyethylene nanocomposites characterized by SEM, Raman spectroscopy and AFM" Polymer, 48, 4751–4764, 2007; [5] Alamo, R. G., et al. "Crystallization of Polyethylenes Containing Chlorines: Precise vs. Random Placement:" Macromolecules, 41, 7141–7151, 2008.

Name: Gheorghe Virgil Aldica
Country: Romania
Date and place of birth: 1949, Brasov, Romania
Present position and address: 1st degree senior scientist, Atomistilor street 105 bis, Magurele 077125, Ilfov, Romania (work place)
Email: aldica@infim.ro, aldica2000@yahoo.com
Website: www.infim.ro
Researcher ID/ORCID: C-2165-2011/0000-0002-3131-7698
Education and scientific degrees: Bucharest University, Faculty of Physics (1967–1972), Ph.D. (1997)
Workplaces: National Institute of Materials Physics (NIMP), Magnetism and Superconductivity Laboratory (1976-to date)
Main fields of interest: superconductivity, thermal analysis, spark plasma sintering
Relevant categories in thermal analyses: fields (inorganic, materials, nano, minerals, complex, organic, polymer, glass, ceramics); methods (TG, DTA, DSC, kinetics, cryo, extremely high temperature (above 1,000 °C) specific heat, mass spectrometry)
Awards and acknowledgments: Romanian Academy Award for Physics—"Constantin Miculescu"—1989; Certificate de apperceive, World Congress on Superconductivity, 1992
Professional activities: Head of Thermal Analysis Center (2004-to date); member of National Organizing Committee of CEEC-TAC1/2011
Publication record: papers (149), books (2), patents (4), citation (467), h-index (11), sum of impact factors (134)
Equipments: TGA-DSC Setaram Sestys Evolution+QMS 200, Netzsch DSC Apparatus DSC 204 F1+LN2 cooling CC200F1, Diamond TG-DTA PerkinElmer Instruments
5 most important publications: [1] G. Aldica, S. Popa, M. Enculescu, P. Badica, Scripta Materialia 68(6), (2013) 428; [2] G. Aldica, S. Polosan, J. Non-Crystalline Solids 358 (2012) 1221; [3] G. Aldica, M. Secu, J. Non-Crystalline Solids 356 (2010) 1631; [4] P. Badica, G. Aldica, A. Crisan, J. Mater. Sci. 37 (2002) 585; [5] S. Mihaiu, S. Scarlat, G. Aldica, M. Zaharescu, J. European Ceram. Soc. 21(10–11) (2001) 1801.

Name: Antonio S. Araujo
Country: Brazil
Date and place of birth: 1963, Natal—RN State, Brazil
Present position and address: Titular Professor, Federal University of Rio Grande do Norte, Institute of Chemistry, 59078-970, Natal RN, Brazil
Email: araujo.ufrn@gmail.com, antonio.araujo@pq.cnpq.br
Website: http://lattes.cnpq.br/9770622597949866
ORCID: 0000-0002-1223-2939
Education and scientific degrees: Bachelor in Chemistry, Federal University of Rio Grande do Norte (1983–1986); Master in Chemistry, Federal University of Paraiba (1986–1988); Doctor in Science, Institute of Chemistry—University of Sao Paulo (1989–1992); Post-Doctoral: Kent State University, Ohio, USA (1998–1999).
Workplaces: Petrochemical Pole of Camacari, BA, Brazil (1988–1989); Titular Professor at Federal University of Rio Grande do Norte (1992-actual)
Main fields of interest: hydrothermal synthesis of microporous, mesoporous and hybrid materials; nanostructured materials; adsorption and catalysis; characterization of heterogeneous catalyst; petroleum and petrochemistry
Relevant categories in thermal analyses: fields (inorganic, materials, nanoscience and nanotechnology); methods (TG, DTA, DSC, kinetics)
Awards and acknowledgments: Fellow of Research Productivity of the Brazilian Concil of Science and Technology (1993-actual); Honorary membership of the Thermoanalytical Group of the Hungarian Chemical Society (2007, 2010); Member of the International Advisory Committee of International Symposium on Nanoporous Materials (2011, 2014)
Professional activities: Membership of the Brazilian Association of Thermal Analysis and Calorimetry; Brazilian Society of Catalysis; organizing committee of national and regional meetings of catalysis, thermal analysis and adsorption
Publication record: papers (122), books (1), patents (3), citations (1260), h-index (18)
Equipments: SDTA 851 Mettler; STD Q600 TA (TG-DTA), PDSC 204 (High Pressure) Netsch.
5 most important publications: [1] Morgado, E., de Abreu, M. A. S., Pravia, O. R. C., Marinkovic, B. A., Jardim, P. M., Rizzo, F. C., Araujo, A. S., A study on the structure and thermal stability of titanate nanotubes as a function of sodium content. Solid State Science, 8 (2006) 888–900; [2] Morgado, E., de Abreu, M. A. S., Moure, G. T., Marinkovic, B. A., Jardim, P. M., Araujo, A. S., Characterization of nanostructured titanates obtained by alkali treatment of TiO_2-anatases with distinct crystal sizes, Chemistry of Materials, 19 (2007) 665–676; [3] Polli, H., Pontes, L. A. M., Araujo, A. S., Application of model-free kinetics to the study of thermal degradation of polycarbonate, Journal of Thermal Analysis and Calorimetry, 79 (2005) 383–387; [4] Souza, M. J. B., Silva, A. O. S., Aquino, J. M. F. B., Fernandes, V. J., Araujo, A. S., Kinetic study of template removal of MCM-41 nanostructured material, Journal of Thermal Analysis and Calorimetry, 75 (2004) 693–698; [5] Araujo, S. A., Ionashiro, M., Fernandes, V. J., Araujo, A. S., Thermogravimetric investigations during the synthesis of silica-based MCM-41, Journal of Thermal Analysis and Calorimetry, 64 (2001) 801–805.

Name: Giuseppe Arena
Country: Italy
Date and place of birth: 1949, Enna, Italy.
Present position and address: Chair of Analytical Chemistry, Department of Chemical Sciences, University of Catania, Viale A. Doria 6, 95125 Catania, Italy.
Email: garena@unict.it
Website: http://www.dipchi.unict.it/personale/
Education and scientific degrees: Laurea in Industrial Chemistry
Workplaces: Research assistant (1974). Post-doctoral fellow, Chemistry Dept., University of St. Andrews, Scotland (1976). Lecturer, University of Catania (1977–1982). Associate professor University of Catania (1982–1985). Analytical Chemistry Chair, University of Messina (1985). Analytical Chemistry Chair, University of Catania (1990–). NATO Senior Scientist, Thermochemical Institute, Brigham Young University (BYU), Provo, Utah, USA (1984). Lecturer, Chemistry Dept., BYU, USA (1984). NATO Senior Scientist, BYU, USA (1989). Visiting Professor, BYU, USA (1998).
Relevant categories in thermal analyses: fields (equilibria in solution, supramolecular chemistry, bio-inorganic chemistry); methods (micro- and Nano-ITC)
Awards and acknowledgments: Sunner Memorial Award, The Calorimetry Conference (USA), Oak Ridge, Tennessee (USA, 1989); Canneri Medal, Italian Chemical Society-Division of Analytical Chemistry (Italy, 2013).
Professional activities: Member elected of the Board of Directors of the Calorimetry Conference (1990); Member of the Editorial Board of Thermochim. Acta (1992–2001), J. of Inclusion Phenomena and Molecular Recognition and Journal of Supramolecular Chemistry. Regional Editor of Journal of Thermal Analysis and Calorimetry; Deputy-Chairman of the Division of Analytical Chemistry, Italian Chemical Society (SCI) for two terms (1997–2000 and 2000–2003); Member of the Board of Directors of the Italian Association of Calorimetry and Thermal Analysis (AICAT) and President of the Italian Group of Calorimetry and Thermal Analysis. Chairman elected of AICAT (2014–); Coordinator of the Ph.D. Chemistry Program, University of Catania, (1997–2003); Chairman of the Division of Analytical Chemistry-SCI (2010–2012) and member of the Board of Directors of the same Institution (2013–); Chairman elected of AICAT.
Publication record: papers (103), books (5), citations (2250), h-index (29), sum of impact factors (317)
Equipments: Isoperibol and isothermal calorimeters, nano-ITC calorimeters
5 most important publications: [1] Anal Bioanal Chem., 2013, 405, 1085–1094; [2] Chem. Commun., 2010, 46, 7139–714; [3] Chem. Commun., 2011, 47 (21), 6117–6119; [4] J. Am. Chem. Soc., 2010, 132(3), 1005–1009; [5] Chem. Eur. J., 1999, 5, 738–744.

Name: Hakan Arslan
Country: Turkey
Date and place of birth: 1972, Çorum, Turkey
Present position and postal address: Department of Chemistry, Faculty of Arts and Science, Mersin University, Mersin, 33343, Turkey
Email: hakan.arslan.acad@gmail.com
Website: http://www.mersin.edu.tr/apbs/arslanh
Researcher ID/ORCID: B-1081-2008/0000-0003-0046-9442
Education and scientific degrees: BSc Degree: Erciyes University, Turkey (1989–1993); M.Sc. Degree: Erciyes University, Turkey (1993–1995); Ph.D. Degree: Nigde University, Turkey (1995–1998). Assist. Prof. (1998–2006), Assoc. Prof. (2006–2012), Prof. (2012–).
Workplaces: Nigde University (1995–1998), Mersin University (1998–).
Main fields of interest: Synthesis and characterization of novel thiourea derivatives, N-heterocyclic carbene derivative ligands and redox active ligands and their metal complexes, thermal behavior and decomposition kinetic studies, single crystal and powder X-ray diffraction studies, theoretical molecular spectroscopy, and vibrational spectroscopy
Relevant categories in thermal analyses: Fields (inorganic, materials, complex, organic, pharmaceutical); methods (TG, DTA, DSC, kinetics)
Awards and acknowledgements: Second place in the undergraduate degree (1993); Second place in the research project competition sponsored by TUBITAK in 1999–2000 (2000); Scholarship for oversee graduate study in biochemistry (1994).
Professional activities: Editor: European Journal of Chemistry (2009–). Associate Editor: Journal of Medicinal Plants Research (2010–2011) (Isi Journal-Sci), International Journal of Chemical Research (2009–). Editorial Board Member: E-Journal of Chemistry (ISI Journal-Sci) (2009–2012), Journal of Chemistry (Isi Journal-Sci) (2012–), Natural Science (2009–), International Journal of Inorganic Chemistry (2008–), etc.
Publication record: papers (114), citations i (1143), h-index (21)
Equipments: (1) Shimadzu model DT-40 simultaneous TG, DTA thermal analysis system (2) Shimadzu model DTG/60H TG, DTA combined system
5 most important publications: [1] Arslan, H., Ozpozan, N., Ozpozan, T. Thermochimica Acta, 329, (1), 57–65, (1999); [2] Arslan, H., Kulcu, N., Florke, U., Transition Metal Chemistry, 28, (7), 816–819, (2003); [3] Arslan, H., Mansuroglu, D. S., VanDerveer, D., Binzet, G., Spectrochimica Acta Part A-Molecular and Biomolecular Spectroscopy, 72, (3), 561–571, (2009); [4] Akbay, C., Hoyos, Y., Hooper, E., Arslan, H., Rizvi, S. A. A., Journal of Chromatography A, 1217, (32), 5279–5287, (2010); [5] Ozdemir, I., Arslan, H., Demir, S., VanDerveer, D., Cetinkaya, B., Inorganic Chemistry Communications, 14, (5), 672–675, (2011).

Name: Ramón Artiaga
Country: Spain
Date and place of birth: 1963, Rio de Janeiro (Brasil)
Present position and address: Profesor titular, Escola Politécnica Superior, University of A Coruña, Avda Mendizábal s/n, 15403 Ferrol (Spain)
Email: ramon.artiaga@udc.es
Website: http://complexmaterials.wikispaces.com/Ramon+Artiaga
Researcher ID/ORCID: G-4312-2010/0000-0003-2506-7263
Education and scientific degrees: Pharmacy, University of Santiago de Compostela (1987), Ph.D. Universidad Complutense de Madrid (1995)
Workplaces: University of A Coruña
Main fields of interest: thermal properties of complex materials
Relevant categories in thermal analyses: fields [inorganic, materials, nano, minerals, complex, organic, polymer, food, other (metal organic frameworks, multiferroic)]; methods (TG, DSC, kinetics, specific heat, calorimetry, other [DEA, DMA, PDSC, TMDSC, pressure-TMDSC, photocuring)]
Professional activities: Teaching and researching at the University of A Coruña; Coordinator on the Spanish side of the double Master in Complex Materials: Thermal Analysis and Rheology (Univ of A Coruña)/Physics of Soft Matter (Univ. Paris 7-Paris Diderot)
Publication record: papers (52), books (1), citations (380), h-index (11)
Equipments: TMDSC, PDSC, TG, STA (DSC+TG), DMA, DEA
5 most important publications: [1] Thermal analysis. Fundamentals and applications to material characterization/Ramón Artiaga Díaz (ed.) A Coruña: Universidade da Coruña, 2005. ISBN: 84-9749-100-9; [2] López-Beceiro, J., Gracia-Fernández, C., Artiaga, R. A kinetic model that fits nicely isothermal and non-isothermal bulk crystallizations of polymers from the melt (2013) European Polymer Journal, 49(8), pp. 2233–2246; [3] López-Beceiro, J., Gracia-Fernández, C., Gómez-Barreiro, S., Castro-García, S., Sánchez-Andújar, M., Artiaga, R. Kinetic study of the low temperature transformation of Co(HCOO) 3[(CH 3) 2NH 2] (2012) Journal of Physical Chemistry C, 116(1), pp. 1219–1224; [4] C. Gracia-Fernández, J. Tarrío-Saavedra, J. López-Beceiro, S. Gómez-Barreiro, S. Naya, R. Artiaga. Temperature modulation in PDSC for monitoring the curing under pressure. Journal of Thermal Analysis and Calorimetry. Journal of Thermal Analysis and Calorimetry 106(2011) pp. 101–107; [5] C. A. Gracia Fernández, S. Gómez Barreiro, J. López Beceiro, J. Tarrío Saavedra, S. Naya, R. Artiaga. Comparative study of the dynamic glass transition temperature by DMA and TMDSC. Polymer Testing 29(2010) pp. 1002–1006.

Name: Aline Auroux
Country: France
Date and place of birth: 1946, Melay 71340, France
Present position and address: Directeur de recherche, IRCELYON, 2 Avenue Einstein, 69626 Villeurbanne, France
Email: aline.auroux@ircelyon.univ-lyon1.fr
Website: http://www.ircelyon.univ-lyon1.fr/
ORCID: 0000-0002-3063-3445
Education and scientific degrees: Chemical Engineer (1968), Ph.D. (1973) University Lyon1, Director of research (2000)
Workplaces: INSA de Lyon (1968–1975), CNRS—IRCELYON (1975–)
Main fields of interest: catalysis, clean and renewable energies, hydrogen production and storage, heat measurements, acid-base properties of solids, water and air depollution; determination of surface properties of solid catalysts by adsorption calorimetry of probe molecules; scales of acidity/basicity in number and strength for numerous oxide and zeolite catalysts have been established; clean and renewable energies: relationships are established between acid/base properties of catalysts (as determined by adsorption calorimetry both in liquid phase or gas phase) and selectivity or conversion
Relevant categories in thermal analyses: fields (materials, inorganic, catalysis); methods (microcalorimetry, DSC, TG)
Awards and acknowledgments: "I.G. Murgulescu" prize of the Romanian Academy of Sciences (2003); Calvet Prize in calorimetry (2007)
Professional activities: Organization of 3 international congresses (CTEC2000, CTEC2004, CTEC2012), one national congress (JCAT2013) and 7 international summer schools in calorimetry and thermal analysis (from 2007 to 2013); Active member of the Société Chimique de France. Guest editor of 3 issues of Thermochimica Acta, Editor of a calorimetry book (Spinger-2013)
Publication record: papers (288), books (1 book and 16 chapters in 13 books), patents (1), citations (6918), h-index (40)
Equipments: heat flow calorimeters: one MS80, two C80, two HT, two Titrys; TG-DSC: one TG-DSC111-MS, one Labsys, two TA
List of 5 most important publications: [1] A. Auroux, Calorimetry and thermal methods in catalysis, A. Auroux Ed., Springer series in Materials Science, 2013, vol 154, 561 pp; [2] A. Auroux, Microcalorimetry, Molecular sieves—Science and technology: Acidity and Basicity, Vol. 6, Springer Ed., 2008, 45–152; [3] L. Damjanovic, A. Auroux, Heterogeneous catalysis on solids, Handbook of Thermal Analysis and Calorimetry, Further advances, Techniques and applications, M. E. Brown, P. K. Gallagher Eds, Elsevier, Amsterdam, vol 5, Chap 11 (2008) 387–438; [4] Microcalorimetric studies of the acidity and basicity of metal oxide surfaces. A. Auroux, A. Gervasini, J. Phys. Chem., 94 (1990) 6371; [5] Acidity characterization by microcalorimetry and relationship with reactivity. A. Auroux, Acidity in Aluminas, Amorphous and Crystalline Silica-aluminas, G. Somorjaï and J.M. Thomas Eds., Topics in Catalysis, 4 (1997) 71–89.

Name: Isak Avramov
Country: Bulgaria
Date and place of birth: 1946, Sofia, Bulgaria
Present position and address: Institute Physical Chemistry, 1113 Sofia, Bulgaria
Email: avramov@ipc.bas.bg
Website: http://ipc.bas.bg/PPages/Avramov/issak.htm
Education and scientific degrees: Prof.; Ph.D.; D.Sci
Workplaces: Institute Physical Chemistry, Sofia, Bulgaria
Main fields of interest: glasses, kinetics of phase transitions, thermal analyses
Relevant categories in thermal analyses: fields (DTA, DSC, photoaccoustics)
Publication record: papers (150), books (chapters in 3 books), patents (5)
Equipments: DTA/TG, DSC
5 most important publications: [1] I. Avramov, Thermochimica Acta 280 (1996) 363–382 "Kinetics of Structural relaxation of glass-forming melts"; [2] I. Avramov J. Non-Cryst. Sol. 351 (2005) 3163–3173 "Viscosity in disordered media"; [3] I. Avramov Journal of Volcanology and Geothermal Research, Volume 160, Issues 1–2, 1 February 2007, Pages 165–174; [4] I. Avramov, Nanoscale Res Lett (2007) 2:235–239 "Kinetics of growth of nanowhiskers (nanowires and nanotubes)"; [5] I. Avramov, C. Rüssel Chapter 18 " Controlled Nucleation and Crystallization for Nanoparticle Synthesis" Springer, Ed. J. Sestak, P. Simon; "Thermal Analysis of Micro, Nano- and Non-Crystalline Materials Transformation, Crystallization, Kinetics and Thermodynamics", Springer (2013) ISSN 1571–3105.

Name: Elena Badea
Country: Romania
Date and place of birth: 1967, Craiova, Romania
Present position and address: Senior Lecturer, Department of Chemistry, Faculty of Sciences, University of Craiova, Str. Calea Bucuresti 107 I, 200512 Craiova, Romania
Email: elena.badea@unito.it
ORCID: 0000-0003-1437-2844
Education and scientific degrees: Degree in Food Chemistry and Engineering (1992) and Ph.D. in Chemical Engineering (2001), University Dunarea de Jos of Galati
Workplaces: Faculty of Sciences, University of Craiova, tenured position (1994–); Department of Chemistry, University of Turin, fellowships and research contracts (2002–2010); National Research and Development Institute for Textiles and Leather, Senior Research Scientist Appointments (2007–2008, 2012–)
Main fields of interest: plant biochemistry; solution thermodynamics of non-electrolytes; thermodynamics of phase transition; thermodynamics of model molecules; physical-chemical properties of collagen-based heritage materials
Relevant categories in thermal analyses: fields (biology, life, cultural heritage); methods (isothermal microcalorimetry, DSC)
Awards and acknowledgments: NATO Outreach Fellowship granted by the Italian National Research Council (2001)
Professional activities: president and founder of the the Romanian Association Science and Cultural Heritage in Connection (i-CON); representative of the Romanian Chemistry Society to the EuCheMS Working Party on Chemistry for Cultural Heritage (WP CCH); chair of scientific committee: 1st and 2nd International Conference Matter and Materials in/for Heritage Conservation (MATCONS 2009 and MATCONS 2011), Craiova, Romania; conference chair: 1st and 3rd International Seminar and Workshop Cultural Heritage–Emerging Technology and Innovation, Bucharest, Romania (2012 and 2013) and Sibiu, Romania (2014)
Publication record: papers (73), books (2), Romanian patent pending (2), h-index (9)
5 most important publications: [1] E. Badea, D. D'Angelo, B. Brunetti, Z. Rečkova, G. Della Gatta, Odd-even effect in melting properties of twelve alkane-α, ω-diamides, J. Chem. Thermodyn. 38 (2006) 1546; [2] G. Della Gatta, E. Badea, M. Jóźwiak, P. Del Vecchio, Thermodynamics of solvation of urea and some mono-substituted *N*-alkylureas in water at 298 K, J. Chem. Eng. Data 52 (2007) 419; [3] E. Badea, L. Miu, P. Budrugeac, M. Giurginca, A. Mašić, N. Badea, G. Della Gatta, Study of deterioration of historical parchments by various thermal analysis techniques, complemented by SEM, FTIR, UV-Vis-NIR and unilateral NMR investigations, J. Therm. Anal. Calorim. 91, (2008) 17; [4] E. Badea, G. Della Gatta, T. Usacheva, Effects of temperature and relative humidity on fibrillar collagen within parchment: a micro Differential Scanning Calorimetry (micro DSC) study, Polym. Degrad. Stabil., 97 (2012) 346; [5] E. Badea, B. Nowicka, G. Della Gatta G., Thermodynamics of fusion and sublimation for a homologous series of eleven alkane-α, ω-diols $HO–(CH_2)_n–OH$: Structure-related odd–even effect, J. Chem. Thermodyn. 68, (2014) 90.

Name: Elena Mihaela Badea
Country: Romania
Date and place of birth: 1964, Bordeni, Romania
Present position and address: assisted professor, Faculty of Chemistry, University of Bucharest, 90 Panduri Str., Sect. 5, Bucharest, Romania
Email: e_m_badea@yahoo.com
Education and scientific degrees: M.Sc. Physical Chemistry (1988) Polytechnic Institute of Bucharest; Ph.D. Chemistry (1990) University of Bucharest
Workplaces: 1988–1990 Petrochemical Factory Brazi; April 1990–November 1990 Institute of Physics and Technology of Radiation Devices—IFA Magurele; November 1990—up to the present Faculty of Chemistry, University of Bucharest
Main fields of interest: the main research domain is coordination chemistry namely complexes with potential biologic activity: synthesis strategies, chemical analyses, physico-chemical and biologic characterization as well as synthesis of materials with special properties
Relevant categories in thermal analyses: fields (inorganic, complex, pharmaceutical, materials, nano, minerals); methods (TG, DTA, DSC, kinetics)
Professional activities: editorial board member of Biointerface Research in Applied Chemistry; scientific vice-president of Central and Eastern European Committee for Thermal Analysis and Calorimetry (CEEC-TAC); member in organizing committees of CEEC-TAC1 and CEEC-TAC2 conferences; member in local committee of Romanian Chemistry Society; member of International Confederation for Thermal Analysis and Calorimetry (ICTAC) since 2004
Publication record: papers (68), books (5), patents (1), citations (327), h-index (11), sum of impact factors (129,695)
Equipments: TG, DTA LABSYS 1200 from SETARAM Instrumentation; DSC 550 from CAHN Instrument Corp.
5 most important publications: [1] M. Badea, P. Budrugeac, A. Cucos, E. Segal, Thermal decomposition kinetics of bis(pyridine)manganese(II) chloride, J. Therm. Anal. Calorim., 115 (2014) 1999-2005; [2] I.D. Vlaicu, M. Constand, R. Olar, D. Marinescu, M. N. Grecu, V. Lazar, M. C.Chifiriuc, M. Badea, Thermal stability of new biologic active copper (II) complexes with 5,6-dimethylbenzimidazole, J. Therm. Anal. Calorim., 113 (2013) 1369–77; [3] M. Badea, R. Olar, V. Uivarosi, D. Marinescu, V. Aldea, Synthesis and characterization of some vanadyl complexes with flavonoid derivatives as potential insulin-mimetic agents, J. Therm. Anal. Calorim., 107 (2012) 279–285; [4] M. Badea, R. Olar, D. Marinescu, V. Uivarosi, T. O. Niculescu, D. Iacob, Thermal study of some new quinolone ruthenium (III) complexes with potential cytostatic activity, J. Therm. Anal. Calorim., 99 (2010) 829–834; [5] M. Badea, R. Olar, D. Marinescu, G. Vasile, B. Jurca, A. M. Madalan, M. Andruh, a Michael-type reaction between acrylato ions and ethylenediamine coordinated to Ni(II). Synthesis, crystal structure and magnetic properties of $[Ni_2(EDDP)_2(H_2O)_2]\cdot 2H_2O$ (H_2EDDP = ethylenediamine-N, N-dipropionic acid), Inorg. Chem. Comm., 12 (2009) 555–557.

Name: Harvey E. Bair
Country: USA
Date and place of birth: 1936, Williamsport, Pennsylvania
Present position and address: Consultant, 17 Seminole Ct., Newton, NJ 07860
Email: hebair@embarqmail.com
Education and scientific degrees: Dickinson College, Carlisle, PA, BS Chemistry (1958); Penn State University, State College, PA, MS Chemistry (1964); General Electric Research Laboratory, Schenectady, NY, Research Training Program in Polymers, mentor: Prof. F.E. Karasz; (1962–64)
Workplaces: John's Hopkins Univ. Applied Physics Laboratory, Silver Spring, MD (1964–65); Bell Labs, Lucent Technologies (formerly AT&T), Murray Hill, NJ (1965–2001)
Main fields of interest: innovating TA techniques for the solution of plastics engineering problems in areas such as performance, processing and reliability in engineering applications, i.e. after a device's laser is aligned and secured in front of an optical fiber with an epoxy, completion of cure can only be carried out in its glassy state or alignment will be lost; DSC enthalpy relaxation studies determined the maximum heating rate that will not exceed T_g and, hence, remain rigid and aligned is about 24°C/h (US Patent 4,978,712)
Relevant categories in thermal analysis: fields (materials, organic, polymers, glasses); methods (DSC, TG, TMA, DMA, cryo, specific heat, calorimetry and instrument development)
Awards and acknowledgments: Mettler Award (1987); SPE Conley Award for engineering technology (1998); SPE International Award for "outstanding contributions in the field of plastics" (2000); NATAS Lifetime Achievement Award (2008); NATAS Fellow (1984) and APS Fellow, High Polymer Physics Division (1987)
Professional activities: NATAS—Symp. Organizer, 1980; Notes Editor, 1981, President, 1984, Program Co-chair, 1985, Chairman, Gordon Conf. on Thermosetting Polymers, 1990; Chairman, ACS Symp. on "TA of Polymers", Boston 1990; Co-chair on "Chemical and Physical Effects on the Long Term Performance of Polymers" Boston 1995; Co-chair, Int. Congress of TA on "Polymer Blends and Blocks", Philadephia, 1996; Polymer Webinars-"Small Molecules, Fast-Scans and Blocks and Blends" with Harvey Bair at jgotro@innocentrix.com, 2011
Publication record: papers (188), book chapters (7), patents (7)
List of 5 most important publications: [1] H. E. Bair, "Glass Transition Measurements by DSC," in Assignment of the Glass Transition, ASTM STP 1249, R.J. Seyler, ed., ASTM, Philadelphia, pp. 50–74, (1994); [2] F.E. Karasz, H.E. Bair and J.M. O'Reilly, "Thermal Properties of Atactic and Isotactic Polystyrene," J. Phys. Chem. 69, 2657 (1965); [3] F.S. Bates, M.A. Hartney and H.E. Bair, "Block Copolymers near the Microphase Separation Transition" Macromolecules 17, 1987–1993 (1984); [4] F.C. Schilling, H.E. Katz and H.E. Bair, "Structure and Morphology of a Polyether/Polyacrylate Semi-interpenetrating Polymer Network", J. Therm. Anal. Calorim. 58, 83–92 (2000); [5] H.E. Bair, "Thermomechanical Analysis and Thermodilatometry-4.7 Selected Industrial Applications" in Thermal Analysis of Polymers: Fundamentals and Applications, J.D. Menczel and R.B. Prime, eds., Wiley, Hoboken, NJ, pp. 363–386 (2009).

Name: Peter Baláž
Country: Slovakia
Date and place of birth: 1947, Nitra, Slovakia
Present position and address: Leading scientist at Institute of Geotechnics of Slovak Academy of Sciences Košice and profesor at Technical University of Košice; Prof. Peter Baláž, Institute of Geotechnics, SAS, Watsonova 45, 040 01 Košice, Slovakia
Email: balaz@saske.sk
Education and scientific degrees: M.Sc. degree in chemistry, Faculty of Sciences, P.J. Šafárik University Košice; Ph. D. and D.Sc. degrees in mineral processing, Institute of Geotechnics, Slovak Academy of Sciences Košice; Associate Professor and Professor—Berg Faculty, Technical University Košice
Workplaces: Faculty of Sciences, P.J. Šafárik University Košice (1972–1977); Institute of Geotechnics, Slovak Academy of Sciences Košice (1977–)
Main fields of interest: the basic and applied research in mechanochemistry; study of the properties of minerals and advanced materials with various methods including thermoanalytical ones
Awards and acknowledgments: prices of Slovak Literary Fond for books published in Elsevier and Springer including prices for outstanding citations in technical sciences; Gold medal of Slovak Academy of Sciences
Professional activities: member of scientific boards of international conferences on mechanochemistry; invited lectures on scientific conferences; broad international cooperation; elaboration of mechanochemical technology for treatment of non-ferrous ores with application of mechanical and thermal procedures
Publication record: papers (417), books (3 and 14 chapters in books), patents (4), citations (>1200), h-index (21)
List of the 5 most important works: [1] P. Baláž, I. Ebert: Hydrometallurgy 27 (1991) 14; [2] P. Baláž: Extractive Metallurgy of Activated Minerals, Elsevier, Amsterdam 2000, 278 pp; [3] P. Baláž, E. Boldižárová, E. Godočíková, J. Briančin: Materials Letters 57 (2003) 1586; [4] P. Baláž: Mechanochemistry in Nanoscience and Minerals Engineering, Springer, Berlin Heidelberg 2008, 413 pp; [5] P. Baláž, M. Achimovičová, M. Baláž et al.: Chemical Society Reviews 47 (2013) 7521

Name: Wojciech Balcerowiak
Country: Poland
Date and place of birth: 1946. Szamotuły, Poland
Present position and address: Lecturer of Thermal Analysis Lab., Institute of Heavy Organic Synthesis, 9 Energetyków Str., PL-47-232 Kędzierzyn-Koźle, Poland
Email: balcerowiak.w@icso.com.pl
Education and scientific degrees: M.Sc. A. Mickiewicz University, Poznań (1969); Ph.D. Research Institute of Industrial Chemistry, Warszawa (1988)
Workplaces: Institute of Heavy Organic Synthesis in Kędzierzyn-Koźle, Poland (1969 –)
Main fields of interest: determination of composition and thermal behaviour of organic-inorganic materials; cation exchangers; catalysts; heat and kinetic of organic syntheses, calcium lime analysis.
Relevant categories in thermal analyses: fields (organic-inorganic materials, polymers, kinetic of homogenous syntheses); methods (TG, DTA, DSC)
Professional activities: Analyst and R&D in organic syntheses; member and executive council member (1997–2006) of Polish Society of Thermal Analysis and Calorimetry; Lecturer of schools on TA.
Publication record: papers (50), patents (50)
Equipments: Mettler-Toledo TGA/SDTA 851^{e} and DSC 822^{e}
5 most important publications: [1] W. Balcerowiak, W. Jerzykiewicz, H. Szewczyk: Differential thermal analysis using closed pans. The ethoxylation of n-dodecylamine, Tenside Deterg., 21 (1984) 10; [2] W. Balcerowiak: The thermal decomposition of sodium formate in presence of sodium hydroxide, Thermochim. Acta, 92 (1985) 661; [3] W. Balcerowiak: Speciation analysis of sulfonic groups of cation exchanger using thermogravimetry, Reactive and Functional Polymers, 33 (1997) 323; [4] W. Balcerowiak, J. Hetper, B. Kałędkowski, M. Gryta: A new approach to the determination of the heat of phenolic resole synthesis, Thermochim. Acta, 320 (1998) 209; [5] W. Balcerowiak: Phase analysis of high-calcium lime by TG, J. Therm. Anal. Calorim., 60 (2000) 67.

Name: Aparna Banerjee
Country: India
Date and place of birth: India
Present position and address: Associate Professor of Homi Bhabha National Institute and Scientific Officer, Product Development Division, Bhabha Atomic Research Centre, Mumbai, India
Email: aparnab@barc.gov.in; aparna_baner@yahoo.com
Education and scientific degrees: Ph.D. from Mumbai University, India, M.Phil, M.Sc.
Workplaces: Bhabha Atomic Research Centre, Mumbai, India.
Main fields of interest: thermodynamics of oxides and alloys: solid state electrochemistry, calorimetry, vapour pressure measurements, ionic conductivity in solids, thermal expansion.
Relevant categories in thermal analyses: fields (Inorganic, materials, ceramics, alloys); methods (TG, DTA, DSC, Dilatometry, Calorimeters, Galvanic cell, transpiration, KEML)
Awards and acknowledgments: Bronze medal for outstanding contribution to solid state chemistry
Professional activities: Executive council member of Indian Thermal Analysis Society. Editorial board of Society for Materials Chemistry. Presented Invited talk in several International symposia and member of organising committee
Publication record: papers (28 journal + 60 symposia papers)
Equipments: DSC, TGA, Galvanic cells, Dilatometry, Calvet calorimeter, Multi-HTC, Impedance Spectroscopy with HT furnace, HT-XRD, Transpiration, KEML, KEMS
5 most important publications: [1] "The System Yb–Ru–O: HighTemperature studies of the Ternary Oxides $Yb_2Ru_2O_7(s)$ and $Yb_3RuO_7(s)$" Aparna Banerjee, Ziley Singh Chaudhary. Mat. Chem. Phys., 42 (2013) 12; [2] "System Er–Ru–O: High temperature study of the heavy rare earth pyrochlore $Er_2Ru_2O_7(s)$ by electrochemical cell and differential scanning calorimeter" Aparna Banerjee, Solid State Ionics, 253 (2013) 70; [3] "Solid Oxide Electrochemical Cell and Differential Scanning Calorimetry used for Thermodynamic Measurements of the Ternary Oxides: $Nd_2RuO_5(s)$ and $Nd_2Ru_2O_7(s)$" Aparna Banerjee and Ziley Singh Chaudhary, Mat. Chem. Phys., 138 (2013) 417; [4] "The Lu–Ru–O System: Thermodynamic Properties and Impedance Measurements of the Pyrochlore $Lu_2Ru_2O_7(s)$" Aparna Banerjee, R. Mishra and Ziley Singh, Solid State Ionics, 201 (2011) 42; [5] "Heat capacity and Gibbs energy of formation of the ternary oxide $CdRh_2O_4(s)$" Aparna Banerjee, Z. Singh, V. Venugopal, Solid State Ionics, 180 (2009) 1337.

Name: Berta Barta Holló (maiden name: Holló)
Country: Serbia
Date and place of birth: 12th July 1984, Subotica, Serbia
Present position and address: research associate, Department of Chemistry, Biochemistry and Environmental Protection, Faculty of Sciences, 21000 Novi Sad, Trg D. Obradovića 3, Serbia
Email: hberta@uns.ac.rs
Researcher ID/ORCID: 35118890700/0000-0002-5786-442X
Education and scientific degrees: 2007—Graduated chemist-biochemistry, Department of Chemistry, Biochemistry and Environmental Protection, Faculty of Sciences, University of Novi Sad; 2011—Ph.D., Department of Chemistry, Biochemistry and Environmental Protection, Faculty of Sciences, University of Novi Sad
Workplaces: Department of Chemistry, Biochemistry and Environmental Protection, Faculty of Sciences, University of Novi Sad, Serbia
Main fields of interest: complexes, polymers and their thermal analysis
Relevant categories in thermal analyses: fields (materials, complex, polymer); methods (TG, DTA, EGA, DSC)
Publication record: papers (13), citations (29), h-index (3), sum of impact factors (22.336)
Equipments: TAI SDT Q600 Simultaneous TG/DSC
5 most important publications: [1] Holló, B., Tomić, Z. D., Pogány, P., Kovács, A., Leovac, V. M., Mészáros Szécsényi, K., Polyhedron, 28 (2009) 3881; [2] Holló, B., Jašo, V., Leovac, V. M., Divjakovic, V., Kovács, A., Mészáros Szécsényi, K., J.Coord. Chem., 66 (2013) 453; [3] Holló, B., Rodić, M. V., Vojinović-Ješić, L.S., Živković-Radovanović, V., Vučković, G., Leovac, V.M., Mészáros Szécsényi, K., J. Therm. Anal. Cal., 116 (2014) 655; [4] Holló, B., Rodić, M.V., Bera, O., Jovičić, M., Leovac, V. M., Tomić, Z. D., Mészáros Szécsényi, K., Struct. Chem., 24 (2013) 2193; [5] Poręba, R., Špirková, M., Pavličević, J., Budinski-Simendić, J., Mészáros Szécsényi, K., Holló, B., Compos. Part B Eng., 58 (2014) 496.

Name: Parmjit S. Bassi
Country: USA
Date and place of birth: 1938, India
Present position and address: Senior Scientist; Environmental Protection Division Laboratories. 5804 Peachtree Corners East, Norcross, GA 30092 USA
Email: Parmjit.Bassi@dnr.state.ga.us
Education and scientific degrees: M.Sc. (Hons. Sch.) (1960) Punjab University, Chandigarh; Ph.D. (1963) Gorakhpur University, Gorakhpur
Workplaces: Punjabi University, Patiala (1964–1967); Regional Engineering College, Srinagar (1967–1969); Jammu University, Jammu (1969–1981), Guru Nanak University, Amritsar, (1981–1998); Environmental Protection Division Laboratory, Atlanta, GA (2001-to date).
Main Fields of Interest: solid state reactions, thermal decompositions, applications of thermal analysis, ferrites—preparation and characterization, solid-liquid phase equilibria, eutectics—crystallization, micro-structure and strength properties.
Relevant categories in thermal analysis: fields (inorganic materials, decompositions, kinetics); methods (TG, DTA, DTG, DSC, kinetics)
Professional activities: National symposium, 'Reactivity of Solids,' Jammu 1974, Director; National symposium, 'Reactivity of Solids,' Amritsar 1981, Director; National seminar, 'Trends in Structure determination,' Amritsar 1983, Director; Binational Seminar at Novosibirsk 1986, Indian delegate, presented paper; National symposium on thermal analysis, Jammu 1998, Keynote address; President, Indian Association of Solid State Chemists and Allied Sciences ISCAS, 1999–2001, 2001–2008
Publication record: papers (101)
List of the 5 most important publications: [1] P. S. Bassi, P. C. Kalsi and C. M. Khajuria; Thermal decomposition of copper (II) phthalate monohydrate, J. Thermal Anal. 18(1980) 77; [2] P. S. Bassi and G. S. Chopra; Solid state reactivity of organic compounds with inorganic compounds, J. Solid State Chem. 62(1986)253; [3] P. S. Bassi, G. S. Chopra, and Kanwaljit Kaur; EPR and thermal studies on the reaction of copper acetate with 8-hydroxyquinoline in the solid phase, Ind. J. Chem. 29A (1990) 454; [4] P. S. Bassi and G. S. Chopra and Rajinder Singh; Linear free energy relationships in solid state reactions, Indian J. Chem. 35A (1996) 458; [5] P. S. Bassi, Applications of thermal analysis in some solid state reactions, J. Thermal Anal. 49 (1997) 1153.

Name: Manidipa Basu
Country: India
Date and place of birth: 1970, Burdwan, West Bengal, India
Present position and address: Scientific Officer (F), Bhabha Atomic Research Centre, Mumbai, India; Chemistry Division, Bhabha Atomic Research Centre, Mumbai, India-400085
Email: deepa@barc.gov.in
Education and scientific degrees: M.Sc. (Chemistry) (1995), Ph.D. (Chemistry) (2003)
Workplaces: Bhabha Atomic Research Centre, Mumbai, India
Main fields of interest: thermodynamic and transport studies
Relevant categories in thermal analyses: fields (inorganic materials, polymer, glass, ceramics, alloys); methods (TG, DTA, EGA, DSC, thermomechanical analysis, kinetics, specific heat, calorimetry)
Publication record: number of papers (28), books (1 book chapter), citations (160)
5 most important publications: [1] Thermodynamic and transport properties of thoria-urania fuel of Advanced Heavy Water Reactor. M. Basu (Ali), R. Mishra, S.R. Bharadwaj and D. Das, J. Nucl. Mater., 403 (2010) 204; [2] Transport properties of I, Te and Xe in thoria-urania SIMFUEL. N. Shirsat, M. Ali (Basu), S. Kolay, A. Datta and D. Das. J. Nucl. Mater., 392 (2009) 16; [3] Enthalpy increment measurements of $La_2Te_3O_9$ and $La_2Te_4O_{11}$. M. Ali (Basu), S.R. Bharadwaj and D. Das; J. Nucl. Mater., 360 (2007) 99; [4] Volumetric and acoustic properties of binary mixtures of tri-n-butyl phosphate with n-hexane, cyclohexane and n-heptane from T = (298.15–323.15) K, M. Ali (Basu), T. Samanta, D. Das; J. Chem. Thermodyn., 57 (2013) 335; [5] Volumetric and compressibility studies on tri-n-butyl phosphate-phase modifier (1-octanol, 1-decanol and isodecanol) interactions from T = (298.15–323.15) K; M. Ali (Basu), T. Samanta, D. Das; J. Chem. Thermodyn., 70 (2014) 1.

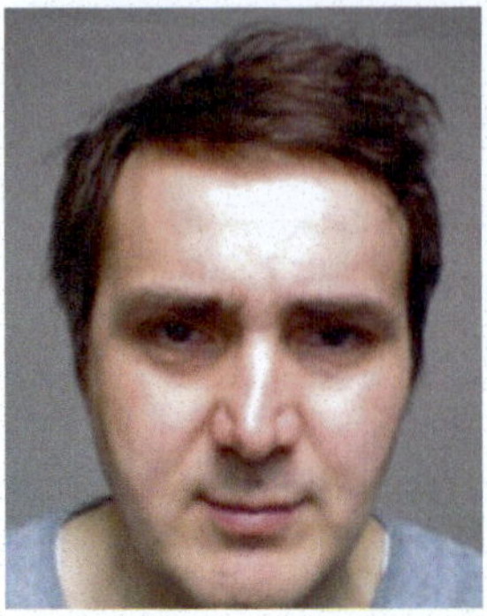

Name: David Bessieres
Country: France
Date and place of birth: 1971, Aureilhan, France
Present position and address: University of Pau and Pays de l'Adour
Email: david.bessieres@univ-pau.fr
Education and scientific degrees: Ph.D. (1999), Assistant Professor (2000), Full Professor (2012)
Workplaces: University of Pau and Pays de l'Adour
Main fields of interest: thermodynamic and interfacial properties, fluids and fluid mixtures, high pressure measurement
Relevant categories in thermal analyses: fields (fluids, fluid mixtures, nanomaterials); methods (Calvet calorimetry, heat capacity, isobaric thermal expansion, high pressure instrument development)
Awards and acknowledgments: AFCAT young resarcher (2004); Stig Sunner Prize Calcon (2009)
Professional activities: Regional Editor Journal of Thermal Analysis and Calorimetry (french speaking language, 2006–2013), co-chairman of the 37th JCAT, Pau 2006
Publication record: papers (49), citations (700), h-index (18)
Equipments: Setaram C80 calorimeter, BT Calvet, High pressure calorimeter, Magnetic suspension balance, Micromeritics Analyzer
5 most important publications: [1] Bessieres D, Saint-Guirons, Daridon, JL, Journal of Thermal Analysis and Calorimetry, 62, 2000, 621–632; [2] Bessieres D, Saint-Guirons, Daridon, JL, Coxam JY, Measurement Science and Technology, 11, 2000, 69–N72; [3] Bessieres D, Lafitte T, Daridon JL, Randzio SL. Thermochimica Acta 428, 2005, 25–30; [4] Bessieres D, Randzio SL. Pineiro MM., Lafitte T, Daridon, JL, Journal of Physical Chemistry B, 110, 2006, 5659–5664; [5] Bessieres D, Pineiro MM., de Ferron G., Plantier F, Journal of Chemical Physics, 133, 2010, 074507.

Name: Ruggero Bettini
Country: Italy
Date and place of birth: 1964, Cremona, Italy
Present position and address: Professor of Pharmaceutical Technology, Department of Pharmacy, University of Parma, Parco Area delle Scienze 27/A, 43124, Parma, Italy
Email: bettini@unipr.it
Website: http://www.unipr.it/ugov/person/15595
Education and scientific degrees: B.Sc. in Pharmacy, University of Parma, Italy (1991); Ph.D. in Pharmaceutical Chemistry and Technology, University of Pavia, Italy (1991–1994); Postdoctoral Fellow (1994–1997); Assistant Professor (1997–2002); Associate Professor (2002–2011); Professor (2011)
Workplaces: Laboratoire de Pharmacie Galénique et Biopharmacie, University of Lille, France (1990); School of Chemical Engineering of the Purdue University, West Lafayette, IN, USA (1993–1994); Department of Pharmacy, University of Parma (1994–)
Main fields of interest: pharmaceutical applications of supercritical fluids; solid-state chemistry of drugs; polymers for regenerative medicine applications and controlled release of drugs
Relevant categories in thermal analyses: fields (pharmaceutical, polymers, cyclodextrins); methods (DSC, TG, hot stage microscopy)
Awards and acknowledgments: The Nagai Foundation Tokyo, 1996; J. Heller Journal of Controlled Release/CRS outstanding paper Award, 1999; First Eurand Award. The Prize for Outstanding Research in Emerging Fields of Oral Drug Delivery, 2000
Professional activities: Director of the Interdepartmental Centre Biopharmanet-TEC, University of Parma, Italy; Member of the Committee for Training and Education (CTE) of the European Federation for Pharmaceutical Sciences (EUFEPS); Member of the Council of the University of Parma (2009–2011)
Publication record: papers (84), books (7 book chapters), patents (15), citations (2717), h-index (26), sum of impact factors (120)
Equipments: DSC; TGA-DSC; Hot Stage Microscope, MicroFT-IR
5 most important publications: [1] R. Bettini et al. J. Pharm. Sci, 99,1855–1870 (2010); [2] S.Nicoli, R. Bettini et al. J. Pharm. Sci., 97, 4830–4839 (2008); [3] I. Pasquali, R. Bettini, F. Giordano Adv. Drug Del. Rev., 60, 300–410 (2008); [4] R. Bettini, et al. J. Therm. Anal. Calorim, 73, 487–497 (2003); [5] R. Bettiniet et al. Eur. J. Pharm. Sci., 13, 281–286 (2001).

Name: Isabelle Beurroies
Country: France
Date and place of birth: 1966, Asnières, France
Present position and address: Lecturer at the Aix-Marseille University;President of the AFCAT association (Association Française de Calorimétrie et d'Analyse Thermique)
Email: isabelle.beurroies@univ-amu.fr
Website: www.lc-provence.fr/madirel/
Education and scientific degrees: Ph.D. in materials science
Workplaces: MADIREL Laboratory AixMarseille University-CNRS at Marseille France
Main fields of interest: characterization of materials, adsorption
Relevant categories in thermal analyses: fields (adsorption, energetic behavior of divided materials, thermal analysis of phase transitions in materials); methods (calorimetry, DSC)
Professional activities: President of the French Association of Calorimetry and Thermal Analysis (AFCAT)
Equipments: Tian-Calvet Micalorimeter, DSC, ATG
5 most important publications: [1] "Microcalorimetric methods for studying vapour adsorption and wetting of powders" R. Dernoyel, I. Beurroies, D. Vincent, J. Therm. Anal. and Calorimetry, 70, 2, 483 2002; [2] "A Comparison between melting-solidification and capillary condensation hysteresis in mesoporous materials : application to the interpretation of thermoporometry data" I. Beurroies, R. Denoyel, P. Llewellyn, J. Rouquerol, Thermochimica Acta 421 (2004) 11; [3] "Comparing the basic phenomena involved in three methods of pore-size characterization: gas adsorption, liquid intrusion and thermoporometry" R. Denoyel, P. Llewellyn, I. Beurroies, J. Rouquerol, F. Rouquerol, L. Luciani, Part. Part. Syst. Charact. 21 (2004) 128; [4] "Detailed in situ XRD and calorimetric study of the formation of silicate/mixed surfactant mesophases under alkaline conditions. Influence of surfactant chain length and synthesis temperature" Beurroies I., Agren P., Buchel G., Rosenholm J.B., Amenitsch H., Denoyel R., Linden M., J. Phys. Chem. B, 2006, 110, 33, 16254; [5] "Using Pressure to Provoke the Structural Transition of Metal-Organic Frameworks" I Beurroies, M Boulhout, P Llewellyn, B Kuchta, G Ferey, C Serre,, R Denoyel, Angewandte chemie (2010), 49, 41, 7526.

Name: Shyamala Rajkumar Bharadwaj
Country: India
Date and place of birth: 1953, Thiruchirappalli, India
Present position and address: Head, Fuel Cell materials and Catalysis Section, Chemistry Division, Bhabha Atomic Research Centre, Mumbai 400085, India
Email: shyamala@barc.gov.in
Education and scientific degrees: B.Sc. (Chemistry) Mumbai University (1976); M.Sc. (Chemistry) Mumbai University (1981); Ph.D. (Physical Chemistry) Mumbai University (1991)
Workplaces: Chemistry Division, Bhabha Atomic Research Centre, Mumbai 400085, India
Main fields of interest: Thermodynamic properties of nuclear and other energy related materials, materials for solid oxide fuel cells, Thermochemical cycles for hydrogen generation
Relevant categories in thermal analyses: Fields (Thermochemical and Thermophysical properties of Materials); Methods (Differential Thermal Analysis, Thermogravimetry, Dilatometry, HTXRD, TPDRO)
Awards and acknowledgments: NETZSCH-ITAS Award (2006); DAE Group Acheivement Award (2010)
Professional activities: Organised several National Symposia and Workshops on Thermal Analysis in India. Regional Editor, Journal of Thermal Analysis and Calorimetry.
Publication record: Papers (142), Books (1), Book Chapters (3)
Equipments: TG-DTA, Calvet Calorimeter
5 most important publications: [1] Nityanand Chaubey, B.N. Wani, S.R. Bharadwaj, M.C. Chattopadhyaya, J. Therm. Anal. Calorim., 112 (2013) 155–164; [2] Pooja Sawant, S. Varma, B.N. Wani and S.R. Bharadwaj, J. Therm. Anal. Calorim., 107 (2012) 185–195; [3] A.M. Banerjee, M.R. Pai, Jagannath, S.R. Bharadwaj, Thermochim. Acta, 516 (2011) 40–45; [4] M. Ali (Basu), B.N. Wani and S.R. Bharadwaj, J. Therm. Anal. Calorim., 96 (2009) 463-468; [5] R.V. Wandekar, B.N. Wani, D. Das and S.R. Bharadwaj, Thermochim. Acta, 493 (2009) 14–18.

Name: Mira R. Bissengaliyeva
Country: Kazakhstan
Date and place of birth: 1955, Karaganda, Kazakhstan
Present position and address: Deputy Director on science. Institute of problems of complex development of mineral resources. 100019, Ippodromnaya str. 5, Karaganda, Kazakhstan.
Email: mirabis@ipkon.kz; 160655@mail.ru
Website: www.ipkon.org
Researcher ID: B-8152-2014
Education and scientific degrees: Chemist, Academician E.A. Buketov Karaganda State University (1972–1977); Ph. D.—Chemistry (1990); Associate Professor (1996); Doctor of Chemical Sciences (2010); Professor (2013).
Workplaces: Academician E.A.Buketov Karaganda State University (1977–1981); The Chemical-Metallurgical Institute named after Zh.N. Abishev affiliated to the RGP "National center on complex mineral processing of the Republic of Kazakhstan" (1981–2004); LLP "The Institute of problems of complex development of mineral resources" (2004–).
Main fields of interest: investigations of low-temperature phase transitions in minerals and inorganic materials; thermodynamics and thermochemistry of natural minerals.
Relevant categories in thermal analyses: fields (inorganic materials, minerals and ceramics); methods (specific heat, adiabatic calorimetry, DSC, DTA, TG, microcalorimetry)
Publication record: papers (110), books (2), patents (1), citations (18), h-index (4), sum of impact factors (28.87)
Equipments: Low-temperature thermophysical device for adiabatic calorimetry ("Termax")
5 most important publications: [1] I. A. Kiseleva, L. P. Ogorodova, L.V. Melchakova, M.R. Bisengalieva, N.S. Becturganov: Physics and Chemistry of Minerals, 19 (1992) 322; [2] I.A. Kiseleva, L.P. Ogorodova, L.V. Melchakova, M.R. Bisengalieva: The Journal of Chemical Thermodynamics, 25 (1993) 621; [3] M.R. Bisengalieva, I.A. Kiseleva, L.V. Melchakova, L.P. Ogorodova, A.M. Gurevich: The Journal of Chemical Thermodynamics, 29 (1997) 345; [4] M.R. Bissengaliyeva, N.S. Bekturganov, D.B. Gogol: Journal of Thermal Analysis and Calorimetry, 101 (2010) 49; [5] M.R. Bissengaliyeva, D.B. Gogol, S.T. Taymasova, N.S. Bekturganov: Journal of Chemical and Engineering Data, 56 (2011) 195.

Name: Ignazio Blanco
Country: Italy
Date and place of birth: 1971, Catania, Italy
Present position and address: Associate Professor, University of Catania, V.le A. Doria 6, 95125 Catania
Email: iblanco@dii.unict.it
Website: http://www.dmfci.unict.it/users/iblanco/
Researcher ID/ORCID/Scopus: F-3406-2010/0000-0001-9252-9435/7006352466
Education and scientific degrees: Industrial Chemistry, University of Catania (1990–1997); Ph.D. Polymeric Materials for Special Uses, University of Catania (2000–2003), Aggregate Professor Science and Technology of Materials (2003); Associate Professor Fundamentals of Chemical Technology (2011)
Workplaces: Department of Physical and Chemical Methodologies for Engineering, University of Catania (2003–2011), Department of Industrial Engineering, University of Catania (2012)
Main fields of interest: physico-chemical characterization of organic molecules; kinetic studies of the thermal degradation of polymers and nanocomposites; thermal analysis of polymers, nanoparticles and nanocomposites
Relevant categories in thermal analyses: fields (materials, nano, organic, polymer); methods (TG, DTA, DSC, thermomechanical analysis, kinetics, specific heat, calorimetry, long-term degradation)
Professional activities: Member of Editorial Board of American Journal of Polymer Science, American Journal of Nanoscience and Nanotechnology, American Journal of Applied Chemistry, Trends in Materials Sciences; Guest Editor of The Scientific World Journal; Member of Italian Association of Calorimetry and Thermal Analysis, Cross-Divisional Group of Calorimetry and Thermal Analysis of the Italian Chemical Society, Italian Association of Chemical for Engineering
Publication record: papers (48), citations (546), h-index (16)
Equipments: Mettler TGA 1 and DSC 1 Star System, Mettler DSC 30, Mettler TC 10 A, Shimadzu DTG-60, Shimadzu DSC 60, TRITEC Dynamic Mechanical Thermal Analyzer
5 most important publications: [1] I. Blanco, G. Cicala, C. Lo Faro, A. Recca. J. Appl. Polym. Sci. 2003, 89(1):268; [2] I. Blanco, G. Cicala, C. Lo Faro, O. Motta, G. Recca. Polym. Eng. Sci. 2006, 46(11):1502; [3] L. Abate, I. Blanco, A. Orestano, A. Pollicino, A. Recca. Polym. Degrad. Stabil. 2003, 80(2):333; [4] L. Abate, I. Blanco, O. Motta, A. Pollicino, A. Recca. Polym. Degrad. Stabil. 2002, 75(3):465; [5] L. Abate, I. Blanco, A. Pollicino, A. Recca. J. Therm. Anal. Calorim. 2002, 70(1):63.

Name: Jerzy Błażejowski
Country: Poland
Date and place of birth: 1945, Wąbrzeźno, Poland
Present position and address: Full professor, University of Gdańsk, Faculty of Chemistry, Wita Stwosza 63, PL-80 308 Gdańsk, Poland
Email: jerzy.blazejowski@ug.edu.pl
Education and scientific degrees: M.Sc. in Chemical Engineering at the Gdańsk University of Technology (1969); Ph.D. (1977) and D.Sc. (1984) at the University of Gdańsk; Professorship in Chemistry (1992)
Workplaces: Gdańsk University of Technology (1969–1970); University of Gdańsk (1970–1979, 1981–1985, 1986–); Pennsylvania State University (1979–1981, 1985–1986, 1989)
Main fields of interest: thermal behaviour and reactivity of solids investigated by thermoanalytical and combined techniques; phenomenological and theoretical description of thermodynamics and kinetics of processes involving solids; X-ray crystallography; crystal lattice energetics and features of ionic and molecular crystals; theory of solid state reaction kinetics
Relevant categories in thermal analyses: fields (inorganic, complex, organic); methods [TG, DTA, DSC, kinetics, other (theoretical)]
Awards and acknowledgments: Scientific Award of the Polish Ministry of National Education (1982, 1987, 1993, 1997–2009); award of the voivode of the province of Gdańsk (1990); The Johannes Hevelius Scientific Award of the City of Gdańsk (2012); The 2012 TA Instruments–ICTAC Award.
Professional activities: Member of the Chemistry Committee of the Polish Academy of Science (1996–2002); member of the Presidium of the Polish Society for Calorimetry and Thermal Analysis (1997–2000) and Polish Chemical Society (1997–2003); member of the Presidium (1998–2001), General Secretary (2001–2013) and President (2013–) of Gdańsk Scientific Society; chairman of the General Council for Higher Education in Poland (2002–2009).
Publication record: papers (200), books (5 ed.), patents (4), citations (1500; 810 independent), h-index (18)
Equipments: DSC 204 Netzsch, TG 209 Netzsch coupled with FTIR IFS 66 Bruker, TG STA 449 F3 coupled with QMS 403 C Netzsch
5 most important publications: [1] J. Błażejowski: Evaluation of kinetic constans for the solid-state reactions under linear temperature increase conditions, Thermochim. Acta, 48 (1981) 109–124; [2] J. Błażejowski: Remarks on the description of reaction kinetics under non-isothermal conditions, Thermochim. Acta, 76 (1984) 359–372; [3] J. Łubkowski, J. Błażejowski: The electrostatic energy in chloride salts of mono-nitrogen organic bases, J. Phys. Chem., 95 (1991) 2311–2316; [4] J. Błażejowski, J. Rak, M. Gutowski: On the possibilities of theoretical analysis of kinetics of the thermal decomposition of solids, J. Therm. Anal., 43 (1995) 45–55; [5] J. Błażejowski, B. Zadykowicz: Computational prediction of the pattern of thermal gravimetry data for the thermal decomposition of calcium oxalate monohydrate, J. Therm. Anal. Calorim. 113 (2013) 1497–1503.

Name: Vladimir Boldyrev
Country: Russia
Date and place of birth: 1927, Tomsk, USSR
Present position and address: Director of the Scientific Education Center "Molecular Design and Ecologically Safe Technologies" at Novosibirsk State University, Professor, Chair of Solid State Chemistry of Novosibirsk State University. Novosibirsk State University, Pirogova, 2, Novosibirsk, 630090, Russia
Email: Boldyrev@solid.nsc.ru
Education and scientific degrees: Chemist, Tomsk State University (1948); Ph.D. (1951); D.Sc. (1961); Professor of Chemistry (1963); Corresponding Member of the Academy of Sciences of the USSR (1979); Full Member of the Russian Academy of Sciences (1991)
Workplaces: Tomsk State University (1951–1957), Tomsk Polytechnical Institute (1957–1963), Institute of Chemical Kinetics and Combustion Acad. Sci. USSR (1963–1975), Institute of Solid State Chemistry and Mechano-chemistry Russ. Acad. Sci. (1975–1998), Novosibirsk State University (1981–)
Main fields of interest: control of the reactivity of solids (components of solid propellents, explosives, non-silver photographic systems, etc.), solid state kinetics, mechanochemistry, dry technology, soft chemistry, solid state pharmaceutical science
Awards and acknowledgments: Order of the Red Banner of Labor (1987), Order "Badge of Honor" (1982), State Prize of Russian Federation for Achievements in Science and Technology (1993), Russian High School Medal (2002), European Chamber of Science and Industry (2012), Order "Badge of Honor" (2013)
Professional activities: ESTAC Committee (1990–94); EUROSTAR (1994–); President of Advisory Committee of International Symposium on the Reactivity of Solids (1992–1994); President of International Mechanochemical Association (IUPAC) (1988–1997); Reactivity of Solids (Elsevier, 1985–1990), Review of Solid State Sci. 1989–1993), J of Material Synthesis and Processing (1991–), Russ. J. "Inorganic Materials" (1994–), Russian J. of Inorganic Chemistry (1994–), Russ. Journal of Physical Chemistry (1994–); Head of Laboratory of Solid State Chemistry, Institute of Chemical Kinetics and Combustion Acad. Sci. USSR (1963–1975); Director of Institute of Solid State Chemistry and Mechanochemistry (1975–1998)
Publication record: papers (900), books (5), patents (80—Russian, 1—German, 1—Japan, 2 —USA), citations (1750)
5 most important publications: [1] V. Boldyrev. On the mechanism of the thermal decomposition of ammonium perhlorate. Combust. Flame. V. 15, N 1, (1970) 71; [2] V. Boldyrev. Influence des defauts cristallins sur la vitesse de decomposition thermique des solides. Bull. Soc. Chim. France. N 4 (1969) 1054–1061; [3] V. Boldyrev. Topochemical reactions. J. Therm. Anal. V. 7 (1975) 685–694; [4] V. Boldyrev. Topochemistry of thermal decomposition of solids. Thermochim. Acta. 100, N 1 (1986) 315–338; [5] V. Boldyrev. Thermal decomposition of silver oxalate. Thermochim. Acta. V. 338, (2002) 63–90.

Name: Elena V. Boldyreva
Country: Russia
Date and place of birth: 1961, Tomsk, USSR
Present position and address: Institute of Solid State Chemistry and Mechanochemistry Siberian Division of Russian Academy of Sciences (chief researcher); Novosibirsk State University (Head of Chair of Solid State Chemistry); ul. Kutateladze, 18, Novosibirsk 630128 Russia
Email: eboldyreva@yahoo.com
Website: www.solid.nsc.ru
SCOPUS Researcher ID: 7006629777
Education and scientific degrees: Graduated from Novosibirsk State University (1982). Ph.D. in Physical Chemistry (1988) and Dr. Sci. in Solid State Chemistry (2000)
Workplaces: 1980–: Institute of Solid State Chemistry (Russian Academy of Sciences, Novosibirsk); 1982–: teaching at Novosibirsk State University. 1990s: research visits in Germany, Italy, UK, France
Main fields of interest: reactivity of solids, Monte Carlo simulations of solid-state reactions, photo- and thermo-mechanical effects, kinetics, polymorphism, effects of pressure, X-ray diffraction, IR-and Raman spectroscopy, thermomicroscopy, thermal analysis and calorimetry, crystallization and dissolution, physical pharmacy, metastable and amorphous states of molecular solids mechanochemistry, API-excipient interactions
Relevant categories in thermal analysis: fields (materials, organic, pharmaceutical, polymorphism, solid drugs, solid-state reactions); methods (DSC, TG, thermomicroscopy, variable-temperature diffraction and spectroscopy in situ, kinetics, calorimetry)
Professional activities: Memberships: International Advisory Committee on the Reactivity of Solids, International Mechanochemical Association, American Nanosociety, COMPRES (USA), American Chemical Society, International Centre for Diffraction Data (USA), Board Member of the Eur. Soc. Appl. Phys. Chem., Eurostar-Science, IUCr Commissions on High Pressure and on Teaching, Executive Committee of the Int. Union of Crystallographers, Advisory Committee of ICCOSS. Editorial boards: Annal. Chim. Sci. Mater., Z. Krist., High Press. Res., J. Struct. Chem. Associate Editor of J. Therm. Anal. Calorim., Co-Editor of Acta Cryst. B., 2012–2013: a member of the Presidential Council of Science and Education of Russia. 2013–: a member of the Scientific Council of Russian Ministry of Science and Education
Awards and acknowledgments: European Society for Applied Physical Chemistry Award (2007); an honorable sign "For the contributions into the development of Novosibirsk" (2003); an honorable sign "For the contributions to the development of the academic research and its applications in Siberia" (2007), "Academina. Women in Science (2012)
Publication record: papers (over 230), books (6 books, 22 book chapters), patents (14)
5 most important publications in JTAC: [1] V. A. Drebushchak, A. G. Ogienko, E. V. Boldyreva, J. Therm. Anal. Calorim., 111 (2013), 2187; [2] V. A. Drebushchak, T. N. Drebushchak, N. V. Chukanov, E. V. Boldyreva, J. Therm. Anal. Calorim., 93 (2008), 343; [3] E.V. Boldyreva, V. A. Drebushchak, I. E. Paukov, Y. A. Kovalevskaya, T. N. Drebushchak, J. Therm. Anal. Calorim., 77 (2004), 607; [4] E. V. Boldyreva, V. A. Drebushchak, T. N. Drebushchak, I. E. Paukov, Y. A. Kovalevskaya, E. S. Shutova, J. Therm. Anal. Calorim., 73 (2003), 409; [5] E. V. Boldyreva, V. A. Drebushchak, T. N. Drebushchak, I. E. Paukov, Y. A. Kovalevskaya, E. S. Shutova, J. Therm. Anal. Calorim., 73 (2003), 419.

Name: Michael Ewart Brown
Country: South Africa
Date and place of birth: 1938, Johannesburg, South Africa
Present position and address: Emeritus Professor, Chemistry Department, Rhodes University, Grahamstown, 6140 South Africa; postal address: 3 Frances Street, Grahamstown, 6139 South Africa
Email: m.brown@ru.ac.za
Education and scientific degrees: B.Sc. (Hons) (Witwatersrand); Ph.D. (Rhodes) (1966); D.Sc. (Rhodes) 2003
Workplaces: Rhodes University Junior Lecturer (1962–1965); Lecturer (1967–1970); Senior Lecturer (1971–1977); Associate Professor (1978–1985); Professor of Physical Chemistry (1986–2003); Research Officer, S.A. Chamber of Mines Research Laboratories (1966); Leverhulme Visiting Research Fellow Queen's University of Belfast (1971); Cavendish Laboratory, University of Cambridge (1980); ICI Explosives, Scotland (1989); Dean of Science, Rhodes University (1986–1991); Acting Dean of Research, Rhodes University (1994); Emeritus Professor 2003
Main fields of interest: decomposition of inorganic solids; kinetics of solid-state reactions; pyrotechnics; thermal analysis
Awards and acknowledgments: Mettler/NATAS Award (1996); the Rhodes University Vice-Chancellor's Distinguished Senior Research Award (1998); the Gold Medal of the South African Chemical Institute (2000); A-rated by the NRF (2002)
Professional activities: Fellow of the Royal Society of South Africa; Fellow of NATAS; Member of the SA Chemical Institute, Member of the Royal Society of Chemistry; Member and past secretary of ICTAC; founder member of the SA Thermal Analysis Society; past member of the Editorial Board of Thermochim. Acta
Publication record: papers (130), books (5)
5 most important publications: [1] M. E. Brown: Steps in a minefield-some kinetic aspects of thermal analysis, J. Therm. Anal. Calorim., 49 (1997) 17; [2] M. E. Brown, R. E. Brown: Kinetic aspects of the thermal stability of ionic solids, Thermochim. Acta, 357 (2000) 133; [3] M. E. Brown, M. Maciejewski, S. Vyazovkin, R. Nomen, J. Sempere et al.: The ICTAC kinetics analysis project. Part 1: Results, Thermochim. Acta, 355 (2000) 125; [4] M. E. Brown: Some thermal studies on pyrotechnic compositions, J. Therm. Anal. Calorim., 65 (2001) 323; [5] M. E. Brown, Beverley D. Glass, Decomposition of solids accompanied by melting—Bawn kinetics, Int. J. Pharm., 254 (2003) 255.

Name: Giovanna Bruni
Country: Italy
Date and place of birth: 1965, Pavia, Italy
Present position and address: Researcher at the School of Pharmacy of the University of Pavia; Department of Chemistry—Physico-Chemical Section, via Taramelli 16, 27100 Pavia, Italy
Email: giovanna.bruni@unipv.it
Researcher ID/ORCID: B-5685-2014/0000-0003-1958-2998
Education and scientific degrees: Degree in Pharmaceutical Chemistry and Technology (1989, University of Pavia); Degree in Pharmacy (1990, University of Pavia); Ph.D. (1994, University of Pavia)
Workplaces: Department of Chemistry—Physico-Chemical Section, University of Pavia
Main fields of interest: polymorphism and pseudo-polymorpism, drug-excipient compatibility, pharmaceutical co-crystals
Relevant categories in thermal analyses: fields (pharmaceutical); methods (DSC, TG)
Professional activities: Associate Editor of J. Therm. Anal. Calorim.
Publication record: papers (87), citations (752), h-index (15), sum of impact factors (181).
Equipments: TGA Q2000 IR, DSC Q2000 and Simultaneous TGA/DSC Q600 apparatus all interfaced with a TA 5000 data station (TA Instruments).
5 most important publications: [1] G. Bruni, V. Berbenni, C. Milanese, A. Girella, A. Cardini, E. Viganò, S. Lanfranconi, A. Marini: J. Pharm. Biomed. Anal., 50 (2009) 764; [2] G. Bruni, V. Berbenni, C. Milanese, A. Girella, A. Marini: J. Therm. Anal. Cal., 102 (2010) 193; [3] G. Bruni, F. Gozzo, D. Capsoni, M. Bini, P. Macchi, P. Simoncic, V. Berbenni, C. Milanese, A. Girella, S. Ferrari, A. Marini: J. Pharm. Sci., 100 (2011) 2321; [4] G. Bruni, V. Berbenni, F. Sartor, C. Milanese, A. Girella, D. Franchi, A. Marini: J. Therm. Anal. Cal., 108 (2012) 235; [5] G. Bruni, M. Maietta, L. Maggi, P. Mustarelli, C. Ferrara, V. Berbenni, M. Freccero, F. Scotti, C. Milanese, A. Girella, A. Marini: J. Phys. Chem. B, 117 (2013) 8113.

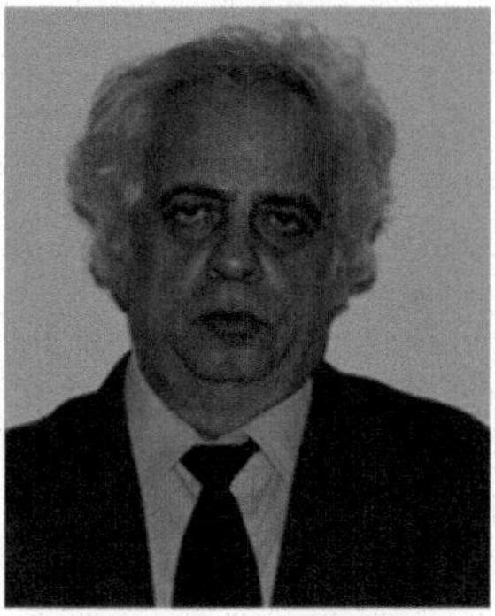

Name: Petru Budrugeac
Country: Romania
Date and place of birth: 1949, Caransebeş, Romania
Present position and address: INCDIE ICPE-CA National Institute for Research and Development in Electrical Engineering, Splaiul Unirii no. 313, sector 3, Bucharest—030138, Romania, Phone: +4021.346.8297
Email: petru.budrugeac@icpe-ca.ro, bp@icpe-ca.ro
Education and scientific degrees: Faculty of Chemistry—Bucharest University (1967–1972); Ph.D. (1981); Senior researcher (2001); Associate Professor—Faculty of Chemistry—Bucharest University (2003)
Workplaces: Institute of Physical Chemistry—Bucharest (1972–1984); College Gheorghe Şincai—Bucharest (1984–1986); INCDIE ICPE-CA National Institute for Research and Development in Electrical Engineering, Bucharest (1986–); Faculty of Chemistry—Bucharest University (2003–)
Main field of interest: non-isothermal and isothermal kinetics; thermal degradation of polymers and polymeric materials; application of thermal analysis methods in cultural heritage; adsorption and catalysis
Professional activities: President of the "Romanian Academy Commission of Thermal Analysis and Calorimetry" (2012), Associated Editor of "Journal of Thermal Analysis and Calorimetry" and of "Chemical Papers", Reviewer at 29 ISI scientific journals
Awards and acknowledgments: Nicolae Teclu Price of the Romanian Academy (1990); Best Reviewer Award 2010, given by Journal of Thermal Analysis and Calorimetry (2010)
Publication record: papers (164), books (3 books and 7 book chapters), patents (3) citations (1297), h-index (21)
5 important publications: [1] P. Budrugeac, E. Segal, Int. J. Chem. Kinet., 33 (2001) 564; [2] P. Budrugeac, J. Therm. Anal. Calorim., 68 (2002) 131; [3] P. Budrugeac, Polym. Degrad. Stab, 89 (2005) 265; [4] P. Budrugeac, L. Miu, J. Cult. Heritage, 9 (2008) 146; [5] P. Budrugeac, Thermochim. Acta, 500 (2010) 30.

Name: Jean-Marc Buisine
Country: France
Date and place of birth: 1952, Wasquehal, North, France
Present position and address: Professor at the University of Lille 1-Sciences and Technologies, Head of Laboratory
Email: jean-marc.buisine@univ-lille1.fr
Website: http://www-udsmm.univ-lille1.fr
Education and scientific degrees: Ph.D.
Workplaces: Dynamics and Structures of Molecular Materiels Laboratory; USTL1-U.F.R. de Physique; University of Lille 1—Sciences and Technologies, F59655 Villeneuve d'Ascq
Main fields of interest: thermodynamics of phase transitions under various pressure, liquid crystals, polymers and their composites, calorimetry, photocalorimetry, photothermal and thermomicroscopical analysis, thermobarometric and piezothermal analysis, use of liquid crystal-polymers composites for ophthalmic applications.
Relevant categories in thermal analyses: fields (organic, polymer, liquid crystals, health); methods (DSC, calorimetry, thermo-optical-microscopy, calorimetry, photo-calorimétry, instrument development: piézothermy—thermobarométry, photo-thermal methods)
Professional activities: Member of the Scientific Board of the annual French Meeting of Calorimetry and Thermal Analysis; President of the north section of the French Physical Society (2006–2010); President of the French Calorimetry and Thermal Analysis Group (2009–2013); Member of the Scientific Board of the Centre Historique Minier North of France; Member of the Scientific Board of the International Workshop on Liquid Crystals for Photonics; Regional Editor of the Journal of Thermal Analysis and Calorimetry (1999–2004).
Publication record: papers (100 journal papers, 167 conference communications), patents (3 patents, 2 licences)
Equipments: Differential scanning calorimeter, adiabatic calorimeter, photocalorimeter, equipments for thermal studies under pressure, equipments for thermo-electrical studies, equipments for photo-thermal studies,
5 most important publications: [1] J. M. Buisine, B. Soulestin, J. Billard, The Metabolemeter I: A New Apparatus to Detect the Phase Transitions of Mesogens; Mol. Cryst. Liq. Cryst. 91 (1983) 115–127; [2] A. Anakkar, J. M. Buisine, C. Alba-Simionesco, L. Ter Minassian, H. T. Nguyen, C. Destrade, Piezothermal analysis of the N_{Re}–S_{Ad} Transition on Small Samples of p-Octylbenzoate-p-cyano benzoyloxyphenyl; J. de Phys. III 2 (1992) 1029–1038; [3] F. Roussel, J. M. Buisine, U. Maschke, X. Coqueret, F. Benmouna, Phase Diagrams and Morphology of Polymer Dispersed Liquid Crystals Based on Nemetic Liquid Crystals/Monofonctional Acrylate Mixtures; Phys. Rev. E 62 (2000) 2310–2316; [4] A. Hadj Sahraoui, S. Longuemart, D. Dadarlat, S. Delenclos, C. Kolinsky, J. M. Buisine, The application of the photopyroelectric method for measuring the thermal parameters of pyroelectric materials. Rev. Sci. Instrum. 73 (2002) 2766–2772; [5] J. M. Buisine, F. Michel, "Accommodative ocular implant"; (2014) Japan Patent; (2012) USA Patent; (2009) EP2120792; (2008) WO08119894.

Name: Donald J. Burlett
Country: USA
Date and place of birth: 1949, Orange, New Jersey, USA
Present position and address: Sr. Research Scientist, Gates Corporation, 2975 Waterview Drive, Rochester Hills, MI 48309
Email: don.burlett@gates.com
Education and scientific degrees: BS—SUNY College at Oswego—Chemistry (1971); Ph.D.—U. Cincinnati—Organic Chemistry (1976); Post-Doctoral—Mass. Inst. Tech.—Organic Chemistry (1977)
Workplaces: Goodyear Tire and Rubber company (1977–2003); Gates Corporation (2003-present)
Main fields of interest: elastomer technology, materials and their interactions within elastomeric materials, thermal analysis and calorimetry of polymers and materials used with them, reactive processing, oxidation and ozonolysis of elastomers, rheology of polymers physical properties of elastomeric systems
Relevant categories in thermal analyses: fields (polymers, material characterization, oxidation and ozonolysis, rheology); methods (DSC, TGA, DMA, TMA)
Awards and acknowledgments: NATAS Fellow (2010)
Professional activities: President, North Coast Thermal Analysis Society (1989–1991); North American Thermal Analysis Society (NATAS)—(Conf. Program Chair 1992, Conference Chair 1994, Vice President 1995, President 1996, Past President 1997, Publications Councilor 1998, Affiliate Councilor to ICTAC 1998–2002); Chair, ICTAC Scientific Commission 2002–2004; ICTAC Vice-President 2004–2006; ICTAC President 2006-2012; ICTAC Past-President 2012-current)
Publication record: number of papers (4), books (2 book chapters), patents (18), citations (277)
5 most important publications: [1] D. J. Burlett, Handbook of Thermal Analysis and Calorimetry, Volume 3—Applications to Polymers and Plastics, Pages 519–586 (2002), Chapter 13 Thermal Analysis and Calorimetry of Elastomers (with M. B. Altman); [2] D. J. Burlett, Handbook of Thermal Analysis and Calorimetry, Volume 5—Recent Advances, Techniques and Applications, Pages 695–732 (2008), Chapter 19 Quality Control; [3] D. J. Burlett, Rubber Chemistry and Technology, March 1999, Vol. 72, No. 1, pp. 165–17, Studies of Elastomer Oxidation via Thermal Analysis; [4] D. J. Burlett and J. T. Lindt, Rubber Chemistry and Technology, July 1993, Vol. 66, No. 3, pp. 411–434, Reactive processing of Rubbers.

Name: Ramón Burriel
Country: Spain
Date and place of birth: 1950, Zaragoza, Spain.
Present position and address: Research Professor, Institute of Materials Science of Aragón, CSIC—University of Zaragoza, Faculty of Science, Plaza San Francisco, 50009 Zaragoza, Spain
Email: burriel@unizar.es
Website: http://www.icma.unizar-csic.es/, http://www.unizar.es/protermat/
Education and scientific degrees: Degree in Physics, University of Zaragoza (1974); Ph.D. (1979)
Workplaces: University of Michigan, Ann Arbor, MI (1980–1981), University of Illinois, Chicago, IL (1981–1983), University of Zaragoza (1983–1987), University of Illinois and Argonne National Laboratory (1988), Centre National de Recherche Scientifique, Grenoble (1992), Institute of Materials Science of Aragón (1987–)
Main fields of interest: calorimetry of solids; phase transitions; magnetic materials; molecular magnetism; permanent magnets; spin crossover transitions; magnetic refrigeration
Relevant categories in thermal analyses: fields (materials); methods (DSC, cryo, specific heat, calorimetry, instrument development)
Awards and acknowledgments: Special award to the best grades in Physics degree (1975).
Professional activities: Director of the Institute of Materials Science of Aragón (2002–2011); Conference Thermo International 2006, Boulder, CO, Symposium organizer; EMMA Conference 1998, Program Chairman; RSEF Conference (Jaca) 1993, Organizing Committee; Junior Euromat Conference (Lausanne) 1992, Organizing Committee; Committee on Professional Qualifications of the EPS, National representative
Publication record: number of papers (157), citations (1736), h-index (21)
Equipments: Adiabatic calorimeter, DSC, TG, AC calorimeter, PPMS HC option.
5 most important publications: [1] L. Tocado, E. Palacios, R. Burriel: Entropy determinations and magnetocaloric parameters in systems with first-order transitions: Study of MnAs, J. Appl. Phys. 105 (2009); [2] F. Palacio, G. Antorrena, M. Castro, R. Burriel, J. Rawson, J. N. B. Smith, N. Bricklebank, J. Novoa and C. Ritter: High-Temperature Magnetic Ordering in a New Organic Magnet, Phys. Rev. Lett. 79 (1997) 2336; [3] J. A. Real, I. Castro, A. Bousseksou, M. Verdaguer, R. Burriel, M. Castro, J. Linares and F. Varret: Spin Crossover in the 2,2′-Bipyrimidine-(bpym-) Bridged Iron (II) Complexes [Fe(L)$(NCX)_2]_2$ (bpym) (L = 2,2′-Bithiazoline (bt) and bpym; X = S, Se). X-ray Absorption Spectroscopy, Magnetic Susceptibility, Calorimetric, and Mössbauer Spectroscopy Studies, Inorg. Chem. 36 (1997) 455; [4] E. Palacios, J. J. Melero, R. Burriel, P. Ferloni: Structural, calorimetric, and Monte Carlo investigation of the order-disorder transition of BF_4 in $(CH_3)_4NBF_4$, Phys. Rev. B 54 (1996) 9099; [5] R. Burriel R., J. Bartolomé, D. González, R. Navarro, C. Ridou, M. Rousseau, A. Bulou: $KZnF_3$ cubic perovskite. Heat capacity and lattice dynamics, J. Phys. C 20 (1987) 2819.

Name: Oana Carp
Country: Romania
Date and place of birth: 1954, Bucharest, Romania
Present position and address: Deputy Director of the Institute of Physical Chemistry "Ilie Murgulescu" of the Romanian Academy, Splaiul Independentei 202, 060021, Bucharest, Romania
Email: ocarp@icf.ro
Education and scientific degrees: Chemist, Faculty of Chemistry, University of Bucharest (1972–1978), Ph.D. (1993); Senior researcher (1997)
Workplaces: Institute of Chemical and Biochemical Energy (1998–1990), Institute of Physical Chemistry "Ilie Murgulescu"of the Romanian Academy, Bucharest
Main fields of interest: synthesis of tailored-made materials (simple and mixed oxides, oxide/C composites, etc.) through controlled thermal treatments; relations between precursors, thermal processing and materials peculiarities
Awards and acknowledgments: Romanian Academy Award (1996) for the papers dedicated to oxides synthesis via thermal decomposition of coordination compounds
Publication record: number of papers (87), books (3 books, 1 book chapters), patents (8), citations (over 2000)
5 most important publications: [1] O. Carp, E. Segal, "Basic program to discriminate among mechanisms of solid-gas decomposition", Thermochimica Acta, 1991, 185, 111–127; [2] O. Carp, L. Patron, E. Segal, R. Barjega, M. Brezeanu, "Nonconventional methods for obtaining hexaferrites. Thermal analysis of some precursors", Journal of Thermal Analysis, 1999, 56, 513–518; [3] O. Carp, L. Patron, L. Diamandescu, A. Reller, "Study of iron oxides obtained by thermal decomposition of the polynuclear coordination compound $[Fe(urea)_6](NO_3)_3$", Thermochimica Acta, 2002, 390, 169–177; [4] C. Paraschiv, B. Jurca, A. Ianculescu, O. Carp, "Synthesis of nanosized bismuth ferrite ($BiFeO_3$) by a combustion method starting from $Fe(NO_3)_3 \cdot 9H_2O$-$Bi(NO_3)_3 \cdot 9H_2O$-glycine or urea systems", Journal of Thermal Analysis and Calorimetry, 2008, 94, 411–416; [5] D. Visinescu, B. Jurca, A. Ianculescu, O. Carp, Starch—A suitable fuel in new low-temperature combustion-based synthesis of zinc aluminate oxides, Polyhedron, 2011, 30, 2824–2831.

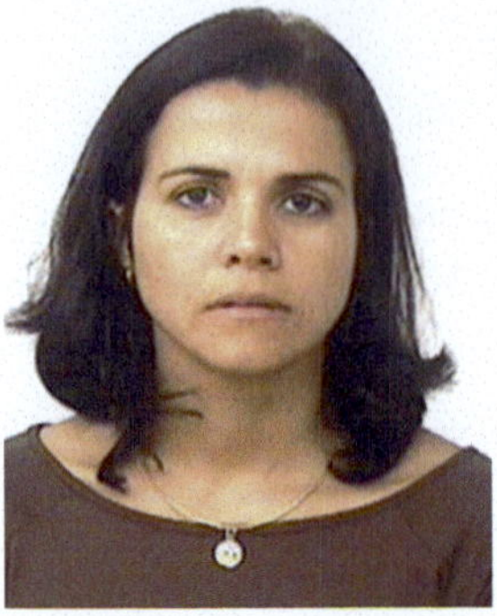

Name: Alessandra Rangel Cassella
Country: Brazil
Date and place of birth: 1971, Niterói, Brazil
Present position: Petroleum Chemistry, Chemistry Division of the Research and Development Center of Petrobras
Postal address: Av. Horácio Macedo, 950, Cidade Universitária—Ilha do Fundão, Rio de Janeiro, RJ, Brasil, CEP 21941-915
Email: arangel@petrobras.com.br
Education and scientific degrees: Industrial Chemical, Fluminense Federal University (UFF), Brazil (1989–1993); M.Sc. in Analytical Chemistry at Chemistry Department/PUC-RIO, Brazil (1994–1996); Dr. in Analytical Chemistry at Chemistry Department/PUC-RIO, Brazil (1996–2000)
Workplace: Research and Development Center (CENPES) of Petrobras, Chemistry Division
Main fields of interest: thermal analysis applied to petroleum products, deposits and biomass characterization and processing.
Relevant categories in thermal analyses: fields (inorganic, organic, materials, industrial products, crude oil, biomass, biodiesel, industrial solid deposits, diesel, wax); methods (TG, DTG, DTA, DSC, TG-FTIR, kinetics, specific heat, calorimetry)
Professional activities: Thermal Analysis Laboratory Coordinator in the Chemistry Department of the Research and Development Center of Petrobras
Publication record: papers (10 journal paper, 50 technical communications, 31 technical reports)
5 most important publications: [1] A. R. Cassella: Evaluation of water in oil emulsions by thermal analysis, Petrobras/Cenpes/RT QM 017/2005; [2] A. R. Cassella: Evaluation of the stability of synthetic drilling fluids by differential scanning calorimetry, Petrobras/Cenpes/CT QM 006/2008; [3] A. R. Cassella: Study of self-ignition diatomaceous phenomenon, Petrobras/Cenpes/CT QM 011/2009; [4] A. R. Cassella: Determination of minerals phases in carbonate rocks by thermal analysis, Petrobras/Cenpes/RT QM 003/2011; [5] A. R. Cassella: Characterization of major components in biomass by thermal analysis, Petrobras/Cenpes/RT/QM 003/2012.

Name: Margarida Lourenço Castelló
Country: Brazil
Date and place of birth: 1966, Rio de Janeiro, Brazil
Present position and address: Professor, Fluminense Federal Institute of Education Science and Technology; Av. Marechal Rondon, 300/307/Bl.2, Maracanã, Rio de Janeiro, RJ, Brasil CEP 20 950-004
Email: mcastello@eq.ufrj.br
Education and scientific degrees: Chemical Engineer and Bachelor in Chemistry, Rio de Janeiro State University (UERJ), Brazil (1985–1990, and 1991–1992); Ed.S. at Center of Philosophy and Human Sciences (2000–2002); M.Sc. and D.Sc. in Chem. And Biochem. Process Technology. at School of Chemistry, Rio de Janeiro Federal University (EQ/UFRJ) (2006–2009, and 2010–2013); Post Doc. Researcher at Inorganic Process Department, EQ/UFRJ (2014–), Brazilian National Council (CNPq) Researcher (2007–)
Workplace: Fluminense Federal Institute of Education Science and Technology, campus Macaé-RJ
Main fields of interest: thermal analysis applied to industrial products and processes, to their effluent and waste thermal characterization and processing, and to catalysts
Relevant categories in thermal analyses: fields (inorganic, organic, catalysts, zeolites, clays, industrial products, coal, biomass, industrial solid wastes, solidified and stabilized wastes); methods (TG, DTG, DTA, DSC, kinetics,, calorimetry, conventional and non-conventional TG and DTA instrument and system development)
Professional activities: Lecturer and Specialist indicated by the National Education Forum in the preparatory meetings of the National Conference on Education CONAE-2014 (2013–). MEC Scientific Awards Committee member for the National Award for Technology and Innovation (2013–). Undergraduate Advisor (2006–)
Publication record: papers (5), citations (16), h-index (2)
5 most important publications: [1] M. L. Castelló, J. Dweck, D. A. G. Aranda: Thermal stability and water content determination of glycerol by thermogravimetry. J. Therm. Anal. Calorim., 97(2) (2009) 627–630; [2] M. L. Castelló, J. Dweck, D A. G. Aranda: Kinetic study of thermal processing of glycerol by thermogravimetry. J. Therm. Anal. Calorim., 105 (3) (2011) 737–746; [3] M. L. Castelló, J. Dweck, D. A. G. Aranda, R. C. L. Pereira, M. J. R. Guimarães Neto: ZSM as a potential catalyst for glycerol pyrolysis. Accepted for publication on J. Sust. Bioen. Syst. 2014; [4] M. L. Castelló: Studying the glycerol through thermal analysis. M.Sc. Monograph. Advisors: Dr. J. Dweck and Dr. D. A. G. Aranda. UFRJ. 2009; [5] M. L. Castelló: Evaluation of the potential use of ZSM5 zeolite for glycerol pyrolysis. D.Sc. Thesis. Advisors: Dr. J. Dweck and Dr. D. A. G. Aranda. UFRJ. 2013.

Name: Éder Tadeu Gomes Cavalheiro
Country: Brazil
Date and place of birth: 1962, Ribeirão Preto/SP, Brazil
Present position and address: Associate professor—Level III, Instituto de Química de Sao Carlos, Universidade de Sao Paulo, Av. Trabalhador São-carlense, 400, 13566-590 São Carlos/SP, Brazil
Email: cavalheiro@iqsc.usp.br
Website: http://www5.iqsc.usp.br/pessoal/docentes/cavalheiro
Researcher ID: D-6339-2012
Education and scientific degrees: Universidade de São Paulo—Faculdade de Filosofia, Ciências e Letras de Ribeirão Preto—Bachelor in Chemistry, 1983; Universidade de São Paulo—Instituto de Física e Química de São Carlos—Máster Degree in Analytical Chemistry, 1987; Universidade de São Paulo—Instituto de Química de São Carlos—Doctoral Degree in Sciences (Analytical Chemistry), 1995
Workplaces: Associate Professor, level III, of Instituto de Química de São Carlos—USP, 2004—present; Assistant Professor of Instituto de Química de São Carlos—USP, 2002–2004; Assistant Professor of Universidade Federal de São Carlos, 1994–2002; Postdoctoral Research Associate with Professor A. Brajter-Toth, Department of Chemistry, University of Florida, Gainesville, EUA. April, 1997–July, 1998; Auxiliary Professor of Instituto de Física e Química de São Carlos, USP, 1987–1989
Main fields of interest: thermal analysis of pharmaceuticals, Schiff bases complexes and biopolymers/modified biopolymers
Relevant categories in thermal analyses: fields (inorganic, materials, complex, pharmaceutical, polymer); methods (TG, DTA, EGA, DSC)
Professional activities: Editorial board member of Journal of Thermal Analysis and Calorimetry. Secretary of the International Confederation on Thermal Analysis and Calorimetry, 2006–2012. Secretary of the Associação Brasileira de Análise Térmica e Calorimetria, 2000–2002 and Vice-president, 2002–2004. Member of the organizing and scientific committee of the Congresso Brasileiro de Análise Térmica e Calorimetria—Editions III, IV and V in 2002, 2004 and 2006, respectively
Publication record: papers (145), books (3 book chapters), patents (1), h-index (19)
Equipments: Simultaneous SDT Q600; Thermobalance Q-50, DSC Q-10, DSC Q-2000 and DMA Q-800 (all from TA Instruments), Nicolet iS10 with TG-FTIR coupling acessory (Thermo Scientific)
5 most important publications: [1] Felicio, R. C., Cavalheiro, E. T. G., Dockal, E. R., Polyhedron, 20 (2001), 261–268; [2] Guinesi, L. S., Cavalheiro, E. T. G., Thermochim. Acta, 444 (2006), 128–133; [3] Guinesi, L. S., Cavalheiro, E. T. G., Carbohydr. Polym., 65 (2006), 557–561; [4] Santos, J. E., Dockal, E. R., Cavalheiro, E. T. G., J. Therm. Anal. Calorim. 79 (2005) 243–248; [5] Simencio-Otero, R. L., Galvao, R. K. H., Araujo, M. C. U., Cavalheiro, E. T. G., Thermochim. Acta, 526 (2011) 200–204.

Name: Stefan Cebulak
Country: Poland
Data and place of birth: 1932, Skarszewy, Poland
Email: scebulak@wnoz.us.edu.pl
Education and scientific degrees: Gimnazjum and Liceum General 1945–1950; University of Warsaw 1950–1955; Dr Sc 1972
Workplaces: National Geological Istitute—Uper Silesian Division 1955–1980; Silesian University Geological Division 1980–1997 (to my retirement); Scientific works in University as volunteer for today
Main fields of interest: study in applying the oxyreactive thermal analysis (OTA) in geochemical explorations of all types of coals; applying this methods (OTA) in the study of the influence of chemo-physical conditions in transformation processes of all types of organic matter
Awards and acknowledgments: Silver Cross of Merit—resolution of the State Council, 1973; Gold Cross of Merit—resolution of the State Council, 1983; the First Degree State Award, 1972; Award for the complement of discovery of the Lublin coal basin and testification of the coal lode in category C_2
Professional activities: Member in the chairman committee in works on testification of the coal lode in the Lublin coal basin (1965–1972). Chairman in the study of the influence of chemo-physical conditions in transformation processes of all types of organic matter (permanent till today)
Publication record: papers (125), books (2)
5 most important publications: [1] S. Cebulak, J. Szwed-Lorenz, Group components of main brown-coal lithotypes: thermal reactivity and microscopic characterization, Elsevier Science Publishers B.V., Amsterdam 1990; [2] S. Cebulak, M. Pliński, A new application of oxyreactive thermal analysis in marine algological studies, Oceanology, No. 38 (1), 1996; [3] S. Cebulak, A. Karczewska, A. Mazurek, A. Langier-Kuźniarowa, Kerogen as a geochemical marker of the thermal history of rocks, Application of thermal analysis, Journal of Thermal Analysis, Vol. 48, 1997; [4] S. Cebulak, A. Langier-Kuźniarowa, I. Grotek, The application of the OTA of kerogen to the evaluation of the rock mass heating, Journal of Thermal Analysis and Calorimetry, Vol. 69, 2002; [5] S. Cebulak, B. Śmieja-Król, S. Duber, M. Misz, A.W. Morawski, Oxyreactive thermal analysis, A good tool for the investigation of of carbon materials, Journal of Thermal Analysis and Calorimetry, Vol. 77, 2004.

Name: Peggy Cebe
Country: USA
Date and place of birth: 1949, Erie, Pennsylvania USA
Present position and address: Professor of Physics, Tufts University, Physics Department, STC Rm. 208, 4 Colby Street, Medford, MA, 02155, USA
Email: peggy.cebe@tufts.edu
Website: http://www.tufts.edu/~pcebe/
Education and scientific degrees: BSEd, Edinboro State College of PA, 1970; MS, Mathematics, Edinboro State College of PA, 1976; MS, Physics, Cornell University, 1981; Ph.D., Physics, Cornell University, 1984
Workplaces: Caltech-NASA Jet Propulsion Laboratory (1984–1987); Massachusetts Institute of Technology (1988–1995); Tufts University (1995–)
Main fields of interest: use of advanced methods of thermal analysis to investigate transitions between the solid and liquid states in semicrystalline polymers and biopolymers
Relevant categories in thermal analyses: fields (polymer, biopolymer), methods (TG, DSC, fast scanning calorimetry)
Awards and acknowledgments: Fellow, North American Thermal Analysis Society (NATAS); Fellow, American Physical Society; Recipient of the 2010 Presidential Award for Excellence in Science, Mathematics, or Engineering Mentoring; NATAS Outstanding Achievement (Mettler-Toledo) Award 2013
Professional activities: Chair, Committee on the Status of Women in Physics, American Physical Society, 1998; Chair, American Chemical Society Division of Polymeric Materials: Science and Engineering, 2001; NATAS Conference Organizer, 2012; Guest Editor of JTAC Special Issue, 2013; Vice-President of NATAS, 2014
Publication record: number of papers (141), citations (3400); books (2), h-index (34), citation report (247)
Equipments: TA Instruments: Q100 DSC, Q500 TGA, Ares Rheometer, RSA III
5 most important publications: [1] Peggy Cebe and Su-Don Hong. "Crystallization Behavior of Poly(etheretherketone)," Polymer, 27(8), 1183–1192 (1986); [2] Xiao Hu, David Kaplan, and Peggy Cebe. "Determining Beta Sheet Crystallinity in Fibrous Proteins by Thermal Analysis and Infrared Spectroscopy." Macromolecules, 39, 6161–6170 (2006); [3] Jennifer Buckely, Peggy Cebe, Daniel Cherdack, Jennifer Crawford, B. Seyhan Ince, Matthew Jenkins, Jingjing Pan, Matthew Reveley, Niesha Washington, and Natalie Wolchover. "Nanocomposites of Poly(vinylidene Fluoride) with Organically Modified Silicate." Polymer, 47(7), 2411–2422 (2006); [4] Peggy Cebe, Xiao Hu, David Kaplan, Evgeny Zhuravlev, Andreas Wurm, Daniella Arbeiter, Christoph Schick. "Beating the Heat—Fast Scanning Melts Beta Sheet Crystals." Scientific Reports, 3, 1130; (2013); [5] Guokui Qin, Xiao Hu, Peggy Cebe and David L. Kaplan. "Molecular Mechanism of Resilin Elasticity." Nature Communications, 3, 1003 (2012).

Name: Eduard Cesari
Country: Spain
Date and place of birth: 1951, Barcelona, Spain
Present position and address: Professor, Dept. de Física, Universitat de les Illes Balears, Crtra. de Valldemossa, km 7.5 E-07122 Palma de Mallorca, Spain
Email: eduard.cesari@uib.cat
Website: www.uib.cat
ORCID: 0000-0002-5992-4369
Education and scientific degrees: Ph.D. in Physics, Univ. de Barcelona, 1979
Workplaces: Univ. de Bardelona, 1974–1986; Univ. Politècnica de Catalunya, 1976–1983; Univ. de les Illes Balears, 1986-present
Main fields of interest: solid-solid phase transformations; thermodynamics of solids; martensitic transformations; shape memory alloys
Relevant categories in thermal analyses: fields (materials); methods (TG, DTA, DSC, thermomechanical analysis, kinetics, specific heat, calorimetry)
Awards and acknowledgments: "Premio Extraordinario de Licenciatura", Univ. de Barcelona; "Premio Extraordinario de Doctorado", Univ. de Barcelona
Professional activities: European Symposium on Martensitic Transformations; Europ. Advisory Committee member (2002–); International Conference on Martensitic Transformations; Internat. Advisory Committee (2003–) member
Publication record: papers (290), patents (3), h-index (30)
Equipments: DSC, DMA (low and high frequency), Thermomechanical anal., DC and AC electr. resistance
5 most important publications: [1] V. A. Chernenko, E. Cesari, V. V. Kokorin, I. N. Vitenko: "The development of new ferromagnetic shape memory alloys in Ni-Mn-Ga system", Scripta Met. et Mater., 33 (1995) 1239–1244; [2] V. A. Chernenko, C. Seguí, E. Cesari, J. Pons, V. V. Kokorin: "Sequence of martensitic transformations in Ni–Mn–Ga alloys", Phys. Rev. B, 57 (1998) 2659–2662; [3] J. Pons, V. A. Chernenko, R. Santamarta, E. Cesari: "Crystal structure of martensitic phases in Ni-Mn-Ga shape memory alloys", Acta Materialia, 48 (2000) 3027–3038; [4] V. A. Chernenko, J. Pons, C. Seguí, E. Cesari: "Premartensitic phenomena and other phase transformations in Ni–Mn–Ga alloys studied by dynamical mechanical analysis and electron diffraction", Acta Materialia, 50 (2002) 53–60; [5] J. Pons, E. Cesari, C. Seguí, F. Masdeu, R. Santamarta: "Ferromagnetic shape memory alloys: Alternatives to Ni–Mn–Ga", Mat.Sci. Eng. A, 481–482 (2008) 57–65.

Name: Attilio Cesàro
Country: Italy
Date and place of birth: 1942, Nola, Italy
Present position and address: Professor, Dept. Chemical and Pharmaceutical Sciences University of Trieste, Via Giorgieri 1, 34127, Trieste, Italy
Email: cesaro@units.it
Website: http://www.dscf.units.it/cesaro
Education and scientific degrees: Dottore in Chimica, University of Naples (1966); Post-doc Researcher, University of California, Irvine (1967); Assistant Professor (1973); Associate Professor (1982); Professor (1990)
Workplaces: University of Naples Institute of Chemistry (1964–1966); Department of Chemistry University of California, Irvine (1967–1968); University of Trieste, Institute of Chemistry (1969–1985); University of Trieste Dept. BBCM (1985–2012) and Dept. Chemical and Pharmaceutical Sciences (2013–)
Main fields of interest: thermodynamics; structure, conformation, dynamics and functional properties of industrial biomacromolecules and biomolecules
Relevant categories in thermal analyses: fields ((bio)polymer, organic glasses, nano-pharmaceutical); methods (DSC, bio-DSC, isothermal microcalorimetry, kinetics)
Awards and acknowledgments: Fulbright (1975); Award Price of Biothermodynamics by Schweiz.Gesellschaft Thermoanal. Kalor. (Basel 1989); Award Prize AICAT-SETARAM (Porto, 2011)
Professional activities: Chairman and organizer of the ESTAC-6 in Grado, of the 5th (1983) and 32nd (2010) National AICAT Meetings (Trieste); President of AICAT (1990–1993), Chairman ESTAC (1994–1998); AICAT: member of the Board (1980–1984); AIMacromolecules: member of the Board (1985–1987); Regional Editorial, J. Thermal Analysis and Calorimetry (1998–2002); member of the Editorial Board: Acta Chemica Slovenica (1996–); Carbohydrate Research (2001–2008); Food Biophysics (2006–); editor and co-editor of several special issues (JTAC, TA, PAC, CR, FB)
Publication record: papers (200), books (1 book, 4 book chapters)
List of the 5 most important publications: [1] A. Cesàro, B. Bellich, M. Borgogna, Biophysical functionality in polysaccharides: from Lego-blocks to nano-particles, Eur. Biophys J 2012, 41, 379–395 (review); [2] L. Tavagnacco, U. Schnupf, P.E. Mason, M.-L. Saboungi, A. Cesàro, J. W. Brady, Molecular Dynamics Simulation Studies of Caffeine Aggregation in Aqueous Solution, J. Phys. Chem. B 2011, 115, 10957–10966; [3] A. Cesàro, Bioprotection: All dried up, Nature Materials 2006, 5, 593–594 (perspective); [4] A. Cesàro, F. Sussich, L. Navarini.Order-disorder conformational transitions of carbohydrate polymers. In "The Nature of Biological Systems as Revealed by Thermal Methods" (D. Lorinczy ed.), Kluwer Academic Publisher, 2004, Chapt 1; [5] F. Sussich, R. Urbani, F. Princivalle, A. Cesàro, Polymorphic amorphous and crystalline forms of trehalose, J. Am. Chem Soc., 31 (1998) 7893.

Name: Dorina-Rodica Chambre
Country: Romania
Date and place of birth: 1967, Chisineu-Cris, Arad, Romania
Present position and address: Associate Professor at University "Aurel Vlaicu" Arad; D.Chambre, 310272 Arad, Barzava str., no.105, sc.B, ap.17, Romania
Email: dorinachambree@yahoo.com
Researcher ID: B-9386-2013
Education and scientific degrees: Chemical Engineer, "Traian Vuia" Polytechnic Institute of Timisoara (1985–1990); Ph.D. (2004); University Assist. (1991–1997); University Lecturer (1997–2006); Assoc. Prof. (2006-present); Post university Fellowship from Italian Government, University of Trieste (2000); Post university Fellowship from Romanian Government, University of Milano (2002).
Workplaces: University Aurel Vlaicu of Arad, Faculty of Food Engineering, Tourism and Environmental Protection, Department of Techical and Natural Sciences
Main fields of interest: thermal stability, non-isothermal degradation kinetics and catalytically activity of synthetic cationites
Relevant categories in thermal analyses: fields (complex, organic, polymer, food); methods (TG, DTA, EGA, DSC, kinetics, microcalorimetry)
Awards and acknowledgments: "Start Cup Udine 2005" Award, University of Udine (2005); CNCSIS Awards (2008, 2009); "Virtute e Sapientia" Award, University of Bucharest (2008); Honorary degree of Romanian Chemistry Society (2011)
Professional activities: President of "Romanian Chemistry Society" Arad branch; Member of "Romanian Thermal Analysis Committee" from Romanian Academy; Member of the Association for Multidisciplinary Research in Western Romania; Main organizer of the Italian-Romanian conference "Strategie di internazionalizzazione e cooperazione allo sviluppo delle filiere agro-energetiche", 2–3 Feb. 2007, Arad, Romania
Publication record: papers: (68); books: (6); number of citations: (60)
Equipments: STA 409 Luxx system (Netzsch)
5 most important publications: [1] D. Chambre, A. M. Bodescu, C. Sîrghie, J. Therm. Anal. Cal., 112(2) (2013) 851; [2] D. Chambre, M. R. Szabo, C. Popescu, C. Idiţoiu, J. Therm. Anal. Cal., 94(2) (2008) 417; [3] D. Chambre, C. Idiţoiu, E. Segal, J. Therm. Anal. Cal. 88(3) (2007) 673; [4] D. Chambre, C. Idiţoiu, M. R. Szabo, J. Therm. Anal. Cal., 88(3) (2007) 681; [5] D. Chambre, C. Idiţoiu, E. Segal, A. Cesàro, J. Therm. Anal. Cal., 82(3) (2005) 803.

Name: Edward Leonard Charsley
Country: UK
Date and place of birth: 1939, London, UK
Present position and address: Emeritus Professor, School of Applied Sciences, University of Huddersfield, Queensgate, Huddersfield HD1 3DH, UK.
Email: e.l.charsley@hud.ac.uk
Website: http://www.hud.ac.uk
Education and scientific degrees: B.Sc. Special Honours Chemistry, University of London (1962). Ph.D., University of Salford (1990)
Workplaces: British Paper and Board Industry Res.Assocn. (1962–1963), Battersea College of Technology (1963–1966), Stanton Redcroft Ltd (1967–1987), Leeds Metropolitan University (1988–1998), University of Huddersfields (1998-present)
Relevant categories in thermal analyses: fields (reaction mechanisms of pyrotechnic systems, thermal analysis instrumentation, sample controlled thermal analysis, standardisation and calibration methods)
Awards and acknowledgments: Kurnakov Medal, Soviet Academy of Sciences (1991), Honorary Lifetime Membership ICTAC (2006), NATAS Mettler-Toledo Award (2007), NATAS Fellowship Award (2008), International Heat Flow Calorimetry Symposium on Energetic Materials Award (2012)
Professional activities: Chairman UK Thermal Methods Group Committee (1979–1981), Chairman ESTAC Committee (1981–1985), Chairman ICTAC Standardisation Committee (1988–1992), President ICTAC (1996–2000), Editorial Board Member of Thermochimica Acta (1994–2011), UK ICTAC Affiliate Councillor (1982–1988 and 2010-present), Chairman ICTAC Scientific Awards Committee (2010-present)
Publication record: papers (136), books (1), patents (2)
5 most important publications: [1] A New High Sensitivity Thermobalance, E. L. Charsley and A. C. F. Kamp, J. Therm. Anal. 7 (1975) 173; [2] DTA and temperature profile analysis of pyrotechnic delay systems: mixtures of tungsten and potassium dichromate, E. L. Charsley, M. C. Ford, D. E. Tolhurst, S. Baird-Parker, T. Boddington and P. G. Laye, Thermochim. Acta, 25 (1978) 131; [3] Quantitative studies on the zirconium-potassium perchlorate-nitrocellulose pyrotechnic system using DSC and chemical analysis, B. Berger, E. L. Charsley, J. J. Rooney and S. B. Warrington, Thermochim. Acta, 255 (1995) 227; [4] An investigation of strontium nitrite and its role in the ageing of the magnesium-strontium nitrate pyrotechnic system using isothermal microcalorimetry and thermal analysis techniques, I. M. Tuukkanen, E. L. Charsley, S. J. Goodall, P. G. Laye, J. J. Rooney, T. T. Griffiths and H. Lemmetyinen, Thermochim. Acta, 443 (2006) 116; [5] Development and Applications of a Sample Controlled DSC System, E. L. Charsley, P. G. Laye, G. M. B. Parkes and J. J. Rooney, J. Therm. Anal. Calorim. 105 (2011) 699.

Name: Richard P. Chartoff
Country: USA
Date and place of birth: 1939, Rochester, New York USA
Present position and address: Research Professor, Director, CAMCOR Polymer Characterization and Thermal Analysis Laboratory, University of Oregon, Materials Science Institute, Department of Chemistry and Biochemistry, 1253 Franklin Blvd., Eugene, Oregon 97403 USA
Email: chartoff@uoregon.edu
Website: http://Camcor.uoregon.edu
Education and scientific degrees: BS Chemical Engineering, Case-Western Reserve University, 1961; Ph.D. Princeton University Chemical Engineering/Polymer Engineering, 1968
Workplaces: University of Cincinnati, 1968–1978; University of Dayton 1978–2008; University of Oregon 2009 to present
Main fields of interest: use of thermal analysis in applications and processing of polymers and polymer composites, polymer science and engineering, polymer structure-property relationships; rheology; polymer matrix nano-composites
Awards and acknowledgments: Outstanding Engineering Award from Engineering Society of Dayton Ohio, Fellow of Royal Norwegian Council for Scientific and Industrial Research, Fellow of North American Thermal Analysis Society (NATAS), Research Grantee of the Bill and Melinda Gates Foundation
Publication record: papers (127), books (3 book chapters), patents (10)
Equipment: DSC, TG, TMA, DMA, DEA
5 most important publications: [1] "The Effect of Oxygen on Cross-linking and Mechanical Properties of a Thermoset Formed by Free Radical Photocuring", J. Appl. Polymer Sci. 119 (4), 2359-2370 (2011); [2] "Thermal Characteristics of Thermosets Formed by Free Radical Photocuring", Journal of Thermal Analysis and Calorimetry, 85 (1), 213–217 (2006); [3] "Particle Surface Treatment for Nanocomposites Containing Ceramic Particles", Composite Interfaces, special issue honoring Professor Bela Pukanszky, 13(8–9), 801–817 (2006); [4] "The Influence of Cure Conditions on the Morphology and Phase Distribution in a Rubber Modified Epoxy Resin Using Scanning Electron Microscopy and Atomic Force Microscopy", Polymer, 46, 785–798 (2005); [5] "Polymerization and Viscoelastic Behavior of Networks from a Dual-Curing, Liquid Crystal Monomer", J. Polym. Sci.: Part B, Polym. Phys., 37, 1183–1190 (1999).

Name: Stephen Z. D. Cheng
Country: USA
Date and place of birth: 1949, Shanghai, China
Present position and address: Robert C. Musson and Trustees Professor and Dean, College of Polymer Science and Polymer Engineering, The University of Akron
Email: scheng@uakron.edu
Education and scientific degrees: Methamatics (Diploma, 1977), Polymer Engineering (M.S., 1981), Chemistry (Ph.D. 1985, Rensselaer Polytechnic Institute)
Workplaces: The University of Akron, Akron, Ohio, USA
Main fields of interest: his research interests are in the area of chemistry, physics and engineering of polymers and advanced functional materials including ordered structure, morphology, phase transition thermodynamics, kinetics, and molecular motions; his recent interests in particular, are focusing on nano-hybrid materials with different molecular chemical structures and physical topologies, architectures and interactions and their assemblies in the bulk, solution and thin films; he is also active in developing researches of conducting polymers, photovoltaics, polymer optics and photonics
Relevant categories in thermal analyses: fields (polymers, materials); methods (DSC, TG, DTA, thermomechanical analysis)
Awards and acknowledgments: Presidential Young Investigator Award (1991), John H. Dillon Medal (APS, 1995), NATAS Outstanding Achievement (Mettler-Toledo) Award (1999), TA-Instrument Award (ICTAC, 2004), PMSE Cooperative Research Award (ACS, 2005), Polymer Physics Prize (APS, 2013), Fellow of AAAS and APS and an Honorable Fellow of Chinese Chemical Society. Member of the National Academic of Engineering of US (2008)
Publication record: papers (450), books (2 and many chapters), patents (10), citations (12500), h-index (60)
5 most important publications: [1] Li, C. Y.; Cheng, S. Z. D.; Ge, J. J.; Bai, F.; Zhang, J. J.; Harris, F. W.; Chien, L.-C.; Yan, D.; He, T.; Lotz, B. Double twist in helical polymer "soft" crystals (featured in APS Physical Forces). Physical Review Letter, 1999, 83, 4558-4561; [2] Zhu, L.; Cheng, S. Z. D.; Calhoun, B. H.; Ge, Q.; Quirk, R. P.; Thomas, E. L.; Hsiao, B. S.; Yeh, F.; Lotz, B. Crystallization temperature-dependant crystal orientations within nanoscale confined lamellae of a self-assembled crystalline-amorphous diblock copolymer. Journal of the American Chemical Society, 2000, 122, 5957–5967; [3] Cheng, S. Z. D. Phase transitions in polymers: The role of metastable states. pp. 324, Elsevier: Amsterdam 2008; [4] Yu, X.; Zhong, S.; Li, X.; Tu, Y.; Yang, S.; Van Horn, R. M.; Ni, C.; Pochan, D. J.; Quirk, R. P.; Wesdemiotis, C.; Zhang, W.-B.; Cheng, S. Z. D. A giant surfactant of polystyrene-(carboxylic acid-functionalized polyhedral oligomeric silsesquioxane) amphiphiles with highly stretched polystyrene tails in micellar assemblies. Journal of American Chemical Society, 2010, 132, 16741–16744; [5] Yu, Y.; Yue, K.; Hsieh, I.-F.; Li, Y.; Dong, X.-H.; Liu, C.; Xin, Y.; Wang, H.-F.;Shi, A.-C.; Newkome, G. R.; Ho, R.-M.; Chen, E.-Q.; Zhang, W.-B.; Cheng, Z. D. C. Giant surfactants provide a versatile platform for sub-10-nm nanostructure engineering. Proceedings of the National Academy of Sciences of the United States of America, 2013, 110, 10078–10083.

Name: Yi Cheng
Country: P.R. China
Date and place of birth: 1952, P.R. China
Present position and address: Nanjing Unicersity of Science and Technology, Nanjing 210094, P.R. China
Email: chengyi20@aliyun.com
ORCID: 0000-0002-9001-8655
Education and scientific degrees: Chemical professor
Workplaces: Nanjing Unicersity of Science and Technology
Main fields of interest: Energetic materials, kinetics of thermal analysis
Relevant categories in thermal analyses: Fields (kinetics); methods (differential kinetic equations, integral kinetic equation)
Awards and acknowledgments: Second Award for Science and Technology Progress, Ministry of Industry and Imformayion Technology of P.R. China, 2009; Third Award for Science and Technology Progress, Ministry of Industry and Imformayion Technology of P.R. China, 2008; Third Award for Science and Technology Progress (Yunnan Province of P.R. China, 1994
Professional activities: Vice chairman of the Committee on thermal analysis, Jiangsu Province, P.R. China; Vice chairman, 2010 Nanjing International Thermal Analysis Kinetics forum (2010)
Publication record: papers (81), books (1), patents (11)
Equipments: Netzsch STA449C TG-DSC, Netzsch QMS403C,Nisolet 6700-FTIR
5 most important publications: [1] Yi Cheng, A New Kinetic Equation for Non-Isothermal Reaction Process. Chinese Journal of Inorganic Chemistry, 2006, 2, 22(2): 287–292; [2] Yi Cheng, Yanchun Li, Shi Yan, Chuan Huang, Deviation of activation energy caused by neglecting a temperature term in Ozawa Equation. J. Math Chem, 2010, 48: 704–713; [3] Yi Cheng, A kinetic method in simultaneous thermal analysis, Analytical Methods, 2010, 2: 1255–1257; [4] Yi Cheng, Li Yanchun, Hui Yunlong. The correctional kinetic equation for the peak temperature in the differential thermal analysis. Journal of Thermal Analysis and Calorimetry. 2008.07, 93(1): 111–113; [5] Chuan Huang, Xinliang Mei, Yi Cheng, Yanchun Li, Xuqiang Zhu, A model free method for evaluating theoretical error of Kissinger equation Journal of Thermal Analysis and Calorimetry. 2014, doi:10.1007/s10973-013-3624-2.

Name: Ionel Chicinaş
Country: Romania
Date and place of birth: 1955, Micesti-Cluj County, Romania
Present position and address: Professor, Head of Department of Materials Science and Engineering, Technical University of Cluj-Napoca, 103–105 Muncii Ave., 400641 Cluj-Napoca, Romania
Email: ionel.chicinas@stm.utcluj.ro
Website: http://www.imm.utcluj.ro/research/matmagnano/
Researcher ID: A-5139-2009
Education and scientific degrees: Physicist—Babeş-Bolyai Univ. (1974–1979), Engineer—Technical Univ. of Cluj-Napoca (1986–1991), Ph.D. (1998), Professor (2004)
Workplaces: Inst. Nuclear Power Reactors—Pitesti (1979–1983), Department of Materials Science and Engineering, Technical University of Cluj-Napoca (1983–)
Main fields of interest: Magnetic and nanocrystaline/nanosized materials, mechanosynthesis
Relevant categories in thermal analyses: Fields (materials, nano, ceramics); methods (DTA, DSC, TG)
Awards and acknowledgments: Excelence Award of Technical University of Cluj-Napoca
Professional activities: Invited professor/researcher—Université Joseph Fourier and Néel Institute, Grenoble, France, 5 stages: 2004, 2006, 2008, 2010, 2013; Editorial Board member —Adv. Mater. Research; Evaluating expert for Czech Science Foundation and for Hungarian Scientific Research Fund (OTKA); Int. Committee member of the 6th Int. Powder Metall. Conf., Oct. 2011, Ankara, Turkey, Sci. Committee—2nd C.E. European Conf. Thermal Anal. Calorim., Vilnius, August. 2013, Invited lecture at European School on Magnetism, Sept. 2007, Cluj-Napoca; Keynote at Materiaux 2010 Conf., Oct. 2010, Nantes, France; Peer to peer regular reviewer for 12 ISI journals, conference organiser: 2nd Int. Conf. (MATEHN '02 şi MATEHN '06, Cluj-Napoca)—General Secretary, 4th Int. Conf. Powder Metallurgy (RoPM 2009, Craiova)—Chairman
Publication record: published papers (190, 65 ISI ranked journals), published book (4), Editor of books (2), patents (1), citations (451), h-index (11), sum of impact factors (89.711)
Equipments: DTA—Labsys Setaram, TG+DSC—Labsys Setaram
5 most important publications: [1] I. Chicinaş, V. Pop, O. Isnard, J. M. Le Breton, J. Juraszek, J. Alloys Compd. 352 (2003) 34; [2] I. Chicinaş, J. Optoelectron Adv. Mater. 8 (2006) 439—review; [3] B. V. Neamţu, O. Isnard, I. Chicinaş, C. Vagner, N. Jumate, P. Plaindoux, Mater. Chem. Phys. 125 (2011) 364; [4] T. F. Marinca, I. Chicinaş, O. Isnard, V. Popescu, J. Am. Ceram. Soc. 96 (2013) 469; [5] B. V. Neamţu, I. Chicinaş, O. Isnard, et al., J. Magn. Magn. Mater. 353 (2014) 6.

Name: Benoy B. Chowdhury
Country: USA
Date and place of birth: 1930, Calcutta, India
Present position and address: Retired, Consultant. 6 Lotus Terrace, Lake Ariel, PA 18436, USA
Email: matech2@verizon.net
Website: www.askmatech.webs.com
Education and scientific degrees: Chemistry/Materials Science and Engineering, Univ. of London, University of Waterloo, Calif. Western University. B.Sc., M.Sc., Ph.D., C. Chem., MRSC (ret.)
Workplaces: Kay Laboratories, Bristol Laboratories, Schering Corporation, M&T Chemicals, Matech Associates (A wholly owned Technology Laboratory for Fee-For Service Materials Testing and Research)
Main fields of interest: thermal and thermo-electric behavior of materials
Relevant categories in thermal analyses: fields (inorganic, synthetic materials, polymers, organic, composites, ceramics, cement, cotton); methods [TG, DSC, TMA, sp. heat, thermal conductivity, thermally stimulated current measurement (TSC)]
Awards and acknowledgments: Invited speaker: Brookhaven National Laboratories, Long Island, NY; Delaware Valley TA Forum at the Univ. of Pennsylvania, Philadelphia, PA; Stevens Inst. of Technology Graduate School, Hoboken, NJ.; Slovak Chemical Society/ Slovak Technical Univ., Bratislava, Slovakia
Professional activities: Chairman, ACS North Jersey Thermal Analysis Topical Group; Program, Session and Publications Chairman, NATAS; Advisory Board Member, Test and Measurement World; R&D Council Member of Research and Development Journal; Advisory Team Member of International Superconductor Corporation; Presenter at ICTAC Conferences; Referee JTAC
Publication record: papers (29), books (1), patents (1)
Equipments: TG, TMA, DMA (Cryogenic and High Temp.), DSC (Standard, High Temp. and High Pressure), High Temp. Liquid and Solid Cells coupled to high Capacitance Bridges for High Temp. Studies of Dielectrics, Laboratory-Designed Instrument for Thermo-Electric Polarization Studies.
5 most important publications: [1] "Thermal Studies of Compositional Variations of Some Novel Silicone Polyimides" in "Synthesis, Characterization and Applications," K. L. Mittal, ed., Plenum Press (N.Y.) 1984, 401–415; [2] "Thermal Evaluation of Transgenic Cotton containing Polyhydroxybutyrate," Thermochim.Acta, 313 (1998) 43–53; [3] "The role of Thermal analysis in Chemistry and Materials science" in "Challenges for Coordination Chemistry in the New Century," M. Melnik and A. Sirota, eds., Slovak Technical University Press, Bratislava, Slovakia, 2001; [4] "Thermal and Polarization studies of carbonaceous Cement," Trans Tech Publications (Switzerland) Vols. 90–91 (2003) 25–32; [5] "Scope of Electron Transport Studies by Thermally Stimulated Discharge Current Measurement," J. Therm. Anal. Calorim., vol. 73 (2003) 53–57.

Name: Konstantinos Chrissafis
Country: Greece
Date and place of birth: 1957, Thessaloniki, Greece
Present position and address: Prof. Chrissafis K., Solid State Physics Dept., School of Physics, Aristotle University of Thessaloniki, 54124 Thessaloniki, Greece
Email: hrisafis@physics.auth.gr
Website: http://ftirlab.physics.auth.gr/Chrissafis.html
Education and scientific degrees: Physicist, Ph.D. in Physics, Professor
Workplaces: Solid State Physics Department, School of Physics, Aristotle University of Thessaloniki
Main fields of interest: crystallization of amorphous materials, phase transitions and degradation of materials, thermal characterization of materials in the temperature range 100 till 1,750 °C, kinetic study of phase transformation and decomposition
Relevant categories in thermal analyses: fields (materials, nano, minerals, pharmaceutical, polymer, glass, ceramics, cement); methods (TG, DTA, DSC, thermomechanical analysis, kinetics, extremely high temperature (above 1,000 °C); specific heat)
Professional activities: member of the Editorial Board of Thermochimica Acta, President of the Hellenic Society of Thermal Analysis (HSTA), national representative in the ESTAC committee
Publication record: papers (135), citations (1245), h-index (22)
Equipments: SETARAM DSC 141, SETARAM SetSys TG-DTA 16/18, PerkinElmer Diamond DMA, PerkinElmer Pyris-1 Diamond DSC with a PerkinElmer Intracooler II
5 most important publications: [1] Chrissafis, K., Antoniadis, G., Paraskevopoulos, K. M., et al., Comparative study of the effect of different nanoparticles on the mechanical properties and thermal degradation mechanism of in situ prepared poly(E-caprolactone) nanocomposites, Composites Science andTechnology, 67(10), 2007, 2165–2174; [2] Bikiaris, D.; Vassiliou, A.; Chrissafis, K.; et al., Effect of acid treated multi-walled carbon nanotubes on the mechanical, permeability, thermal properties and thermo-oxidative stability of isotactic polypropylene, Polymer Degradation and Stability, 93 (5), 2008, 952–967; [3] Chrissafis, K; Paraskevopoulos, K. M., Bikiaris, D. N., Thermal degradation mechanism of poly(ethylene succinate) and poly(butylene succinate): Comparative study, Thermochimica Acta, 435(2), 2005, 142–150; [4] Chrissafis, K., Paraskevopoulos, K.M., Bikiaris, D. N., Thermal degradation kinetics of the biodegradable aliphatic polyester, poly(propylene succinate), Polymer Degradation and Stability 91(1), 2006, 60–68; [5] Chrissafis, K., Kinetics of thermal degradation of polymers, Journal of Thermal Analysis and Calorimetry, 95(1), 2009, 273–283.

Name: Mária Chromčíková
Country: Slovakia
Date and place of birth: 1967, Považská Bystrica, Czechoslovakia
Present position and address: Institute of Inorganic Chemistry of the Slovak Academy of Sciences, VILA Glass Center, Študentská 2, SK-911 50 Trenčín
Email: maria.chromcikova@tnuni.sk
Website: www.uach.sav.sk
Education and scientific degrees: Ing.—2002, A. Dubček University of Trenčín; Ph.D.—2007, Institute of Inorganic Chemistry SAS, Bratislava
Workplaces: Institute of Inorganic Chemistry, Slovak Academy of Sciences, Bratislava (1986–); A. Dubček University of Trenčín, VILA Glass Center, Head of the laboratory of thermal methods.
Main fields of interest: physical chemistry, glass science and technology
Relevant categories in thermal analyses: fields (inorganic, glass, ceramics, materials); methods (DTA, DSC, TMA, physical properties, crystallization kinetics)
Professional activities: Member of the Technical Committee TC03 of the International Commission on Glass
Equipments: TMA Q400 EM (TA Instrument), STA 449 F1 Jupiter (Netzsch), TMA 402 F1 Hyperion (Netzsch), TMA 402 (Netzsch), DILATOMETER 402 EP (Netzsch), Renishaw inVia Reflex Raman spectrometer with LINKAM TS1500 and DSC600 heating stages.
Publication record: papers (49), books (1 book chapter), citations (38)
5 most important publications: [1] M. Chromčíková, M. Liška: Viscosity and structural relaxation of $15Na_2O{\cdot}xMgO{\cdot}(10-x)CaO{\cdot}75SiO_2$ glasses, J. Therm. Anal. Calorimetry, 90, 421–429 (2007); [2] M. Chromčíková, M. Liška: Stress strain testing of the strand of E-glass fibers, Adv. Materials Research, 39–40, 165–168 (2008); [3] M. Liška, M. Chromčíková: Thermal properties and related structural and thermodynamic studies of oxide glasses. In: Šesták J, Holeček M, Málek J, Editors. Glassy, amorphous and nanocrystalline materials: thermal physics, analysis, structure and properties, Chap. 11. New York: Springer; 2011. pp. 179–97; [4] M. Chromčíková, M. Liška, J. Macháček, J. Šulcová: Thermodynamic model and structure of CaO-P2O5 glasses. J. Therm. Anal. Calorimetry, 144, 757–789 (2013); [5] M. Chromčíková, M. Liška, J. Holubová, Z. Černošek: Structure of As_2S_3–Sb_4S_4 Glasses by Combined Raman Spectroscopy and Thermodynamic Modeling Approach. J. Non-Cryst. Solids (2013) doi:10.1016/j.jnoncrysol.2013.12.015.

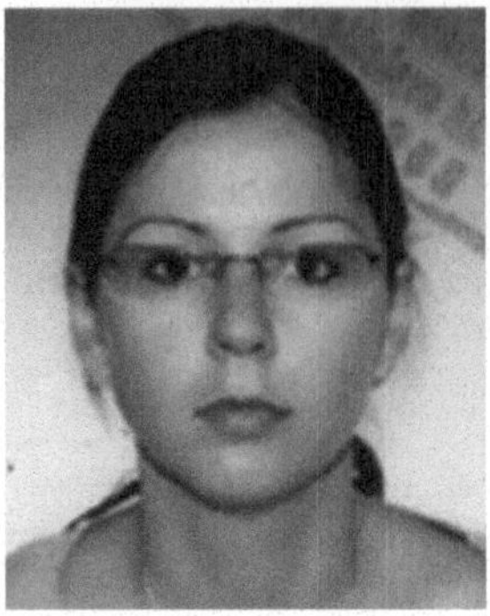

Name: Zuzana Cibulková
Country: Slovakia
Date and place of birth: 1980, Levice, Slovak Republic
Present position and address: Institute of Physical Chemistry and Chemical Physics, Faculty of Chemical and Food Technology, Slovak University of Technology, Bratislava, Slovakia
Email: zuzana.cibulkova@stuba.sk
Education and scientific degrees: Chemical Engineer, Faculty of Chemical and Food Technology, Slovak University of Technology, Bratislava (1998–2003), Ph.D. (2006)
Workplaces: Istitute of Physical Chemistry and Chemical Physics, Faculty of Chemical and Food Technology, Slovak University of Technology, Bratislava, Slovakia (2007–)
Main fields of interest: thermodegradation, thermooxidation and stabilization of polymers and organic materials; kinetics and thermodynamics of the processes in solid state; kinetics of the reactions exhibiting the induction period.
Relevant categories in thermal analyses: fields (materials, polymers, food); methods (DSC, DTA, TG)
Publication record: papers (21), citations (120)
Equipments: PerkinElmer DSC 7, Shimadzu DSC-60, Seiko EXSTAR 6300
5 most important publications: [1] Z. Cibulková, P.Šimon, P. Lehocký, J. Balko: Antioxidant activity of p-phenylenediamines studied by DSC. Polym. Deg. Stab. 87, 479–486 (2005); [2] P. Šimon, Z. Cibulková, P. Thomas: Accelerated thermooxidative ageing tests and their extrapolation to lower temperatures. J.Therm. Anal. Cal. 80, 381–385 (2005); [3] Z. Cibulková, P.Šimon, P. Lehocký, J. Balko: Antioxidant activity of 6PPD derivatives in polyisoprene matrix studied by non-isothermal DSC measurements. J. Therm. Anal. Cal. 80, 357–361 (2005); [4] A. Gregorová, Z. Cibulková, B. Košíková, P. Šimon: Stabilization effect of lignin in polypropylene and recycled polypropylene. Polym. Deg. Stab. 89, 553–558 (2005); [5] Šimon P., Hynek D., Malíková M., Cibulková Z.: Extrapolation of accelerated thermooxidative tests to lower temperatures applying non-arrhenius temperature functions. J. Therm. Anal. Cal., 93, 817–821(2008).

Name: Gülbanu Koyundereli Çılgı
Country: Turkey
Date and place of birth: 1980, İzmir-Türkiye
Present position and address: Assistant Professor Doctor, Pamukkale University Technology Faculty, Materials Science and Engineering Department, Denizli/TURKEY PO 286 20020
Email: gkoyundereli@pau.edu.tr, gkcilgi@gmail.com
Website: http://www.pau.edu.tr/gkoyundereli/tr
ORCID: 0000-0002-0016-019X

Education and scientific degrees: B.Sc.: Karadeniz Technical University Chemistry Department—2001; MS: Pamukkale University Chemistry Department—2004; Ph.D.: Pamukkale University Chemistry Department—2012
Workplaces: Research Assistant, Pamukkale University, Chemistry Department, 2001–2012; Dr. Research Assistant, Pamukkale University, Chemistry Department, 2012–2013; Assistant Professor Doctor (as a Head of Department), Pamukkale University, Materials Science and Engineering Department, 2013–
Main fields of interest: thermogravimetry, thermal behaviour of organic compounds and textile fibers, decomposition kinetics, modeling of decomposition reactions
Relevant categories in thermal analyses: fields (organic and inorganic compounds, polymer, textile fibers, marbles, ceramics); methods (TG, DTA, DSC)
Professional activities: Scietific Committee Member in 2nd Central and Eastern European Conference on Thermal Analysis and Calorimetry CEEC-TAC2; Organization Committee Member in 1st International Symposium on Secondary Metabolites Chemical, Biological and Biotechnological Properties; Organization Committee Member in 4th Physical Chemistry Congress
Publication record: papers (13), citation index (56), h-index (5)
Equipments: Simultaneous TG-DTA (Shimadzu DTG-60H) and Shimadzu DSC-60
5 most important publications: [1] H. Deligöz, Ö. Özen, G. K. Çılgı and H. Çetişli, Thermochimica Acta, 426, 33–38, 2005; [2] G.K Çılgı, H. Cetişli, J Therm Anal Calorim., 98, 855–861, 2009; [3] H. Cetişli, G. K. Çılgı, R. Donat, J Therm Anal Calorim., 108, 1213–1222, 2012; [4] G. K. Çılgı, H. Cetişli, R. Donat, J Therm Anal Calorim., 110, 127–135, 2012; [5] M. Ak, G. K. Çılgı, F. D. Kuru, H. Cetişli, J Therm Anal Calorim., Vol 111, 1627–1632, 2013.

Name: Marie-Vanessa Coulet
Country: France
Date and place of birth: 1974, France
Present position and address: CNRS Researcher—Aix Marseille University, MADIREL Laboratory, Centre de St Jérôme, 13397 Marseille Cedex 20, France
Email: vanessa.coulet@univ-amu.fr
Website: www.madirel.fr
Education and scientific degrees: Physics, University of Provence (1991–1996), Ph.D. in Material Science (1996–1999)
Workplaces: CNRS Research Center for Thermodynamics and Microcalorimetry, Marseille (1996–1999)—Cristallography Laboratory (Université Joseph Fourier and CNRS), Grenoble (1999–2000)—TECSEN (CNRS and Université Paul Cezanne), Marseille (2000–2008), IM2NP (CNRS and Aix Marseille Université), Marseille (2008–2013), MADIREL (CNRS and Aix Marseille Université) Marseille, (2013–)
Main fields of interest: phase transitions and structure in non-crystalline systems (glasses and liquids); thermal stability, reactivity and structure in metallic nanopowders
Relevant categories in thermal analysis: fields (inorganic, nanopowders, phase change materials, liquids, glasses); methods (DSC, TGA, combined DSC-TGA, combined DSC-EXAFS)
Professional activities: Leader of a project funded by the French National Agency (ANR-ASTRID-ANPA) on "Aluminium nanoflakes for Propulsion applications". Member of a project funded by the French National Agency (ANR-Blanc-TEAM) on "New functionalities of tellurium based amorphous materials"
Publication record: papers (32)
Equipments: Differential scanning calorimeters, Thermogravimetric apparatus, Combined DSC-ATG apparatus (up to 1,600 °C)
5 most important publications: [1] B. Rufino, F. Boulc'h, M.-V. Coulet, G. Lacroix, R. Denoyel, Influence of particles size on thermal properties of aluminium powder, Acta Mat. 55 (8) pp 2815–2827 (2007); [2] B. Rufino, M.-V. Coulet, R. Bouchet, O. Isnard, R. Denoyel, Structural changes and thermal properties of aluminium micro- and nanopowders, Acta Mat., 58, 4224–4232 (2010); [3] P. Zalden, G. Aquilanti, C. Prestipino, O. Mathon, B. André, M. Wuttig, M.-V. Coulet, Simultaneous calorimetric and quick-EXAFS measurements to study the crystallization process in phase-change materials, J. Synchr. Radiation, 19, 806–813 (2012); [4] B. André, M.-V. Coulet., M. Dumont, J. Rogez, V. Heresanu, B. Rufino, R. Bouchet, R. Denoyel, Morphology and reactivity of aluminium nanocrystalline powders, Int. J. Nanotechnology, 9, pp 618–629 (2012); [5] B. André, M.-V. Coulet, P.-H. Esposito, B. Rufino, R. Denoyel, High-energy ball milling to enhance the reactivity of aluminum nanopowders, Materials letters, 110 pp 108–110 (2013).

Name: José M. Criado
Country: Spain
Date and Place of birth: 1944, Sevilla, Spain
Present position and address: Professor and Head of the group of "Solid State Chemistry" at the Institute of Material Sciences of Sevilla, C.S.I.C.-University of Sevilla
Email: jmcriado@icmse.csic.es
Education and scientific degrees: Master in Chemistry, University of Sevilla 1966; Ph.D. in Chemistry, University of Sevilla, 1970
Workplaces: Department of Inorganic Chemistry of the University of Sevilla (1966–1972); Institute of Physics and Chemical Researches (1972–1985); Visiting Professor University of Stanford (1975); Institute of Material Sciences of Sevilla (ICMSE) (1986–)
Main fields of interest: heterogeneous catalysis; kinetics ofsolid state reactions; synthesis of materials with controlled texture and structure; mechanochemistry
Relevant categories in thermal analyses: fields (nanomaterials, polymers and ceramics); methods (TG, DTA, DSC, EGA, XRD, SEM, TEM, mechanosynthesis from high energy grinding mills)
Awards and acknowledgments: National Spanish Award "Alfonso X el Sabio" (1972); Honor Medal of the Pardubice University, Pardubice (Czech Republic) (2002)
Professional activities: ICTAC Councilor at the ICTAC council for the periods 1989-1996. He has served as member of the ESTAC Board and the Spanish Group of Calorimetry and Thermal Analysis (GECAT). Regional Editor of J. Thermal Anal. Calorim. from 1984 to 2002. Member of the Honorary Board of J. Thermal Anal. Calorim. from 2003 and member of the Editorial Board of Thermochim. Acta since 2003. He has served as member of the Organizing and Scientific Committees of a large number of International Congresses connected with his fields of interest
Publication records: papers (271), books (2), citation index (223), h-index (37)
Equipments: Homemade TG systems equipped with Mass and IR spectrometers able to work from high vacuum up to 20 bars at temperatures up to 1,500 °C. Simultaneous TG-DTG TA Instruments, model 6000. Thermobalance (TA Instruments 2000). DSC (TA Instruments 200). High pressure DSC (TA Instrument 20). Thermomechanical Analyser Lynseis. Home made dilatometer operating up to 1,600 °C under high vacuum and controlled atmosphere
List of the 5 most important publications: [1] J. M. Trillo, G. Munuera and J. M. Criado: Catalytic decomposition of formic acid on metal oxides; Catal.Rev. 7 (1972) 51; [2] J. M. Criado, A. Ortega and F. Gotor: Correlation between the shape of Controlled Rate Thermal Analysis curves and the kinetics of Solid State Reactions; Thermochim. Acta 157 (1990) 171; [3] P. E. Sánchez-Jiménez, L. A. Pérez-Maqueda, A. Perejón and J. M. Criado: Quantitative characterization of Multicomponents Polymers by Sample Controlled Thermal Analysis; Anal. Chem. 82 (2010) 8875; [4] S. Vyazovkin, A. K. Burnham, J. M. Criado, L. A. Pérez-Maqueda, C. Popescu, and N. Sbirrazzuoli: ICTAC Kinetics Committee recommendations for performing kinetic computations on thermal analysis data; Thermochim. Acta 526 (2011) 1; [5] P. Balaz, M. Achimovocova, M. Balaz, P. Billik, Z. Cherkezova-Zheleva, J. M. Criado, F. Delogu, E. Dutkova, E. Gaffet, F. J. Gotor, R. Kumar, I. Mitov, T. Rojac, M. Senna, A. Streletskii and K. Wieczorek-Ciurowa; Chem Soc. Rev.42 (2013) 7571.

Name: Zsuzsanna Czégény
Country: Hungary
Date and place of birth: 1975, Budapest, Hungary
Present position and address: senior research fellow, Institute of Materials and Environmental Chemistry, Research Centre for Natural Sciences, Hungarian Academy of Sciences, 1117 Budapest, Magyar tudósok körútja 2, Hungary
Email: czegeny.zsuzsanna@ttk.mta.hu
Website: www.ttk.mta.hu
ORCID: 0000-0001-9636-9547
Education and scientific degrees: Chemist (M.Sc.), Eötvös Loránd University, Faculty of Natural Sciences, Budapest, Hungary (1998); Ph.D. Eötvös Loránd University, Faculty of Natural Sciences, Budapest, Hungary (2003)
Workplaces: Institute of Material and Environmental chemistry, Research Centre for Natural Sciences, Hungarian Academy of Sciences, Budapest, Hungary (1998–)
Main fields of interest: thermal decomposition of polymers, biomass materials and their mixture; effect of additives and role of the components under the thermal decomposition process of mixtures
Relevant categories in thermal analyses: fields (polymer, biomass, additives); methods (TG, TG-MS)
Publication record: papers (25), citations (256), h-index (11), sum of impact factors (42)
Equipments: TG-MS; pyrolysis-GC/MS
5 most important publications: [1] Zs. Czégény, E. Jakab, M. Blazsó: Pyrolysis of wood, cellulose, lignin–brominated epoxy oligomer flame retardant mixtures, J. Anal. Appl. Pyrol. 103: pp. 52–59. (2013); [2] Zs. Czégény, E. Jakab, M. Blazsó, T. Bhaskar. Y. Sakata: Thermal decomposition of polymer mixtures of PVC, PET and ABS containing brominated flame retardant: Formation of chlorinated and brominated organic compounds, J. Anal. Appl. Pyrol. 96: pp. 69–77. (2012); [3] Zs. Czégény, M. Blazsó: Effect of phosphorous flame retardants on the thermal decomposition of vinyl polymers and copolymers, J. Anal. Appl. Pyrol. 81(2): pp. 218–224. (2008); [4] J. Bozi, Zs. Czégény, M. Blazsó: Conversion of the volatile thermal decomposition products of polyamide-6,6 and ABS over Y zeolites, Thermochimica Acta 472(1–2): pp. 84–94. (2008); [5] Blazso, M., Czegeny, Z., Csoma, C., Pyrolysis and debromination of flame retarded polymers of electronic scrap studied by analytical pyrolysis, J. Anal. Appl. Pyrol. 64(2): pp. 249–261. (2002).

Name: Dorin Nicolae Dadarlat
Country: Romania
Date and place of birth: 1953, Cluj-Napoca, Romania
Present position and address: Senior Scientist I, Head of Photothermal Phenomena Group, National RD Institute for Isotopic and Molecular Technologies, Donath Str. 65-103, Cluj-Napoca, Romania
Email: ddadarlat@gmail.com, dadarlat@itim-cj.ro
Website: webmail.itim-cj.ro
Education and scientific degrees: "Babes-Bolyai" University, Cuj-Napoca, (1972–1977) MS in Physics; Ph.D. in physics in 1982. Specializations in laboratories of other countries: Lebedev Institute of Physics, Moscow; Tor Vergata University, Rome; Agricultural University, Wageningen; University of Reims; Universite du Littoral, Dunkerque; University of Campinas
Workplaces: Tehnofrig Enterprize Cluj-Napoca (1977–1981), National RD Institute for Isotopic and Molecular Technologies Cluj-Napoca (1981-present)
Main fields of interest: photothermal calorimetry and spectroscopy; phase transitions
Relevant categories in thermal analyses: fields (materials, nano, food, phase transitions); methods [calorimetry, instrument development, photothermal techniques (PPE, PTE, PTR)]
Awards and acknowledgments: Prize «Constantin Miculescu», awarded by the Romanian Academy of Science, in 1984 for contributions in the field of the detection of radiation
Professional activities: Referee at J. Food Engn., Meas. Sci. Technol., Int. J. Thermophys., J. Phys. D Appl. Phys., Instr. Sci. Techn., Food Biophysics, J. Physics: C, Appl. Phys. A, Central European J. Phys., Int. J. Mol. Sci., J. Mat. Sci., Thermochimica Acta, Int. J. Thermal Sci, etc.
Publication record: papers (140), books chapters (3), citations (all: 620; without self citations: 340), h-index (14)
Equipments: Experimental setups for PPE, PTR and PTE calorimetry (lasers, IR and pyroelectric detectors, light modulators, lock-in amplifiers, various detection cells)
5 most important publications: [1] D. Dadarlat, M. Chirtoc, R. Candea, I. Bratu, Infrared Phys., 24, 469 (1984); [2] D. Dadarlat, M.Chirtoc, C. Neamtu, R. Candea, D. Bicanic, Phys. Stat. Sol., (a) 121, K231 (1990); [3] M. Marinelli, F. Mercuri, U. Zammit, R. Pizzoferrato, F. Scudieri, D. Dadarlat, Phys. Rev. B49, 9523 (1994); [4] D. Dadarlat, Laser Physics, 19, 1330 (2009); [5] D. Dadarlat, M. Streza, R. Chan Yu King, F. Roussel, M. Kuriakose, M. Depriester, E. Guilmeau, A. Hadj Sahraoui, Meas. Sci. Technol. 25, 015603 (2014).

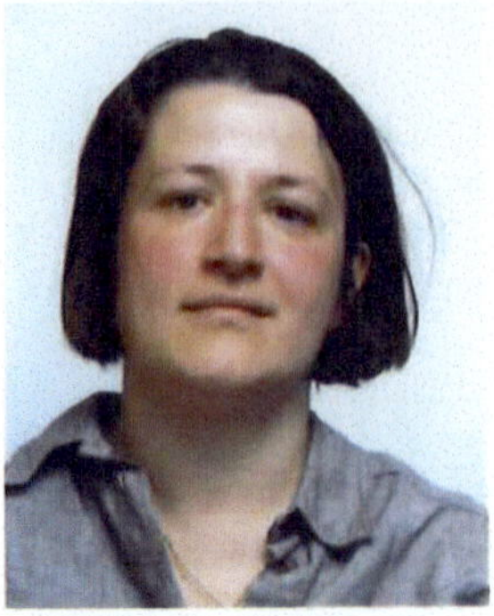

Name: Christine Dalmazzone
Country: France
Date and place of birth: 1965, Reims, France
Present position and address: Expert Research Engineer, IFPEN, 1&4, Av. Bois Préau, 92852 Rueil Malmaison Cedex, France
Email: christine.dalmazzone@ifpen.fr
Website: http://www.ifpenergiesnouvelles.fr
Education and scientific degrees: Chemical Engineer, Ph. D. from the University of Technology of Compiègne (UTC), Habilitation to supervise research
Workplaces: Rueil Malmaison (France)
Main fields of interest: physical chemistry of complex fluids, colloids, emulsions, foams, gas hydrates in the oil and gas industry
Relevant categories in thermal analyses: fields (complex fluids and materials (emulsions), crude oil); methods (DSC, High Pressure DSC, TG/DTA)
Professional activities: Secretary of AFCAT (Association Française de Calorimétrie et d'Analyse Thermique), member of the SPE (Society of Petroleum Engineers)
Publication record: papers (39 journal papers, 63 conference proceedings), patents (15), h-index (10)
Equipments: High Pressure micro DSCVII SETARAM, DSC1 Mettler Toledo, TG/DTA Pekin Elmer
5 most important publications: [1] Le Parlouër, P., Dalmazzone, C., Herzhaft, B., Rousseau, L., Mathonat, C., (2004). Characterisation of gas hydrates formation using a new high pressure micro-DSC, Journal of Thermal Analysis and Calorimetry, 78, 165–172; [2] Clausse, D., Gomez, F., Dalmazzone, C., Noik, C., (2005). A method for the characterization of emulsions: Thermogranulometry. Application to water-in-crude oil emulsion, Journal of Colloid and Interface Science, 287, 694–703; [3] Dalmazzone, D., Hamed, N., Dalmazzone, C., (2009). DSC measurements and modelling of the kinetics of methane hydrate formation in water-in-oil emulsion. Chemical Engineering Science, 64, 2020–2026; [4] Dalmazzone, C., Noïk, C., Clausse, D., (2009). Application of DSC for emulsified system characterization, OGST-Rev. IFP, 64(5), 543–555; [5] Dalmazzone, C., Noïk, C., Argillier, J.F., (2012). Impact of Chemical EOR on the Separation of Diluted Heavy oil Emulsions, Energy and Fuels, 26 (6), 3462–3469.

Name: Kim Dam-Johansen
Country: Denmark
Date and place of birth: 1958, Frederiksvaerk, Denmark
Present position and address: Head of Department, Professor DTU Chemical Engineering
Email:kdj@kt.dtu.dk
Website: http://www.kt.dtu.dk/english
Education and scientific degrees: Ph.D. Technical University of Denmark (1987); M.Sc. Technical University of Denmark (1983)
Workplaces: Head of Department, Professor DTU Chemical Engineering (2000-present); Group Vice President, Hempel A/S (1997); Professor (1993), Associate Professor (1990), Assistant Professor (1988) at DTU Chemical Engineering; Assistant Professor DTU/Engineering Academy (1986)
Main fields of interest: experimental and theoretical reaction engineering applied in combustion, high temperature processes, emissions control, product design and pharma production
Relevant categories in thermal analyses: fields (inorganic, materials, minerals, organic, polymer, cement); methods [TG, DTA, kinetics, extremely high temperature (above 1,000 °C)]
Awards and acknowledgments: **2011**: Honorary Professor, Institute of Process Engineering, China. 2011; Einstein Processor, Chinese Academy of Sciences. 2011; Doctor honoris causa, Aabo Academy University, Finland. World most cited researcher in combustion and 5th most cited in Energy and Fuels (1998–2008). A number of awards from (e.g. ASME, Lim Prize, IFV Jubilee Foundation, Fælleskemikernes Jubilee Prize)
Professional activities: Member of the board of Hempel Holding A/S and Hempel Foundation, VGB Scientific Advisory Board, Danish Academy of Technical Sciences, Danish Technical Chemical Foundation (chairman), The Board of Appeal for Patents and Trademarks (DK), Steering groups and boards of many Danish and international journals, research programs, conferences and evaluation committees. Head of SDC Education in Chemical and Biochemical Engineering
Publication record: papers (212), books (1), patents (13), citation index (5842), h-index (44)
Equipment: Can be seen at http://www.chec.kt.dtu.dk/Our-Research/Facilities
5 most important publications: [1] Telschow, Samira; Frandsen, Flemming; Theisen, Kirsten; Dam-Johansen K.: Cement Formation—A Success Story in a Black Box: High Temperature Phase Formation of Portland Cement Clinker. Industrial and Engineerin Chemistry Research, 51, 34, pp 10983–11004. 2012; [2] Andric, Pavle; Meyer, Anne S; Jensen, Peter A., Dam-Johansen K.: Reactor design for minimizing product inhibition during enzymatic lignocellulose hydrolysis: I. Significance and mechanism of cellobiose and glucose inhibition on cellulolytic enzymes. Biotechnology Advances, 28, 3, pp 308–324. 2010; [3] Sorensen, P. A., Kiil, S., Dam-Johansen, K., Weinell, C. E.: Anticorrosive coatings: a review. Journal of Coatings Technology and Research, 6, 2, pp 135–176. 2009; [4] Yebra, D.M.; Kiil, S.; Dam-Johansen, K.: Antifouling technology—past, present and future steps towards efficient and environmentally friendly antifouling coatings. Progress in Organic Coatings, 50, 2, pp 75–104, 2004; [5] Lin, W. G., Dam-Johansen, K., Frandsen, F.: Agglomeration in bio-fuel fired fluidized bed combustors. Chemical Engineering Journal, 96, 1–3, pp 171–185. 2003.

Name: Dasarathi Das
Country: India
Date and place of birth: 1951, Arambagh, West Bengal, India
Present position and address: Raja Ramanna Fellow, Department of Atomic Energy (DAE), Chemistry Group (Mod. Lab. 2-214S), Bhabha Atomic Research Centre, Trombay, Mumbai 400085, India
Email: dasd1951@gmail.com
Education and scientific degrees: Master degree in Chemistry (1973), University of Calcutta (Kolkata), Post Graduation Training at BARC (1974–1975), Doctor of Philosophy (Physical Chemistry) in 1984, University of Bombay (Mumbai). Joined BARC as Scientific officer in 1975, retired from BARC as Outstanding Scientist
Workplaces: Chemistry Division, BARC, Trombay, Mumbai 400 085
Main fields of interest: high temperature thermochemistry of nuclear materials, solid oxide fuel cell materials
Relevant categories in thermal analyses: fields (materials inter-compatibility study, phase diagram and thermodynamic property evaluations); methods (XRD, SEM, EDAX, TG-DTA-EGA, DSC, solution calorimtery, Calvet calorimetry, Knudsen effusion and transpiration methods for vapor pressure measurements at high temperatures)
Publication record: number of papers (100), books (1), patents (1)
5 most important publications: [1] Thermodynamics of liquid uranium vaporisation, D. Das, S.R. Dharwadkar, and M.S. Chandrasekharaiah, J. Nucl. Mater., 130(1985) 217–222; [2] Platinum point reference in the calibration of (visual) opt. pyrometer, D. Das, and M. S. Chandrasekharaiah, High Temperature Science, 21 (1987) 161–164; [3] Temperature Its Measurement and Control in Science and Ind., M. S. Chandrasekharaiah, J. L. Margrave, D. Das, vol.6, Part 1, American Institute of Physics, New York, 1992, pp 373–378, J. F. Schooley (Ed.); [4] Vapor species over SOFC interconnect materials and their potential, K. Hilpert, D. Das, M. Miller, D. H. Peck, and R. Wei, J. Electrochem. Soc., 143 (1996) 3642–3647; [5] Transport properties of iodine and tellurium in a thoria-2 mol% urania matrix, K. N. G. Kaimal, A. S. Kerkar, A. N. Shirsat, D. Das, A. Datta, A. G. C. Nair, and S. B. Manohar, J. Nucl. Mat. 317 (2003) 189–194.

Name: Smruti Dash
Country: India
Date and place of birth: 1961, Bhubaneswar, India
Present position and address: Scientific officer, Product Development Division, Bhabha Atomic Research Centre, Trombay, Mumbai, India, PIN: 4000085
Email: smruti@barc.gov.in
Website: http://www.barc.gov.in
Researcher ID/ORCID: 11400/BARC
Education and scientific degrees: Master in Chemistry, Ph.D. (Chemistry)
Workplaces: Bhabha Atomic Research Centre
Main fields of interest: thermodynamic studies of materials and phase diagrams; carried out experimental as well as theoretical studies to find out thermodynamic properties, particularly heat capacity, enthalpy of formation and Gibbs free energy of formation of technologically important compounds
Relevant categories in thermal analyses: fields (thermodynamic measurements and phase diagram); methods (solution calorimeter and differential scanning calorimeter)
Awards and acknowledgments: Science and Technical Excellence Award, 2007 (Bhabha Atomic Research Centre)
Professional activities: Life member of a number of scientific organizations
Publication record: number of papers (70)
Equipments: DSC, High temperature calorimeter, Solution Calorimeter, Quadrupole mass spectrometer, Knudsen effusion and transpiration
5 most important publications: [1] Smruti Dash, Ziley Singh, "Calorimetric investigations of stoichiometric compounds of barium and uranium oxides", Journal of Nuclear Materials, 404 (2010) 9–18; [2] Smruti Dash, S. C. Parida, Ziley Singh, B. K. Sen, V. Venugopal, " Thermodynamic investigations of ThO_2–UO_2 mixed oxide solid solutions", Journal of Nuclear Materials, 393 (2009) 267–281; [3] Smruti Dash, Z. Singh, S. C. Parida, V. Venugopal, "Thermodynamic studies on $Rb_2ThO_3(s)$", Journal of Alloys and Compounds 398 (2005) 219–227; [4] Smruti Dash, Ziley Singh, R. Prasad and V. Venugopal, "Computer calculation of the 700 K isothermal section of the Ba–Mo–O system from thermodynanic data", High Temperatures and High Pressures., 32; [5] Smruti Dash, D. D. Sood and R. Prasad, "Phase diagram and thermodynamic calculations of alkali and alkaline zirconates", Journal of Nuclear Materials, 228 (1996) 83–116.

Name: Wim De Klerk
Country: The Netherlands
Date and place of birth: 1962, Alblasserdam, The Netherlands
Present position and address: Program Manager, TNO Defence, Safety and Security, P.O. Box 45, 2280 AA Rijswijk, The Netherlands
Email: wim.deklerk@tno.nl
Website: www.tno.nl
Education and scientific degrees: Chemistry (Analytical and Physical), Technical College, Rotterdam, Bachelor of Science
Workplaces: Rijswijk, The Netherlands
Main fields of interest: Thermal analysis, energetic materials, microcalorimetry, test procedures, materials
Relevant categories in thermal analyses: Fields (organic, inorganic, polymers, composites, energetic materials, toxicity); methods (TG, DTA, EGA, DSC, DMA, thermomechanical analysis, kinetics, calorimetry, microcalorimetry, instrument development)
Professional activities: President of ESTAC, Vice-preseident of ICTAC, Secretary of Dutch Thermal Analysis Society—TAWN, Dutch representative in NATO Ammunition groups, chairman of the ICTAC—scientific group "lifetime studies", Member of editorial board of Journal of Thermal Analysis and the Journal Propellants, Explosives and Pyrotechnics, Chairman/organizer of ESTAC-10
Publication record: papers (>40)
Equipments: DSC, TG, DMA, TMA, HFC, HPDSC
5 most important publications: [1] W. De Klerk, C. Popescu, A. E. D. M. van der Heijden, Study on the decomposition kinetics of FOX-7 and HNF, 8th ESTAC Barcelona (Spain), Journal of Thermal Analysis and Calorimetry, Vol. 72, pp 955–966 (2003); [2] W. de Klerk, M. Boers, Sample geometry as critical factor for stability research, 3rd International Heat Flow Calorimetry for Energetics Symposium, French Lick (USA), April 2002, Thermochimica Acta nr 401, pp 43–52 (2003); [3] W. de Klerk, Overview of HFC applications, ESTAC-10 symposium, Rotterdam, The Netherlands (2010); [4] W. de Klerk, P. van Harmelen, Sustainable ammunition Safety, PARARI Symposium 2013, Canberra, Australia (2013).

Name: Pompea Del Vecchio
Country: Italy
Date and place of birth: 1958, Latronico (Potenza), Italy
Current position and address: Associate Professor of Physical Chemistry, Department of Chemical Sciences, University of Naples Federico II, Via Cintia—80126 Naples—Italy
Email: pompea.delvecchio@unina.it
Website: www.docenti.unina.it/pompeagiuseppinagrazia.del_vecchio
Education and scientific degrees: Degree in Chemistry (1983); Ph.D. in Chemical Sciences (1988); Assistant Professor of Physical Chemistry (1991); Associate Professor of Physical Chemistry (2003)
Workplaces: Department of Chemical Sciences—University of Naples Federico II; Visiting Appointments: Visiting researcher at CNRS of Marseille studying the thermodynamics of sublimation of N-acetyl amides of some amino acids (1992). Visiting researcher at ETH in Zurich studying the self-reproduction properties of vesicles enclosed by fatty acid membranes (1994)
Professional activities: Member of: Italian Chemical Society (S.C.I.), Division of Physical Chemistry, Division of Chemistry of Biological Systems (D.C.S.B.)
Main fields of interest: biophysical chemistry: thermodynamic stability of biological macromolecules and their complexes
Relevant categories in thermal analyses: fields (thermodynamics, biology, life); methods (DSC and ITC)
Publication record: papers (63), citations (597)
Equipments: Nano ITC—TA Instruments and Nano DSC—TA Instruments
5 most important publications: [1] F. Foglia; L. Mandrich; M. Pezzullo; G. Graziano; G. Barone; M. Rossi; G. Manco; P. Del Vecchio. Role of the N-terminal region for the conformational stability of esterase 2 from Alicyclobacillus acidocaldarius. Biophysical Chemistry, 127, (2007) 113; [2] P. Del Vecchio, P. Carullo, G. Barone, B. Pagano, G. Graziano, A. Iannetti, R. Acquaviva, A. Leonardi, S. Formisano. Conformational stability and DNA binding energetics of the rat thyroid transcription factor 1 homeodomain. Proteins, 70 (2008) 748; [3] C. Esposito, P. Carullo, E. Pedone, G. Graziano, P. Del Vecchio, R. Berisio. Dimerisation and structural integrity of Heparin Binding Hemagglutinin A from Mycobacterium tuberculosis: Implicatios for bacterial agglutination. FEBS Letters, 584 (2010) 1091; [4] L. M. Pavone, P. Del Vecchio, P. Mallardo, F. Altieri, V. De Pasquale, S. Rea, N. M. Martucci, C. S. Di Stadio, P. Pucci, A. Flagiello, M. Masullo, P. Arcari, E. Rippa. Structural characterization and biological properties of human gastrokine 1. Molecular Biosystems, 9 (2013) 412; [5] O. Ortona, V. Vitagliano, D. Fessas, P. Del Vecchio, G. D'Errico. Inhomogeneities in sodium decylsulfate doped 1,2-dipalmitoylphosphatidylcholine bilayer. Journal of Colloid and Interface Science, 343 (2010) 401.

Name: Giuseppe Della Gatta
Country: Italy
Date and place of birth: 1935, Turin, Italy
Present position and address: Retired, Full Professor of Physical Chemistry, Department of Chemistry, University of Turin, via Pietro Giuria 9, 10125, Turin, Italy
Email: giuseppe.dellagatta@unito.it
ORCID: 0000-0002-9770-1457
Education and scientific degrees: Degree Dottore in Chimica, Faculty of Sciences, University of Turin (1961); Libero docente in Physical Chemistry (1970)
Workplaces: Assistant Professor and Lecturer (1962–1981) and Associate and Full Professor of Physical Chemistry (1982–2007) at the Faculty of Pharmacy, University of Turin. Research grants: CNRS, Vitry/Seine, France (1961–1962); CNRS, Marseilles, France (1965–1966).
Main fields of interest: physical chemistry of high-purity metals; surface chemistry; solution thermodynamics of non-electrolytes; thermodynamics of phase transitions; thermodynamics of model molecules; collagen-based heritage materials (parchment and leather)
Relevant categories in thermal analyses: fields (inorganic, organic, biology, materials, cultural heritage); methods (isothermal microcalorimetry, DSC, specific heat, instrument development)
Awards and acknowledgments: N. S. Kurnakov Medal, Russian Academy of Sciences, (1990); Honorary Medal 50th Anniversary of the University of Łódź, Poland (1998); Mettler-Toledo Award—NATAS (2003); Gold Medal of the Italian Society of Chemistry (2007)
Professional activities: president and founder, Associazione Italiana di Calorimetria e Analisi Termica (1979–1981); member: Council of International Confederation for Thermal Analysis (1980–1985), IUPAC Commission on Thermodynamics (1982–1997), Board of Directors, Associazione Italiana di Chimica Fisica (1985–1986); committee member, European Symposium on Thermal Analysis and Calorimetry (1985–1989); organiser: 2nd and 4th Meetings, Section Méditerranéenne, Société Française de Chimie Physique (1968 and 1970), 9èmes Journées de Calorimétrie et d'Analyse Thermique, Turin (1978) and 11th IUPAC Conference on Chemical Thermodynamics, Como (1990); co-organiser: ESTAC 6 Conference, Italy (1994) and MEDICTA'95, Italy (1995); chair/organiser, IUPAC 41st World Chemistry Congress/IUPAC General Assembly, Turin (2007)
Publication record: papers (135), editor of special issues (12), h-index (28)
5 most important publications: [1] G. Della Gatta, B. Fubini, G. Ghiotti, C. Morterra, The chemisorption of carbon monoxide on various transition aluminas, J. Catal., 43 (1976) 90; [2] G. Barone, G. Della Gatta, D. Ferro, V. Piacente, Enthalpies and entropies of sublimation, vaporization and fusion of nine polyhydric alcohols, J. Chem. Soc., Faraday Trans., 86 (1990) 75; [3] G. Della Gatta, M. Jóźwiak, B. Brunetti, Enthalpies and entropies of fusion and of sublimation at the temperature 298.15 K of thiourea and seven *N*-alkylthioureas, J. Chem. Thermodyn., 32 (2000) 979; [4] G. Della Gatta, M. J. Richardson, S. M. Sarge, S. Stølen, Standards, calibration, and guidelines in microcalorimetry—Part 2. Calibration standards for differential scanning calorimetry—(IUPAC Technical Report), Pure Appl. Chem., 78 (2006) 1455; [5] E. Badea, G. Della Gatta, B. Pałecz, Thermal properties of some small peptides (*N*-acetyl-amino acid-*N*′-methylamides) with non-polar side groups, J. Chem. Thermodyn., 63 (2014) 178.

Name: Costas Demetzos
Country: Greece
Present position and address: Professor, Faculty of Pharmacy, University of Athens
Email: demetzos@pharm.uoa.gr
Education and scientific degrees: Ph.D. in Pharmaceutical Sciences (1990) University of Athens. 8/1985—12/1988 Faculty of Pharmacy, University of Athens, 1/1989—7/1989 University Paris V, France; B.Sc. in Pharmaceutical Sciences (1983) Faculty of Pharmacy, University of Athens
Workplaces: 2012—today: Professor in Pharmaceutical Nano-Technology, Faculty of Pharmacy, University of Athens
Main fields of interest: pharmaceutical technology and nanotechnology, colloidal science, thermal analysis and its applications in drug development; advanced drug delivery nano systems; chimeric and hybridic liposomal and dendrimeric drug delivery nano systems; bio-inspired advanced drug delivery nano systems, physical stability, morphology and fractal analysis; regulatory aspects in drug development
Professional activities: Reviewer in more than 30 International Scientific Journals; Member and/or President of Scientific Committee of several National (Greek) and International Congresses (i.e. MEDICTA 2013); President in the Hellenic Pharmaceutical Society (HPS) (2008—today); Member of the Management Board of the National Organization for Medicines (EOF) (2010-today); Member of the Management Committee in the COST action (TD0802) on dendrimeric technologies for drug delivery (2009–2013).; Delegate of the Hellenic Pharmaceutical Society (HPS) in European Federation of Pharmaceutical Sciences (EUFEPS) (2012-today)
Publication record: papers (160 journal paper, 200 conference proceedings), books (3), patents (9), citations (>2000)
5 most important publications: [1] "PEO-b-PCL/DPPC chimeric nanocarriers: self-assembly aspects in aqueous and biological media and drug incorporation" N. Pippa, E. Kaditi, S. Pispas, and C. Demetzos, Soft Matter, 9, 4073–4082, 2013; [2] "The fractal hologram and elucidation of the structure of liposomal carriers in aqueous and biological media." N. Pippa, S. Pispas, C. Demetzos, Int. J. Pharm. 430, 65–73, 2012; [3] "New drug delivery nanosystem combining liposomal and dendrimeric technology (liposomal locked-in dendrimers) for cancer therapy", K. Gardikis, S. Hatziantoniou, M. Bucos, D. Fessas, M. Signorelli, T. Felekis, M. Zervou, C. Skrettas, B. Steele, M. ionov, M. Micha-Skretta, B. Klajnert, M. Bryszewska, C. Demetzos, J. of Pharm. Sc., 99(8), 3561–71, 2010; [4] "Atomic force microscopy (AFM): a tool to study the structure, dynamics and stability of liposomal drug delivery systems", E. Spyratou, E. Mourelatou, M. Makropoulou, C. Demetzos, Expert Opinion on Drug Delivery Systems, 6, 305–317, 2009; [5] "Differential scanning calorimetry (DSC): A tool to study the thermal behavior of lipid bilayers and the liposomal stability", C. Demetzos, J. Liposome Res., 18, 159–173, 2008.

Name: Renaud Denoyel
Country: France
Date and place of birth: 1957
Present position and address: Director of MADIREL Laboratory
Email: renaud.denoyel@univ-amu.fr
Website: www.madirel.fr
Workplaces: MADIREL Laboratory, Marseille
Main fields of interest: high surface area materials
Relevant categories in thermal analyses: fields (thermal behaviour of materials); methods (calorimetry, TGA, DSC)
Publication record: papers (140), books (5 book chapters), patents (1), h-index (26)
Equipments: TG, DSC, adsorption calorimetry, liquid intrusion calorimetry
5 most important publications: [1] Thermal methods in the synthesis of new ordered mesoporous adsorbents. R. Denoyel, M. J. Keene, P. L. Llewellyn, J. Rouquerol, J. Thermal Analysis and Calorimetry, 1999, 56, 261; [2] Thermodynamic (including microcalorimetry) study of the adsorption of nonionic and anionic surfactants onto silica, kaolin and alumina. R. Denoyel, J. Rouquerol, J. Colloid Interface Sci. 1991, 143, 555; [3] Determining the contact angle of a non-wetting liquid in pores by liquid intrusion Calorimetry. F. Gomez, R. Denoyel, J. Rouquerol, Langmuir, 2000, 16, 3474; [4] Influence of particles size on thermal properties of aluminium powder, B. Rufino, F. Boulc'h, M. V. Coulet, G. Lacroix, R. Denoyel, Acta Materialia 55(8): 2815–2827 2007; [5] Structural changes and thermal properties of aluminium micro- and nano-powders. B. Rufino, M. V. Coulet, R. Bouchet, O. Isnard, R. Denoyel, Acta Materialia, 2010, 58, 4224–4232.

Name: Nicolae Doca
Country: Romania
Date and place of birth: 1946, Sînandrei, Timiş, Romania
Present position and address: Since 2011 retired from West Univ. of Timişoara, Cornelia Sălceanu str., no. 9, 300561, Timisoara, Romania
Email: docanicolae@yahoo.com
Researcher ID: A-5612-2014
Education and scientific degrees: Chem. eng., Ph.D., Professor emeritus
Workplaces: West Univ. of Timişoara, Research Center Thermal Analysis in Environmental Problems
Main fields of interest: catalysis, kinetics, environmental friendly technologies
Relevant categories in thermal analyses: fields (pharmaceutical, polymer); methods (TG, DTA, EGA, DSC, kinetics)
Awards and acknowledgments: "Gheorghe Spacu" Award in chemistry given by Romanian Academy (2005)
Professional activities: ICTAC counselor, Manager of the Research Center Thermal Analysis in Environmental Problems
Publication record: papers (56), books (14), patents (11), citation index (467), h-index (12), sum of impact factors (108)
Equipments: TG/DTG/DTA Diamond PerkinElmer, DSC Diamond PerkinElmer, TMA Diamond PerkinElmer, EGA (FT-IR Spectrum 100 with Multiscope FT-IR Microscope PerkinElmer)
5 most important publications: [1] Doca, N., Vlase, G., Vlase, T., Perţa, M., Ilia, G., Plesu, N., TG, EGA and kinetic study by non-isothermal decomposition of a polyaniline with different dispersion degree., (2009) Journal of Thermal Analysis and Calorimetry, 97 (2), pp. 479–484; [2] Doca, N., Vlase, G., Vlase, T., Ilia, G., Thermal behavior of Cd^{2+} and Co^{2+} phenyl-vinyl-phosphonates under non-isothermal condition., (2008) Journal of Thermal Analysis and Calorimetry, 94(2), pp. 441–445; [3] Doca, N., Segal, E. Kinetics of dehydrogenation of low aliphatic alcohols on copper catalysts containing small amounts of Cr, Mn, Fe and Ni., (1985) Reaction Kinetics and Catalysis Letters, 28(1), pp. 123–129; [4] Pop, N., Mogoş, A.M., Vlase, G., Vlase, T., Doca, N., Theoretic analysis and experimental evidence for relationships between the derivative thermogravimetric curves and the Gramm-Schmidt profiles (2013) Journal of Thermal Analysis and Calorimetry, 113(1), pp. 113–119; [5] Segal, E., Budrugeac, P., Carp, O., Doca, N., Popescu, C., Vlase, T., Thermal Analysis-Fundamentals and applications. Kinetic analysis of heterogeneous transformations (in romanian), Ed. Romanian Academy, Bucarest, 2013.

Name: Žaneta Dohnalová
Country: Czech Republic
Date and place of birth: 1976, Karvina, Czech Republic
Present position and address: researcher and lecturer, University of Pardubice, Faculty of Chemical Technology, Department of Inorganic Technology, Technological Pavvilion, Doubravice 41, 532 10 Pardubice, Czech Republic
Email: zaneta.dohnalova@upce.cz
Education and scientific degrees: Chemical engineer, University of Pardubice (1994–1999); Ph.D. (2004).
Workplaces: University of Pardubice, Faculty of Chemical Technology, Department of Inorganic Technology,
Main fields of interest: chemistry and synthesis of inorganic materials, especially the research of inorganic pigments and powder materials, their application possibilities for ceramic glazes, organic binders and building materials; study of high-temperature syntheses of pigments and evaluation of their colour properties, granulometric composition, thermal behaviour and chemical and light stability
Relevant categories in thermal analyses: fields (inorganic materials, pigments); methods (TG, DTA, heating microscope)
Professional activities: member of audit committee of Czech Chemical Society, organiser of the of 4th Joint Czech-Hungarian-Polish-Slovak Thermoanalytical Conference in Pardubice (2013)
Publication record: papers (16), citations (51), h-index (7)
Equipments: STA 449C Jupiter (NETZSCH, Germany); heating microscope EM201-15 (Hesse Instruments, Germany)
5 most important publications: [1] Dohnalová Ž., Šulcová P., Trojan M.: Preparation and selected properties of pigments on base of Ln-doped $CaSnO_3$, J. Therm. Anal. Calorim. 93 (2008) 3, 857–861; [2] Dohnalová Ž., Šulcová P., Trojan M.: Effect of Er^{3+} substitution on the quality Mg–Fe spinel pigments, Dyes and Pigments 80 (2009) 22–25; [3] Dohnalová Ž., Šulcová P., Trojan M.: Synthesis and colour properties of pigments based on terbium-doped Mg_2SnO_4, J. Therm. Anal. Calorim. 101/3 (2010) 973–978; [4] Dohnalová Ž., Vontorčíková M., Šulcová P.: Characterization of metal oxide-doped lutetium orthoferrite powders from the pigmentary point of view, J. Therm. Anal. Calorim. 113 (2013) 1223–1229; [5] Gorodylova N., Dohnalová Ž., Šulcová P. Effect of the Synthesis Conditions on the Formation of $MgSrP_2O_7$ and its Characterization for Pigmentary Application, J. Therm. Anal. Calorim. 113 (2013) 147–155.

Name: Valeri A. Drebushchak
Country: Russia
Date and place of birth: 1958 Rudnoe, Kazakhstan
Present position and address: Senior researcher—V. S. Sobolev Institute of Geology and Mineralogy, SB RAS, Pr. Ak. Koptyuga, 3, 630090 Novosibirsk, Russia; Associated professor—(1) Department of Natural Sciences and (2) Physical Department of the Novosibirsk State University, Pirogova St., 2, 630090 Novosibirsk, Russia.
Email: dva@igm.nsc.ru; dva@xray.nsu.ru
Researcher ID: B-6763-2008
Education and scientific degrees: physicist (1980) Novosibirsk State University; Ph.D. in Chemistry (1993) United Institute of Geology, Geophysics and Mineralogy, SB RAS.
Workplaces: (1) V. S. Sobolev Institute of Geology and Mineralogy, SB RAS; (2) Novosibirsk State University.
Main fields of interest: thermal physics, condensed matter, materials sciences
Relevant categories in thermal analyses: fields (minerals, molecular crystals, pharmaceuticals, thermoanalytical instrumentation); methods (DSC, TG, TMA)
Awards and acknowledgments: Young Scientist in Applied Researches from the Siberian Branch of Russian Academy of Sciences (1987); Scientist in Applied Researches from the Siberian Branch of Russian Academy of Sciences (1989); Young Scientist in Basic Researches from the Siberian Branch of Russian Academy of Sciences (1990); Best Reviewer Award from the Journal of Thermal Analysis and Calorimetry (2014).
Professional activities: Regional Editor in the Journal of Thermal Analysis and Calorimetry
Publication record: papers (101), books (5), citation index (821), h-index (14) (strictly speaking, this is an update, not a correction)
Equipments: DSC-111 Setaram; DSC-30, TG-50, TMA-40 Mettler; DSC-204, TG-209, TMA-202 Netzsch
5 most important publications: [1] Drebushchak VA, Kravchenko TA, Pavlyuchenko VS. Synthesis of pure pentlandite in bulk. J Cryst Growth. 1998;193:728–31; [2] Drebushchak VA. Calibration coefficient of a heat-flow DSC—Part II. Optimal calibration procedure. J Therm Anal Calorim. 2005; 79: 213–8; [3] Drebushchak VA, Mylnikova LN, Drebushchak TN. The mass-loss diagram for the ancient ceramics. J Therm Anal Calorim 2011; 104: 459–66; [4] Drebushchak VA, Ogienko AG, Boldyreva EV. Polymorphic effects at the eutectic melting in the H_2O–glycine system. J Therm Anal Calorim. 2013; 111:2187–94; [5] Drebushchak VA. Thermoelectricity in thermal analysis and calorimetry. Novosibirsk State University: Novosibirsk, 2009, 166 p. (in Russian).

Name: Zofia Drzazga
Country: Poland
Date and place of birth: 1948, Sosnowiec, Poland
Present position and address: Head of Department of Medical Physics; Silesian Intercollegiate Center for Education and Interdisciplinary Research, ul. Pulku Piechoty 1A, 41-500 Chorzow, Poland
Email: zofia.k.drzazga@gmail.com; zofia.drzazga@us.edu.pl
Education and scientific degrees: Master of Science, University of Silesia 1980; Doctor of Physics 1980, University of Silesia; Assistant professor 1991, Institute of Low Temperature and Structure Research, PAS, Wroclaw; Professor of Physical Science 2000.
Workplaces: University of Silesia, A. Chelkowski Institute of Physics in Katowice, Silesian Intercollegiate Center for Education and Interdisciplinary Research in Chorzow, Poland,
Main fields of interest: application of non-ionizing radiation in diagnostics and therapy in vitro and in vivo (microcalorimetry and thermal imaging, respectively)
Relevant categories in thermal analyses: fields (organic, pharmaceutical, life, effect of external factors on organic); methods (microcalorimetry, thermovision, TG-DTA)
Awards and acknowledgments: Award of the Minister of Higher Education and Science 1988
Professional activities: organization of education in medical physics at University of Silesia, Head of Department of Medical Physics (1996-present), organization and chief person of International Symposium on Medical Physics 2001, 2003, 2006, 2009, early Symposium on Medical Physics (1996, 1998), member of the Polish Society of Medical Physics (president of Silesian Branch 2004–2008), member correspondence of the American Association of Physicist in Medicine (1999–2001), member of the Committee on Specialization of Medical Physics (2006–), member of the Committee of Medical Physics, Radiobiology and Diagnostic of Polish Academy of Science (2007–2012), Member of American Society for Photobiology (2010–), member of the Polish Society of Calorimetry and Thermal Analysis (2012–)
Publication record: papers (125), books (2 monographs, editor or co-editor of 6 books and 15 chapters in books), citations (511), h-index (11)
Equipments: VP DSC MicroCal instrument, Thermovision Camera A40M
5 most important publications: [1] Z. Drzazga, A. Michnik, M. Bartoszek, E. Beck "Thermal stability of haemoglobin solutions under DC and Magnetic field and UV and IR Radiation" J Therm Anal Calorim., 65 (2001), 575–582; [2] A. Michnik, K. Michalik, Z. Drzazga, Stability of bovine serum albumin at different pH, J Therm Anal Calorim, 80 (2005) 399–406; [3] Anna Michnik, Katarzyna Michalik, Zofia Drzazga, Effect of UVC radiation on conformational restructuring of human serum albumin, J. Photochem. Photobiol. B: Biology, 90 (2008) 17, 178; [4] Z. Drzazga, K. Michalik, T. Hałat, A. Michnik, H. Trzeciak, Calorimetric and spectroscopic studies characterization of newborn rat' blood serum after maternal administration of cyclophosphamide J Therm Anal Calorim, 102 (2010) 143–148; [5] A. Michnik, Z. Drzazga, K. Michalik, A. Barczyk, I. Santura, E. Sozańska, W. Pierzchała, "Differential scanning calorimetry study of blood serum in chronic obstructive pulmonary disease", J Therm Anal Calorim, 102 (2010) 57–60.

Name: Celia Duce
Country: Italy
Date and place of birth: 1974, La Spezia, Italy
Present position and address: Permanent Researcher at the Department of Chemistry and Industrial Chemistry, University of Pisa (Unipi), Via Moruzzi, 56124 Pisa, Italy
Email: celia.duce@unipi.it
Website: http://www.dcci.unipi.it/thermolab/
Education and scientific degrees: 2009-present Permanent Researcher, Unipi. 2005 Ph.D. in Biomaterials, Unipi. 2000 Degree in Chemistry, Unipi, Summa cum Laude
Main fields of interest: physico-chemical characterization of materials used in the cultural heritage, polymers, archeological pitch and tar, self-assembly biomaterials; nanoparticles and halloysite nanotube based materials; excess properties of mixtures; cheminformatics.
Relevant categories in thermal analyses: fields (materials, nano, polymer, organic, food, biology, life); methods (TG, DTA, EGA, DSC, ITC, calorimetry, microcalorimetry densimetry, viscosimetry, rheology)
Awards and acknowledgments: Alberto Lucci award (2008)
Professional activities: Elected member of the council board of the Calorimetry and Thermal Analysis Interdivisional Group (GICAT) of Italian Chemical Society (2011–2013) and of Italian Association of Calorimetry and Thermal Analysis (2008–2010). Organizer of: the 1st Winter School "Renewable Energy Systems and Green Nanotechnologies for a Clean Environment", Romania, (2012); the XXXII and XXX National Congress on Calorimetry and Thermal Analysis (Trieste, 2010 and Pisa, 2008); the Workshop on "Nanocalorimetry for Biological and Biomedical applications", Pisa (2008). Member of the scientific committee of: XXXIV National Congress on Calorimetry and Thermal Analysis, Rome, (2012)
Publication record: papers (35), h-index (9)
Equipment: NanoDSC (N-DSC III, CSC); MicroITC (2277 TAM, Thermometrics); DSC (Pyris Diamond, PerkinElmer); TG (Q5000IR, TA Instruments) combined with FT-IR for EGA; Rheometer (HAAKE RheoStress 6000, Thermo Fisher Scientific)
5 most important publications: [1] C. Duce, et al. Dalton Trans., 2013, 42, 5975; [2] C. Duce, et al. Anal Bioanal Chem 2012, 402, 2183; [3] I.Bonaduce; L. Carlyle; M. P. Colombini; C. Duce; C. Ferrari; E. Ribechini; P. Selleri; M. R. Tiné, J Thermal Anal Calorim 2012, 107, 1055; [4] M. R. Tiné, M. Alderighi, C. Duce, L. Ghezzi, R. Solaro, J Thermal Anal Calorim 2011, 103, 75; [5] C. Duce, M. R. Tiné, L. Lepori, E. Matteoli, Fluid Phase Equilibria 2002, 199, 197.

Name: Jo Dweck
Country: Brazil
Date and place of birth: 1946, Istanbul, Turkey
Present position and address: Titular Professor, School of Chemistry, Rio de Janeiro Federal University, Praia de Botafogo 114/403, Rio de Janeiro, RJ, Brasil CEP 22 250-040
E-mail: jodweck@yahoo.com.br, dweck@eq.ufrj.br
Education and scientific degrees: Chemical Engineer, Rio de Janeiro Federal University (UFRJ), Brazil (1965–1969), M.Sc. and Dr. Eng. in Chem. Eng. at COPPE/UFRJ (1970–1974), and São Paulo University, Brazil (USP) (1984–1990), Visiting Scientist, Rutgers University, NJ/US (1993), Adjunct Professor, Louisiana State University, LA/US (2001–), Assistant Prof., Adjunct Prof. and Titular Professor UFRJ (1970–), Thermal Analysis Visiting Professor, USP (2009), Brazilian National Council (CNPq) Researcher (1980–)
Workplace: Rio de Janeiro Federal University: General Physics Dept. at the Physics Institute, Chemical Engineering and Inorganic Process Departments at the Chemical School
Main fields of interest: thermal analysis applied to industrial products and processes and to their effluent and waste thermal characterization and processing
Relevant categories in thermal analyses: fields (inorganic, organic, materials, clays and organoclays, pozzolans, industrial minerals, industrial products, polymers, natural and syntetic oils, ceramics, cement, shale, coal, biomass, biodiesel, bitumen, composite materials, nanocomposites, industrial solid wastes and sludges, foods, construction materials, solidified and stabilized wastes, biomasses, diesel and biodiesel); methods (TG, DTG, DTA, DSC, MDSC, kinetics, specific heat, calorimetry, conventional and non-conventional TG and DTA instrument and system development)
Professional activities: Guest Editor of the J. Therm. Anal. Cal. (JTAC) 67(2), (2002) special issue. JTAC Associate Editor (2010–). ICTAC Scientific Awards Committee member (2011–2012). Annual Thermal Analysis of Materials and Processes Graduate Course Lecturer at UFRJ (1993–). Industry Thermal Analysis applications Consultant (2003–)
Publication record: papers (74 journal papers, 59 conference proceedings), books (2 books, 2 book chapters), patents (4), h-index (9)
Equipments: Has developed 3 prototypes (TGA (1985), DTA (1986) and a non-conventional DTA (NCDTA) system (1987). Two new NCDTA and NCTG systems were developed to analyze cement hydration and carbonation in pastes and mortars in standard molds (2013)
5 most important publications: [1] J. Dweck: Obtaining Modulated Temperature DSC Curves through a Non-conventional DSC Method. J. Therm. Anal. Calorim. 60 (2000) 785–93; [2] J. Dweck, P. F. Ferreira da Silva, R. S. Aderne, M. P. Büchler, F. Cartledge, Evaluating cement hydration by non-conventional DTA—An application to waste solidification, J. Therm. Anal. Calorim. 71 (2003) 821–7; [3] J. Dweck, Qualitative and Quantitative Characterization of Brazilian Natural and Organophilic Clays by Thermal Analysis, J. Therm. Anal. Calorim. 92(1) (2008) 129–35; [4] J. Dweck, M. B. M. Melchert, M. M. Viana, F. K. Cartledge and P. M. Büchler, Importance of quantitative thermogravimetry on initial cement mass basis to evaluate the hydration of cement pastes and mortars J. Therm. Anal. Calorim. 111 (3) (2013) 1–10; [5] A. Neves Junior, R. D. Toledo Filho, E. M. R. Fairbairn, J. Dweck, A study of the carbonation profile of cement pastes by thermogravimetry and its effect on the compressive strength J. Therm. Anal. Calorim. 116(1) (2013) 69–76.

Name: Charles M. Earnest
Country: USA
Date and place of birth: 1941, Goodsprings, Alabama (USA)
Present position and address: Dana Professor of Analytical Chemistry, Berry College, Rome, GA (USA) 30149-0043
Email: cearnest@Berry.edu
Education and scientific degrees: B.Sc. in Chemistry, University of Alabama (1964), Ph.D. in Analytical Chemistry, University of Alabama (1969)
Workplaces: Stillman College (1969–1974), University of Pittsburgh at Johnstown (1974–1976), University of Louisiana- Monroe (1976–1978), PerkinElmer Corporation, Thermal Analysis Product Department (1978–1985), Berry College, Department of Chemistry, Rome, GA (1985-present)
Main fields of interest: thermal analysis instrumentation and quantitatitve aspects of DSC, TGA, and DTA; characterization of clays and minerals as well as inorganic chemicals, fats and oils, polymorphism in both organic and inorganic species; worked on the use of first thin film thermocouples for use in DSC cells as a graduate student under Professor A.F. Findeis at the University of Alabama (1965–1969)
Relevant categories in thermal analyses: fields (Analytical chemistry, geosciences, and material sciences); methods (DTA (high temperature), DSC, TGA, and DTG)
Awards and acknowledgments: North American Thermal Analysis Society (NATAS) Fellow (1988), NATAS Distinguished Service Award (2001), Carden Award (1993), Garrett Award (1998)
Professional activities: Served on editorial board of Thermochimical Acta for 16 years (1981–1997), President of NATAS (two terms) 1990 and 1991. Editor of NATAS Notes 1989–1990. Edited proceedings of the 17th NATAS conference (1988), Member of ASTM E-37 Committee on Thermal Methods (1980–1990). Chairman of "Thermal Analysis Networking" Session 2002, 2013 Pittsburgh Conferences, Past Member of ICTAC Geosciences Committee
Publication record: papers (60), books (3)
Equipment: Power Compensated DSC, High Temperatue DTA, High Temp TGA, Dynamic Mechanical Analyzer, and Thermomechanical Analyzer
5 most important publications: [1] C. M. Earnest, "Modern Thermogravimetry", Anal. Chem., 1984, Vol. 56, pp 1471A-1486A. (Invited Paper); [2] C. M. Earnest, "The Modern Thermogravimetric Approach to the Compositional Analysis of Materials, "Compositional Analysis by Thermogravimetry" C. M. Earnest, Editor, ASTM STP 997, Philadelphia, 1988, pp. 1–18; [3] C. M. Earnest, "Characterization of the Low Temperature Ash Component of the Herrin 6 Seam by Thermal Methods of Analysis, "Thermochim.Acta, 121, 1987, pp. 71–86; [4] C. M. Earnest, "Descriptive Oxidative Profiles for Pyrite in Low Temperature Ash Component of Coals by Differential Thermal Analysis," Thermochim.Acta, 75, 1984, pp. 219–232; [5] C. M. Earnest, 'Thermal Analysis of Selected Illite and Smectite Clay Minerals, Part I and Part II," in Thermals Analysis in the Geosciences, W. Smykatz-Kloss and S. Warne, editors, Springer, 1991, pp. 270–312 and 270–312.

Name: Matko Erceg
Country: Croatia
Date and place of birth: 1976, Vrgorac, Croatia
Present position and address: Associate Professor, Faculty of Chemistry and Technology, Teslina 10/V, 21 000 Split, Croatia
Email: merceg@ktf-split.hr
Website: http://tkojetko.irb.hr/en/znanstvenikDetalji.php?sifznan=12177
Researcher ID: 8691183500
Education and scientific degrees: BSCHE, Faculty of Chemistry and Technology in Split (2000); Ph.D. at Faculty of Chemistry and Technology in Split (2007); Assistant Professor (2008); Associate Professor (2012)
Workplaces: Faculty of Chemistry and Technology, University of Split, Croatia (2001–)
Main fields of interest: preparation and characterization of polymer blends, composites and nanocomposites with the emphasis on their thermal properties and thermal degradation; kinetic analysis of the thermal degradation processes of polymer blends, composites and nanocomposites
Relevant categories in thermal analyses: fields (polymer, materials, nano); methods (TG, DSC, kinetics)
Professional activities: president of the Association of Chemical Engineers and Technologists Split (2012–), member of the executive board of Croatian Society of Chemical Enginners and Technologists (2012–), member of scientific commitee of the 1st and 2nd Central and Eastern European Conference on Thermal Analysis and Calorimetry
Publication record: papers (12), h-index (7), sum of impact factors (14.148)
Equipments: Mettler Toledo 823e Differential Scanning Calorimeter, PerkinElmer Pyris 1 Thermogravimetric Analyser
5 most important publications: [1] M. Erceg, T. Kovačić, I. Klarić, Polymer Degradation and Stability 90 (2005) 86–94; [2] M. Erceg, T. Kovačić, I. Klarić, Polymer Degradation and Stability 90 (2005) 313–318; [3] M. Erceg, T. Kovačić, S. Perinović, Thermochimica Acta 476 (2008) 44–50; [4] M. Erceg, T. Kovačić, I. Klarić, Thermochimica Acta 485 (2009) 26–32; [5] M. Erceg, T. Kovačić, S. Perinović, Polymer Composites 31 (2010) 272–278.

Name: Saeed Farahany
Country: I.R. Iran
Date and place of birth: 1978, Arak, Iran
Present position and address: Postdoctoral Researcher-Department of Materials, Manufacturing and Industrial Engineering.Faculty of Mechanical Engineering. Universiti Teknologi Malaysia (UTM). 81300. Skudai. Johor Bahru. Malaysia
Email: saeedfarahany@gmail.com; fsaeed2@live.utm.my
Website: http://web1.fkm.utm.my/?id=department_MMI&pid=1163
ORCID: 0000-0002-1532-0416
Education and scientific degrees: Ph.D. in Materials Engineering, Universiti Teknologi Malaysia (UTM)
Workplaces: Universiti Teknologi Malaysia (UTM)
Main fields of interest: nucleation and growth; phase transformation during solidification
Relevant categories in thermal analyses: fields (solidification of light alloys (aluminium and magnesium alloys)); methods (computer-aided cooling curve thermal analsysis (CA-CCTA))
Professional activities: Reviewer of the Journal of Thermal Analysis and Calorimetry
Publication record: papers (31), books (1), patents (-), h-index (6), sum of impact factors (40.80)
Equipments: Data acquisition system (DeweSoft 7.5), professional data analysis software (Flexpro.9)
5 most important publications: [1] Saeed Farahany, Ali ourdjini and Mohd Hasbullah Idris. Journal of Thermal Analysis and Calorimetry, 2012, 109(1), p 105–111; [2] Saeed Farahany, Hamid Reza Bakhsheshi-Rad, Mohd Hasbullah Idris, Mohammed Rafiq Abdul Kadir, Amir Fereidoni and Ali Ourdjini. Thermochimica Acta, 2012, 527, p 180–189; [3] Saeed Farahany, Ali ourdjini, Mohd Hasbullah Idris, Saeed Shabestari. Journal of Thermal Analysis and Calorimetry, 2013, 114(2), p 705–717; [4] Saeed Farahany, Ali ourdjini, Mohd Hasbullah Idris, Saeed Shabestari. Thermochimica Acta, 2013, 559, p 59–68; [5] Saeed Farahany, Ali ourdjini, Tuty Asma Abu Bakar, Mohd Hasbullah Idris. Thermochimica Acta, 2014, 575, p 179–187.

Name: Michael Feist
Country: Germany
Date and place of birth: 1953, Halle/Saale, Germany
Present position and address: Senior scientist, Institut für Chemie der Humboldt-Universität zu Berlin, Brook-Taylor-Str. 2, D-12489 Berlin
Email: feistm@chemie.hu-berlin.de
Website: www.gefta.org
Education and scientific degrees: Dr. rer.nat., Chemistry studies at Berlin Humboldt University (1971–1975), Doctoral thesis at Moscow State University (1976–1980)
Workplaces: Institute of Chemistry, Humboldt-Universität zu Berlin
Main fields of interest: heterogeneous halogen exchange reactions on oxides and fluorides, pulse thermal analysis
Relevant categories in thermal analyses: fields (inorganic, materials, nano); methods (TG, DTA, EGA, PulseTA®, Raman spectroscopy, ESR spectroscopy)
Professional activities: Editorial board of J. Therm. Anal. Calorim. (1999–2003), Chairman of the Gesellschaft für Thermische Analyse e.V. (GEFTA) since 1998, ICTAC Council member since 1997, Invited professor at Institut de chimie de la matière condensée de Bordeaux (ICMCB) 2006 and 2010
Publication record: papers (78)
Equipments: Netzsch STA 409 C Skimmer®
5 most important publications: [1] M. Feist, K. Witke, D. Hass, Z. anorg. allg. Chem. 586 (1990) 185–93; [2] M. Feist, S. Trojanov, H. Mehner, K. Witke, E. Kemnitz, Z. anorg. allg. Chem. 625 (1999) 141–6; [3] R. Stößer, M. Feist, J. Phys. Chem. C 112 (2008) 16438–444; [4] M. Gaudon, J. Majimel, J.-M. Heintz, M. Feist, D. Dambournet, A. Tressaud, J. Fluorine Chem. 129 (2008) 1173–9; [5] M. Feist, K. Teinz, S. Robles Manuel, E. Kemnitz, Thermochim. Acta 524 (2011) 170–8.

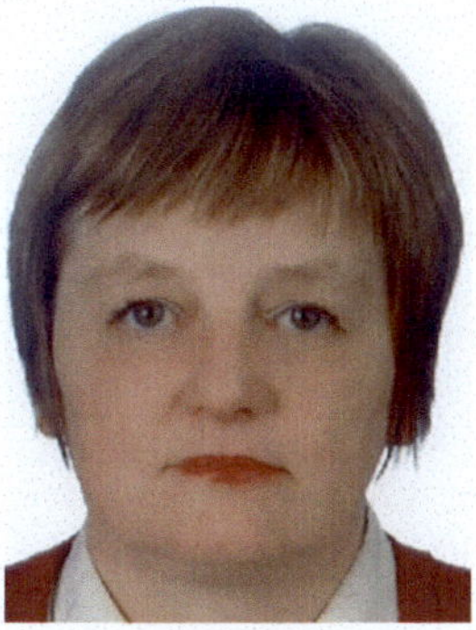

Name: Wiesława Ferenc
Country: Poland
Date and place of birth: 1952, Olsztyn, Poland
Present position and address: Prof. Wiesława Ferenc, Maria Curie-Skłodowska University, Department of General and Coordination Chemistry, Maria Curie-Sklodowska Square 2, 20-031 Lublin, Poland
Email: wetafer@poczta.umcs.lublin.pl
Education and scientific degrees: Master degree (1977); Ph. D. (1982); Assistant Professor (1992); Professor (1999)
Workplaces: Maria Curie-Skłodowska University—Lublin, Poland (1977–)
Main fields of interest: thermal analysis studies (theory and applications); magnetochemistry (determination of magnetic susceptibility by Gouy, Faraday and SQUID-VSM methods); history of chemistry and philosophy of nature sciences.
Relevant categories in thermal analyses: fields (thermal analysis investigation of complexes and inorganic compounds in air and inert atmosphere); methods (TG/DTG, DTA, DSC and TG/FTIR)
Awards and acknowledgments: Maria-Curie Skłodowska medal (2012); National Education Committee medal (2013)
Professional activities: Member of Polish Chemical Society Board (1997–2009); Head of Polish Chemical Society Student Section (1998–2009); Member of Presidium of Polish Chemical Society (2009–2012); Member of Board of Polish Society of Thermal Analysis and Calorimetry (2006–); Member of Lublin Scientific Society
Publication record: papers (154), books (2), citation index (500)
Equipments: Setsys 16/18 (Setaram), TGA Q5000 instrument (TA Instruments) with Nicolet 6700 Spectrophotometer; Magnetometers: Gouy and SQUID-VSM.
5 most important publications: [1] W. Ferenc, B. Bocian: J. Them. Anal. Calorim. 62 (2000) 831; [2] W. Ferenc, A. Walków-Dziewulska: J. Them. Anal. Calorim. 71(2003) 375; [3] M. Lalia-Kantouri, M. Gdaniec, K. Chrisstofis, W. Ferenc, J. Sarzyński, C.D. Papadopoulos: J. Therm. Anal. Calorim. Vol. 109 (1) (2012) 131; [4] W. Ferenc: Na początku była filozofia: od alchemii do chemii (At the begining the philosophy was: from alchemy to chemistry), Lublin 1998, book; [5] W. Ferenc: Wybrane zagadnienia z dziejów alchemii i chemii (The selected problems from the history of alchemy and chemistry), Lublin 1999, book.

Name: Valter José Fernandes Junior
Country: Brazil
Date and place of birth: 1965, São Paulo, Brazil
Present position and address: Vice-Rector for Research and Tilutar Professor of Analytical Chemistry, UFRN-Federal University of Rio Grande do Norte, Institute of Chemistry, Nalat, RN, Brazil, 59078-970
Email: valter.ufrn@gmail.com
Website: http://lattes.cnpq.br/1595902438130772
Education and scientific degrees: Chemist, Federal University of Rio Grande do Norte (1986); D.Sc., University of São Paulo (1991)
Workplaces: (1987–1991) Whirlpool Latin America—BRASTEMP, Laboratory for Research and Development; (1992–2010) Associate Professor in Department of Chemistry, Federal University of Rio Grande do Norte; (2010-to date) Titular Professor of Analytical Chemistry in Institute of Chemistry, Federal University of Rio Grande do Norte
Relevant categories in thermal analyses: fields (fuels, biofuels, lubricants, catalysis, thermal stability, materials); methods (TG, DTA, DSC, EGA, kinetics, high pressure)
Awards and acknowledgments: Award for outstanding contribution, Journal of Thermal Analysis and Calorimetry, 2008
Professional activities: (2000–) Coordinator of the Federal Programm for Fuels Quality in the State of Rio Grande do Norte; (2000–2011) Regional Editor of Journal of Thermal Analysis and Calorimetry for Latin America and Caribean; (2006–) Honorary member of Thermoanalytical Group, Hungarian Chemical Society; (1996–2000) Vice-Presidente of Brazilian Association for Thermal Analysis and Calorimetry
Publication record: papers (118), books (1), patents (2), citations (902), h-index (16)
Equipments: MS/TGA/SDTA 851e Mettler (TG-DTA-MS), STD Q600 TA (TG-DTA), DSC 204 (High Pressure) Netsch
5 most important publications: [1] Analysis of thermal and oxidative stability of biodiesel from Jatropha curcas L. and beef tallow, Journal of Thermal Analysis and Calorimetry, 113, 437–442, 2013; [2] Effect of the AL-MCM-41 catalyst on the catalytic pyrolysis of atmospheric petroleum residue (ATR), Journal of Thermal Analysis and Calorimetry, 106, 759–762, 2011; [3] Thermal investigation of oil and biodiesel from Jatropha curcas L., Journal of Thermal Analysis and Calorimetry, 96, 1029–1033, 2009; [4] Thermo-oxidative stability and cold flow properties of babassu biodiesel by PDSC and TMDSC techniques, Journal of Thermal Analysis and Calorimetry, 97, 611–614, 2009; [5] Properties of Brazilian gasoline mixed with hydrated ethanol for flex-fuel technology, Fuel Processing Technology, 88, 365–368, 2007.

Name: Marta Fernández-García
Country: Spain
Date and place of birth: 1967, Madrid, Spain
Present position and address: Research Scientist at Institute of Polymer Science and Technolology (ICTP-CSIC), Madrid, Spain
Email: martafg@ictp.csic.es
Website: www.ictp.csic.es
Education and scientific degrees: Chemist (1985–1990), Ph.D. in Chemistry (1995), Universidad Complutense de Madrid. Master in Plastics and Rubbers (1991), ICTP-CSIC. Tenured Scientist, (2002–2009), Research Scientist (2009–) at ICTP-CSIC.
Workplaces: ICTP-CSIC (1990–1996), National Institute of Standards and Technology (NIST) (1997–1998), ICTP-CSIC (1999–)
Relevant categories in thermal analysis: fields (synthesis and characterization of polymeric materials, nanomaterials, self-organization, hybrid materials, antimicrobial polymeric materials); methods (DSC, MDSC, TGA, thermomechanical analysis)
Professional activities: Member of the international advisory board of the European Polymer Journal since 2008. Vicepresident of the Spanish Group of Thermal Analisys and Calorimetry (GECAT) since 2013, acting previously as member of the board (2003–2005, 2007–2009), secretary (2005–2007) and treasurer (2009–2013). Member of the Committee of Revista de Plásticos Modernos since 2002. Member of the International Advisory Board of the 43rd and 44rd IUPAC World Polymer Congress, Macro2010 and Macro2012 in UK and US, respectively. Member of Scientific Committee del POLYMAR2013 in Spain
Publication record: papers (120), books (1 book, 6 book chapters), citations (>1,200)
Equipments: DSC, MDSC, TGA, thermomechanical analysis
List of important publications: [1] Polymeric Materials with Antimicrobial Activity: From Synthesis to Applications, Eds. A. Muñoz-Bonilla, M.L. Cerrada, M. Fernández-García; Royal Society of Chemistry (2014), ISBN: 978-1-84973-807-1; [2] A. Muñoz-Bonilla, O. León, V. Bordegé, M. Sánchez-Chaves, M. Fernández-García, Controlled block glycopolymers able to bind specific proteins, Journal of Polymer Science. Part A: Polymer Chemistry, 51, 1337–1347 (2013); [3] A. Muñoz-Bonilla, M. Fernández-García, Polymeric materials with antimicrobial activity, Progress in Polymer Science, 37, 281–339 (2012); [4] M.L. Cerrada, C. Serrano, M. Sánchez-Chaves, M. Fernández-García, F. Fernández- Martín, A. de Andrés, R.J. Jiménez Riobóo, A. Kubacka, M. Ferrer, M. Fernández-García, Self-sterilized EVOH-TiO2 nanocomposites: Interface effects on biocidal properties, Advanced Functional Materials, 18, 1949–1960 (2008); [5] A. Kubacka, C. Serrano, M. Ferrer, H. Lünsdorf, P. Bieleck, M.L. Cerrada, M. Fernández-García, M. Fernández-García, Synthesis high-performance dual-action polymer-TiO_2 nanocomposite films via melting processing, Nano Letters, 7, 2529–2534 (2007)

Name: Dimitrios Fessas
Country: Italy
Date and place of birth: 1961, Rhodes, Greece
Present position and address: Associate Professor of Physical Chemistry, University of Milan, DeFENS, via Celoria 2—20133 Milan Italy
Email: Dimitrios.Fessas@unimi.it
Website: http://users.unimi.it/~thalasa/dimitrios2.htm
Education and scientific degrees: Degree in Physics, University of Naples (1990). CNR, Italy scholar fellow (1991–1992), Ph.D. in Chemical Sciences University of Naples (1995), Research Professor, Univ. of Milan (1996–2005), Associate Professor Univ. of Milan (2006–)
Workplaces: University of Naples, Italy (1991–1996). University of Milan, Italy (1996–)
Main fields of interest: 1. physico-chemical properties of food and pharmaceutical systems. 2. thermodynamics of biological macromolecules in solution. 3. restoring processing and characterization of cultural heritage biomaterials
Relevant categories in thermal analyses: fields (materials, nano, complex, pharmaceutical, bio-polymer, food, biology, life, glass, microbs); methods (TG, DTA, ITC, DSC, Knudsen TG, kinetics, cryo; specific heat, calorimetry, microcalorimetry, instrument development)
Professional activities: Member of AICAT also as member of the Board of Directors (2005–2008, 2011–). Chairman of the "XXVIII National Conference, AICAT-GICAT 2006, Milano, Italy. Member of the National Professional Association of Italian Qualified Experts in Radiological Protection
Publication record: papers (90), books (25), h-index (18)
Equipments: ITC (Calvet) DAM (Setaram); TGA-DSC 111 (Setaram); Micro DSC III (Setaram); DSC C80D (Setaram); DSC 6 (PerkinElmer); DSC 20 (Mettler); Nano-ITC III Model 5300 (CSC USA); Nano-DSC Model 6300 (CSC USA); DSC 2920 (TA Instruments); MASC 2 (CNR—IPCF T-lab Pisa, Italy)
5 most important publications: [1] Fessas, D. et al. Guidelines for buckwheat enriched bread—Thermal analysis approach, Journal of Thermal Analysis and Calorimetry, 91 (2008) 9-16; [2] Fessas, D; Schiraldi, A. Water properties in wheat flour dough II: classical and knudsen thermogravimetry approach. Food Chemistry, 90 (2005) 61–68; [3] Fessas, D; et al. Thermal unfolding of monomeric and dimeric beta-lactoglobulins. European Journal of Biochemistry, 268 (2001) 5439–5448; [4] Fessas, D; et al. Thermal analysis on parchments I: DSC and TGA combined approach for heat damage assessment. Thermochimica Acta, 447 (2006) 30–35; [5] Tripsianes, K; et al. Structural basis for dimethylarginine recognition by the Tudor domains of human SMN and SPF30 proteins. Nature Structural and Molecular Biology, 18 (2011) 1414–1420.

Name: Ludger O. Figura
Country: Germany
Date and place of birth: 1961, Quakenbrueck, Germany
Present position and address: Academic Dean Faculty of Agricultural Sciences, Head of Food Engineering, University of Applied Sciences, Osnabrueck, Germany
Email: l.figura@hs-osnabreueck.de
Website: https://my.hs-osnabrueck.de/al/29271.html
Researcher ID/ORCID: E-9032-2011/0000-0003-4689-0961
Education and scientific degrees: Chemistry, Technical University Braunschweig, Germany (1980–1986), Dr.rer.nat. Physical Chemistry (1989), Professor Food Engineering (1995)
Workplaces: German Institute for Food Technologies, Quakenbrueck, Germany; University of Applied Sciences, Bremerhaven, Germany; University of Applied Sciences, Osnabrueck, Germany
Main fields of interest: physical properties of food
Relevant categories in thermal analyses: fields (food, glass); methods (TG, DSC, calorimetry)
Professional activities: GDL—Gesellschaft Deutscher Lebensmitteltechnologen (Society of German Food Technologists); GEFTA—Gesellschaft für thermische Analyse (Society for Thermal Analysis); ICFP—International Conference of Food Properties
Publication record: papers (28 journal papers, 38 proceedings), books (3 books, 2 book chapters), patents (3)
Equipments: DSC, TGA, Combustion Calorimeter
5 most important publications: [1] Figura, L.O. Quantification of non-crystalline Trehalose by DSC-Devitrification, Proceedings of the 2014 International Conference on Food Properties (ICFP 2014); Kuala Lumpur, Malaysia, January 24–26, 2014; [2] Spanneberg, R., et al. (2012). "Glyoxal modification of gelatin leads to change in properties of solutions and resulting films." Soft Matter 8(7): 2222–2229; [3] Figura, L. O., et al. (2007). Food physics —Physical Properties—Measurement and Application. Springer, Berlin, London, New York, 550 p; [4] Figura, L. O. (2004). Lebensmittelphysik : physikalische Kenngrößen, Messung und Anwendung; mit 195 Tabellen. Springer, Berlin, London, New York; [5] Figura, L. O. and M. Epple (1995). "Anhydrous α-lactose A study with DSC and TXRD." Journal of Thermal Analysis and Calorimetry V44(1): 45–53.

Name: Elżbieta Filipek
Country: Poland
Date and place of birth: 1956, Stalowa Wola, Poland
Present position and address: Professor, West Pomeranian University of Technology, Szczecin (ZUT)—Head of Department of Inorganic and Analytical Chemistry, al. Piastow 42, 71-065 Szczecin, Poland
Email: elafil@zut.edu.pl
Website: www.zut.edu.pl
Education and scientific degrees: M.Sc. (1981—Szczecin University of Technology), Ph.D. (1991—Adam Mickiewicz University, Poznań), D.Sc. (2008—Adam Mickiewicz University, Poznań), Assoc. Professor ZUT (2009–2014), Professor (2014)
Workplaces: Szczecin University of Technology (1981–2009), West Pomeranian University of Technology, Szczecin, Faculty of Chemical Technology and Engineering, Department of Inorganic and Analytical Chemistry (2009–)
Main fields of interest: chemistry and physicochemistry of inorganic solids; synthesis and properties of new compounds and solid solutions formed in the multicomponent metal oxide systems (oxides—ceramic composites, pigments, catalysts, semiconductors, gas sensors); phase equilibria in oxide systems; thermoanalytical methods, XRD, IR, SEM/EDX/WDX
Relevant categories in thermal analyses: fields (inorganic chemistry, new functional materials, phase equilibria, ceramics, minerals catalysts, thermal stability of materials); methods (TG/DTG, DTA, EGA, DSC, high temperature X-ray diffraction)
Awards and acknowledgments: Silver (1998) and gold (2013) Cross of Merit conferred by President of RP (Poland), 15 awards of Rector of the Szczecin University of Technology (1985–2008) and 5 awards of Rector of ZUT (2009–2013)—for the special scientific achievements
Professional activities: Member of the: Board of Polish Society of Calorimetry and Thermal Analysis—PTKAT (2009–), Vice-chairman of Organizing Committee of CCTA 10—Zakopane, Poland (2009), Member of Scientific Committee of CCTA 11—Zakopane, Poland (2012) and ISTAC 10—Plock, Poland (2011) ICTAC Member (2000–2004, 2008–2012), Member of: Senate of Szczecin University of Technology (1996–2005), Senate of ZUT (2012–). Member of Polish Chemical Society—PTCh (1991–2002, 2012–). Reviewer of papers
Publication record: papers (>155), patents (18), citations (386), h-index (11), sum of impact factors ($\sim$ 140)
Equipments: TA Instruments thermoanalyzer (model SDT 2960, USA), TA Instruments microcalorymeter (model DSC 2010, USA), Derivatograph and Q-1500D (MOM)
5 most important publications: [1] E. Filipek: "Homogeneity area of MoO_3 solid solution in $SbVO_5$ in air", Solid State Sciences., 8 (2006) 577–588; [2] E. Filipek, G. Dąbrowska: "Synthesis and Selected Properties of $CrSbVO_6$ and Phase relations in the V_2O_5–Cr_2O_3–α-Sb_2O_4 system in the solid state", Journal of Materials Science, 42 (2007) 4905–4915; [3] E. Filipek, M. Piz: "The reactivity of $SbVO_5$ with T–Nb_2O_5 in solid state in air", Journal of Thermal Analysis and Calorimetry, 101 (2010) 447–453; [4] E. Filipek, G. Dąbrowska: "New solid solution $Fe_{1-x}Cr_xVSbO_6$ with rutile-type structure", Journal of Alloys and Compounds, 523 (2012) 102–107; [5] M. Ziolek, H. Golinska-Mazwa, E. Filipek, M. Piz: "Catalytic properties of new ternary Nb–Sb–V oxide—a comparative study with mechanical mixture of single oxides and binary systems", Catalysis Today, 187 (2012) 159–167.

Name: Henrik Fordsmand
Country: Denmark
Date and place of birth: 1955, Copenhagen, Denmark
Present position and address: Research Scientist. Haldor Topsøe A/S, Nymøllevej 55, DK2800 Kgs. Lyngby, Denmark
Email: hefo@topsoe.dk
Website: www.topsoe.com
Education and scientific degrees: M.Sc. Chemical Engineer, Danish Technical University 1980; Ph.D. equivalent degree from the Industrial Eduactional Reseach Programme under the Danish Academy of Technical Sciences. 1991
Workplaces: Danish Technical University 1980–1981, Sandvik Hard Materials 1981–1989, Danish Technological Institute 1989–199, Faxe Kalk A/S later JM Huber Denmark ApS 1991–2007, Haldor Topsøe A/S 2007–
Main fields of interest: materials science, catalysis, materials chracterization (thermal analysis, porosimetry, particle sizing, rheology, colloid caracterization), materials preparation processes (precipitation, crystalization, filtration, drying, coating, powder handling, calcination)
Relevant categories in thermal analyses: fields (inorganic, materials, nano, ceramics, other (catalysis)); methods (TG, EGA, DSC, thermomechanical analysis, kinetics, high temperature (above 1,000 °C), specific heat, other (temperature programmed reactions))
Professional activities: Member of the Board DaSTA and NoSTAC
Equipments: 1 TG, 2 STA, 1 Dilatometer
5 most important publications: [1] An Explanation of Wettability Problems When Brazing Cemented Carbides. Thorsen K.A., Fordsmand H., Praestgaard P.L., Weld. J., (1984), 65(10), p 308–315; [2] Composite precipitated calcium carbonate/silicate pigment with good combination of optical and mechanical properties that imparts improved printing properties to paper. Haverinen J.P., Seuna E.H., Fordsmand, H., PCT Int. Appl. (2001), WO 2001092422 A1 20011206; [3] A new composite fine paper with high filler loading and functional cellulosic microfines. Subramanian R., Fordsmand H., Paltakari J., Paulapuro H., J. Pulp Paper Sci. (2008), 34(3), 146–152; [4] Role of internal coke for deactivation of ZSM-5 catalysts after low temperature removal of coke with NO_2. Barbera K., Sorensen S., Bordiga S., Skibsted J., Fordsmand H., Beato P., Janssens T.V.W., Catal. Sci. Tech. (2012), 2(6), 1196–1206; [5] High surface area calcite. Schultz L.N., Andersson, M.P., Dalby K.N., Muter D., Okhrimenko D.V., Fordsmand H., Stipp S.L.S., J. Cryst. Growth (2013), 371, 34–38.

Name: Mária Földvári
Country: Hungary
Date and place of birth: 1944, Budapest, Hungary
Present position and address: retired from the Hungarian Geological Institute, Budapest
Email: foldvari@t-online.hu
Education and scientific degrees: Geologist, Eötvös Loránd University, Budapest (1968), Ph.D. (1996)
Workplace: Hungarian Geological Institute, Budapest (1968–2007)
Main fields of interest: thermoanalytical investigation of minerals and complex phase analysis (thermal analysis, XRD, FTIR etc.) of rocks; thermal decomposition of minerals, water bonds and water types in minerals and rocks, crystallinity of minerals, substitution, characterization of geological processed based on thermoanalytical data
Relevant categories in thermal analyses: fields (mineral); methods (TG, DTA)
Awards and acknowledgments: Prize of Hungarian Academy of Science (1983) Vendl Mária medal (1997), Szabó József medal (2012)
Professional activities: member of the TA committee of the Hung. Acad. Sci. (1980–); Geochemistry, Mineralogy and Petrology scientific committee of the Hung. Acad. Sci; secretary (1996–2002); Clay Minerals Division of Hungarian Geological Society: secretary (1981–1994), president (1994–2004); member of ICTA Committee for Thermal Analysis in Geosciences (1988–1996); Hungarian Geological Society member of Board (1994–2012)
Publication record: papers (114), books (4), citations (128), sum of impact factors (51.8)
Equipments: Derivatograph-PC
5 most important publications: [1] Földvári M. 2011:Handbook of thermogravimetric system of minerals and its use in geological practice. Budapest, (Occasional Papers of the Geological Institute of Hungary Vpl. 213.) 180 p; [2] Földvári M. 1991: Measurement of different water species in minerals by means of thermal derivatography. in Series of Lecture Notes in Earth Sciences 38. (Eds.) Smykatz-Kloss W., Warne S.St.J. Thermal Analysis in the Geosciences, Springer Verlag. Berlin-Heidelberg-New York-London-Paris-Tokyo-Hong Kong-Barcelona-Budapest 379. pp. 84–100; [3] Földvári M. 1997: Kaolinite genetic and thermoanalytical parameters. J. Therm. Anal. 48. pp. 107–119; [4] Heide K., Földvári M: 2006: High temperature mass spectrometric release studies of kaolinite decomposition ($Al_2[Si_2O_5(OH)_4]$). Thermochimica Acta, 446. 1–2. pp. 106–112; [5] Kónya P., Földvári M. 2008: Thermoanalytical investigation of cavity filling natrolite group minerals of basalts from Balaton Highland, Hungary. J. Therm. Anal. Calorim. 94. 1. pp. 209–218.

Name: Francisco Fraga-López
Country: Spain
Date and place of birth: 1966, Narón, Galicia, Spain.
Present position and address: Professor of Applied Physics. University of Santaigo of Compostela. Faculty of Sciences. Avenida Alfonso X el Sabio S/N.27002. Lugo. Spain.
Email:francisco.fraga@usc.es
Education and scientific degrees: Professor of Applied Physics (2002–). Ph.D. in Physics (1996).
Workplaces: Faculty of Sciences. University of Santiago de Compostela. Spain.
Main fields of interest: nanotechnology, thermal analysis, DSC, FTIR, thermogravimetric analysis, DMA of thermosets, natural polymer and thermoplastics; in the last years we are working in solar cells with nanocomposites of epoxy resin with gold nanoparticles in my research group
Relevant categories in thermal analyses: fields (materials, nano, organic, pharmaceutical, polymer, food, glass, ceramic, supramolecular chemistry. In the last years we are working in solar cells with nanocomposites of epoxy resin with gold nanoparticles in my research group); methods (TG, DTA, DSC, thermomechanical analysis, kinetics; specific heat, calorimetry, microcalorimetry, DMA, light scattering)
Professional activities: Dean of Faculty of Sciences of University of Santiago de Compostela (2002–2010). Vicepresident of Grupo Especializado de Calorimetría y Análisis Térmico E spañol(2009–2012), President of (2012–) Grupo Especializado de Calorimetría y Análisis Térmico Español
Publication record: papers (70), books (5 book chapters), citation index (450), sum of impact factors (120)
Equipments: DSC, Light Scattering, Rheometer, TGA, Microcalorimeter
5 most important publications: [1] Lisardo Núñez, F. Fraga, M. R. Núñez, and M. Villanueva, Thermogravimetric Study of the Decomposition Process of the System BADGE (n = 0)/1, 2 DCH, Polymer, 41, 4635–4641, 2000; [2] L. Núñez, J. Taboada, F. Fraga and M. R. Núñez, Kinetic Study and Time-Temperature-Transformation Cure Diagram for an Epoxy/Diamine System, Journal of Applied Polymer Science, 66, 1377–1388, 1997; [3] F. Fraga, S. Burgo and E. Rodríguez Núñez, Curing Kinetic of the Epoxy System BADGE n = 0/1, 2 DCH by Fourier Transform Infrared Spectroscopy (FT-IR), Journal of Applied Polymer Science, 82. 3366–3372, 2001; [4] Lisardo Núñez, F. Fraga, A. Castro, M. R. Núñez, and M. Villanueva, TTT Cure Diagram for An Epoxy System Diglycidyl Ether of Bisphenol A/1, 2 Diamine Cyclohexane/Calcium Carbonate Filler, Polymer, 42, 3581–3587, 2001; [5] Lisardo Núñez, F. Fraga, M. R. Núñez and M. Villanueva, Effects of Diffusion on the Kinetic Study and TTT Diagram for an Epoxy/Diamine System, Journal of Applied Polymer Science, 70 (10), 1931–1938, 1998.

Name: Ray L. Frost
Country: Australia
Date of birth: 1943
Present position and address:Professor of Physical Chemistry, QUT PO Box 2434 GPO Brisbane, QLD 4001
Email: r.frost@qut.edu.au/qut.edu.au
Website: http://searching.qut.edu.au/search/search.cgi?query=Ray+Frost&collection=staffsitemeta
Education and scientific degrees: D.Sc., Ph.D., M.Sc., BEd
Workplaces: Queensland University of Technology
Main fields of interest: thermal analysis and spectroscopy of minerals
Relevant categories in thermal analyses: fields (TG, DTG, DSC); methods (CRTA)
Professional activities: RACI
Publication record: papers (>1,200), books (12 book chapters), patents (2), citations (>14,613), h-index (56)
5 most important publications: [1] Frost, R. L; Martens, W.; Ding, Z.; Kloprogge, J. T. DSC and high-resolution TG of synthesized hydrotalcites of Mg and Zn. Journal of Thermal Analysis and Calorimetry (2003), 71(2), 429–438; [2] Frost, R. L. Raman spectroscopy of selected copper minerals of significance in corrosion. Spectrochimica Acta, Part A: Molecular and Biomolecular Spectroscopy (2003), 59A(6), 1195–1204; [3] Frost, Ray L.; Kristof, Janos; Kloprogge, J. Theo; Horvath, Erzsebet. Deintercalation of Hydrazine-Intercalated Kaolinite in Dry and Moist Air. Journal of Colloid and Interface Science (2002), 246 (1), 164–174; [4] Frost, Ray L.; Kristof, Janos; Horvath, Erzsebet; Kloprogge, J. Theo. Separation of Adsorbed Formamide and Intercalated Formamide Using Controlled Rate Thermal Analysis Methodology. Langmuir (2001), 17(11), 3216–3222; [5] Frost, R. L.; Williams, P. A.; Martens, W. Raman spectroscopy of the minerals boleite, cumengeite, diaboleite and phosgenite—implications for the analysis of cosmetics of antiquity. Mineralogical Magazine (2003), 67(1), 103–111.

Name: Patrick Kent Gallagher
Country: USA
Date and place of birth: 1931, IL, USA
Present position and address: Emeritus Prof., The Ohio State Univ. 409 South Way Court, Salem, SC, 29676, USA
Email: p_gallag@bellsouth.net
Education and scientific degrees: BS, MS, and Ph.D. Inorganic Chemistry, Univ. Of Wisconsin
Workplaces: USMC 1954-7, AT&T Bell Telephone Laboratories 1959-89, The Ohio State Univ, 1990-8
Main fields of interest: synthesis, characterization, and reactivity of solids
Relevant categories in thermal analyses: fields (inorganic materials, solid state kinetics, instrument development); methods (TG, DTA, DSC, TMA, calorimetry, spectroscopies, X-ray diffraction)
Awards and acknowledgments: NATAS Outstanding Achievement (Mettler-Toledo) Award (1976), TAI/ICTAC (1982), Netzsh/GEFTA(1996), Kurnikov Medal (1985), Semiconductor International Research (1986), ASTM/Merit (2004), Honorary Membership ICTAC (2000), NATAS Service Award (1985), Fellow-NATAS, AmCerSoc, AmAsAdvSci
Professional activities: ICTAC/Council, VP, Pres, Treas, Tech ProgChair(2), Stand Com Chair; NATAS/ExCom, VP, Pres, Tech ProgChair(3); ASTM E-37/Chair, Editorial/TCA, MRS, AmCerSoc, ACS, ReactSolids, TAAbstracts, TrendsAnalChem
Publication record: papers (>250), books (Series Editor "Handbook of Thermal Analysis", 5 Volumes), patents (5), citations (Most Cited List Materials Chemistry)
5 most important publications: [1] Ceramic Reactions and Phase Behavior, P. K. Gallagher, in "Characterization of Ceramics", R. E. Loehman (ed.), Manning, Pub., Greenwitch, CT, 1993, Chapt. 8, pp. 137–68; [2] Characterization of $Ba_2YCu_3O_X$ as a Function of Oxygen Partial Pressure. Part I: Thermoanalytical Measurements, P. K. Gallagher, Adv. Ceram. Mat., 2, 632–639 (1987). [Am. Cer. Soc., Pittsburg, PA, Apr. 1987]; [3] A New LPCVD Technique of Producing Borophosphosilicate Glass Films by Injection of Miscible Liquid Precursors, R. A. Levy, P. K. Gallagher and F. Schrey, J. Electrochem. Soc., 134, 430–437 (1987); [4] Applications of Evolved Gas Analysis to the Study of Inorganic Materials and Processes, P. K. Gallagher, J. Ther. Anal., 25, 7–20 (1982). [7th Nordic Symposium on Thermal Analysis, Helsinki, Finland, June 1982, Plenary Lecture]; [5] Reactive Powders from Solution, (Invited) D. W. Johnson, Jr. and P. K. Gallagher, "Ceramic Processing Before Firing", G. Onada, Jr. and L. Hench (Eds.), Wiley-Sons, 1978, New York, pp. 125–139. [10th Univ. Conf. on Ceram. Sci., Gainsville, FL, January 27–30, 1975].

Name: Andrew Knox Galwey
Country: United Kingdom
Date and place of birth: 1933, Dublin, Ireland
Present position and address: Retired (1995) Reader in Physical Chemistry, Queen's University of Belfast, Northern Ireland. Currently living in Belfast.
Email: aandk.galwey@talktalk.net
Education and scientific degrees: B.Sc. (Honours Special Chemistry), University of London, 1955; Ph.D. (Physical Chemistry) University of London (Imperial College) 1958; Diploma of Imperial College (DIC) 1958; D.Sc., University of London, 1973.
Workplace: School of Chemistry, The Queen's University of Belfast, 1957–1995
Main fields of interest: kinetics of solid-state thermal decompositions; chemistry of solids
Relevant categories in thermal analyses: fields (mechanisms and reactivity controls in reactions of inorganic and organic solids); methods (literature reviews, theory, vacuum apparatus. microscopy and electron microscopy)
Professional activities: Fellow of the Royal Society of South Africa, Elected 2003
Awards and acknowledgments: 2 journal special issues: (i) Journal of Thermal Analysis, Special Issue for AK Galwey on the occasion of 60th birthday. ME Brown (Ed), 41 (2–3) Feb-Mar 1994, pp. 249–739; (ii) Thermochimica Acta, Special Issue: "Kinetics of Thermally Stimulated Reactions": Invited Papers in Honour of AK. Galwey and ME Brown on their respective 70th and 65th Birthdays. S. Vyazovkin and T. Ozawa (Eds.) 388 (1–2) June 2002, pp. 1–460.
Publication record: papers (163 journal papers, 39 review papers, 22 conference proceedings), books (3 books, 2 edited books)
5 most important publications: [1] Andrew K Galwey and Michael E Brown, Thermal Decomposition of Ionic Solids, Elsevier, Amsterdam, 1999, pp. 597; [2] Boris V L'vov and Andrew K Galwey, Interpretation of the kinetic compensation effect in heterogeneous reactions: thermochemical approach. International Reviews of Physical Chemistry, 32 (2013) 515–557; [3] Andrew K Galwey, Roberto Spinicci and Giulio GT Guarini, Nucleation and growth processes occurring during dehydration of certain alums: the generation, the development and the function of the reaction interface. Proceedings of the Royal Society (London), A378 (1981) 477–505; [4] Andrew K Galwey and Mohamed A Mohamed, The low temperature decomposition of ammonium perchlorate: nitryl perchlorate as the intermediate. Proceedings of the Royal Society (London), A396 (1984) 425–440; [5] Andrew K Galwey, Theory of solid-state thermal decomposition reactions. Scientific stagnation or chemical catastrophe? An alternative approach appraised and advocated. Journal of Thermal Analysis and Calorimetry, 109 (2012) 1625–1635.

Name: Heinz Gamsjäger
Country: Austria
Date and place of birth: 1932, Donawitz, Austria
Present position and address: Professor Emeritus, Montanuniversität Leoben, Franz-Josef-Straße 18, A-8700 Leoben, Austria
Email: gamsjaeg@unileoben.ac.at
Website: https://online.unileoben.ac.at/mu_online/visitenkarte.show_vcard?pPersonenId=4CB74842975D0419&pPersonen Gruppe=3
Education and scientific degrees: 1956: M.Sc. (Dipl.-Ing.) TU Graz1960: Ph.D. (Dr.mont.) MU Leoben, 1969: Docent, MU Leoben, 1971: PD University of Berne, Switzerland, 1974: Assoc. Prof. Uni Berne, 1975: Full Prof. MU Leoben, 2000: Em. Prof. MU Leoben
Workplaces: Institute of Physical Chemistry MU Leoben 1956–1968, Institute of inorganic, anaytical and physical Chemistry Uni Berne (1968–1969), Swiss Federal Institute of Reactor Research (1969–1971), Inst. inorg., anayt. and phys. Chem. Uni Berne (1971–1975), Inst. Phys. Chem. MU Leoben (1975–)
Main fields of interest: investigation of solid-solute phase equilibria in aqueous solutions; computer-assisted calculation and optimization of phase diagrams; modelling of geochemical and metallurgical processes
Relevant categories in thermal analyses: fields (thermodynamic analyses of solubility phenomena); methods (solubility measurements)
Awards and acknowledgments: Schrödinger prize 1995, Austrian Academy of Sciences
Professional activities: Co-chairman and organizer of the 14th ISSP 2010, Leoben, Austria, IUPAC Subcommittee on Solubility and Equilibrium Data: Chairman 2002–2007, IUPAC Commission on Solubility Data V.8: Secretary 1996–2001, Editorial Board "Monatshefte für Chemie—Chemical Monthly" since 1998, Chairman and organizer of the 7th ISSP 1996, Leoben, Austria, Editorial Board IUPAC Solubility Data Series since 1994
Publication record: papers (120), books (2 books, 2 book chapters)
5 most important publications: [1] H. Gamsjäger, Solubility phenomena in science and education: Experiments, thermodynamic analyses, and theoretical aspects. Pure Appl. Chem., 85 (2013) 2059; [2] Gamsjäger, H., Gajda, T., Sangster, J., Saxena, S. K., Voigt, W. Chemical Thermodynamics, vol. 12. In NEA Data Bank, OEC, editor, Chemical Thermodynamics of Tin. Editor, Jane Perrone, OECD NEA, Data Bank, Issy-les-Moulineaux (France), 2013; [3] Gamsjäger, H., Bugajski, J., Gajda, T., Lemire, R.J., Preis, W. ChemicalThermodynamics, vol. 6. In NEA Data Bank, OECD, editor, Chemical Thermodynamics of Nickel. North Holland Elsevier Science Publishers B. V., Amsterdam, The Netherlands, 2005; [4] Gamsjäger, H.; Königsberger, E. In Hefter, G.T.; Tomkins, R.P.T., editor, The experimental determination of solubilities, chapter 4.2 Solubility of sparingly soluble solids in liquids, pages 315–358. Wiley, 2003; [5] Gamsjäger H., Königsberger E., Preis W. Solubilities of metal carbonates. Pure Appl. Chem., **70**:1913–1920, 1998.

Name: Yulai Gao
Country: P.R. China
Date and place of birth: 1975, P.R. China
Present position and address: Professor, School of Materials Science and Engineering, Shanghai University, 149 Yanchang Road, Shanghai 200072, P.O. Box 275 P.R. China
Email: ylgao@shu.edu.cn
Website: http://www.mat.shu.edu.cn/Default.aspx?tabid=2078&ctl=Detail&mid=2755&Id=60869&SkinSrc=[L]Skins/mat_new_con1/mat_new_con1
ORCID: 0000-0001-9732-5490
Education and scientific degrees: Materials Processing Engineering, Shanghai University (2003-present); Assoc. Professor (2005); Professor (2010). Florida State University (2012–2014); Senior visiting professor (2012–2013); Adjunct professor (2013–2014)
Workplaces: Shanghai University (2003-present), Florida State University (2012–2014)
Main fields of interest: nanocalorimetry; metal solidification; nano and amorphous materials
Relevant categories in thermal analyses: fields (metallic materials); methods (DSC, nanocalorimetry)
Publication record: papers (100), citations (520)
5 most important publications: [1] B. G. Zhao, L. F. Li, Q. J. Zhai, Y. L. Gao*. Formation of amorphous structure in Sn3.5Ag droplet by in situ controllable quenching. Applied Physics Letters. 2013, 103: 131913; [2] B. G. Zhao, L. F. Li, B. Yang, G. Wang, Q. J. Zhai, Y. L. Gao*. Structure observation of single solidified droplet by in situ controllable quenching based on nanocalorimetry. Journal of Alloys and Compounds. 2013, 580: 386–391; [3] L. H. Kong, Y. L. Gao*, T. T. Song, G. Wang, and Q. J. Zhai. Non-isothermal crystallization kinetics of FeZrB amorphous alloy. Thermochimica Acta. 2011, 522: 166–172; [4] B. Yang, Y. L. Gao*, C. D. Zou, Q. J. Zhai, A. S. Abyzov, E. Zhuravlev, J. W. P. Schmelzer, C. Schick. Cooling rate dependence of undercooling of pure Sn single drop by fast scanning calorimetry. Applied Physics A. 2011, 104(1): 189–196; [5] Y. L. Gao*, E. Zhuravlev, C. D. Zou, B. Yang, Q. J. Zhai, and C. Schick. Calorimetric measurements of undercooling in single micron sized SnAgCu particles in a wide range of cooling rates. Thermochimica Acta. 2009, 482(1–2): 1–7.

Name: Konstantin S. Gavrichev
Country: Russia
Date and place of birth: 1950, Paris, France
Present position and address: Deputy Director, Head of Laboratory, Kurnakov Institute of General and Inorganic Chemistry of RAS; 31, Leninsky prospect, Moscow, 119991, Russia
Email: gavrich@igic.ras.ru
Website: http://www.igic.ras.ru/
Researcher ID/ORCID: B-6304-2013/0000-0001-5304-3555
Education and scientific degrees: Chemical Engineer, Lomonosov Moscow Institute of Fine Chemical Technology (1973), Ph.D. (1983), Dr.Sci. (2003).
Workplaces: Kurnakov Institute of General and Inorganic Chemistry of the Russian Academy of Sciences (1973-present time)
Main fields of interest: thermal and thermodynamic properties of inorganic compounds; modeling of temperature dependences of the heat capacity in connection with the peculiarities of the structure; design of calorimetric equipment.
Relevant categories in thermal analyses: fields (inorganic, materials, nano, minerals, complex, glass, ceramics); methods (relaxation calorimetry, adiabatic calorimetry, drop-calorimetry, DSC/TG)
Awards and acknowledgments: Medal "In memory of 850 anniversary of Moscow" (1997), Sertificates for the best scienrtific publications in MAIK NAUKA/Interperiodica Publ. (1996, 2010), Kurnakov Medal (2002), Bronze medal of 32nd Geneva Salon of Innovations (2004)
Professional activities: Vice-chairman and organizer of 18th and 19th RCCT (Samara, 2011 and Moscow, 2013) and 14th TAC(Saint-Petersburg, 2013); member of the Mendeleev Russian chemical society and American chemical society, Chairman of the Scientific council on chemical thermodynamics and thermochemistry of Russian Academy of Sciences
Publication record: papers (205), books (2), patents (11), citations index (679), h-index (11), sum of impact factors (195)
Equipments: Self-made IGIC RAS low temperature adiabatic calorimeter, BKT-3 LLC "Termis", MHTC/DSC2000 (Setaram), Netzsch STA 449 F1 Jupiter, DSC Mettler 4000
5 most important publications: [1] K. S. Gavrichev et al., J. of Therm. Anal. Calorim. 102 (2010) 809–811; [2] Zh. V. Dobrokhotova et al., Thermochim Acta 556 (2013) 68–74; [3] K.S. Gavrichev, Inorganic Materials (Supplement 2) 39 (2003) S88–S112; [4] K. S. Gavrichev et al., J. of Therm. Anal. Calorim. 73 (2003) 71–83; [5] K. S. Gavrichev et al., Thermochimica Acta 217 (1993) 77–89.

Name: Konstantin B. Gerasimov
Country: Russia
Date and place of birth: 1955, Barnaul, Russian Federation
Present position and address: Novosibirsk, Russian Federation. Kutateladze-18, Novosibirsk, 630128, Russia
Email: gerasimov@solid.nsc.ru
Website: http://www.solid.nsc.ru
Education and scientific degrees: Chemist, Novosibirsk State University (1972–1977); Ph.D. (1986)
Workplaces: Institute of Solid State Chemistry and Mechanochemistry SB RAS. (1977–)
Main fields of interest: reactivity of solids, mechanism of chemical reactions in solid state, mechanical alloying.
Relevant categories in thermal analyses: fields (inorganic, materials, ceramics); methods (TG, DTA, DSC, dilatometry)
Publication record: papers (53)
Equipments: Simultaneous TG-DTA/DSC Apparatus STA 449 F1 Jupiter with quadrupolle mass spectrometer QMS 403C Aeolos, NETZSCH GmbH, Germany; Dilatometer DIL 402C, NETZSCH GmbH, Germany; Differential Scanning Calorimeter DSC 200 F3 Maia, NETZSCH GmbH, Germany
5 most important publications: [1] K. B. Gerasimov, V. V. Boldyrev: Mat. Res. Bulletin, 31(1991) 1297; [2] K. B. Gerasimov, S. V. Mytnichenko, S. V. Pavlov, V. A. Chernov, S. G. Nikitenko: J. Alloys Comp. 252 (1997) 179; [3] K. B. Gerasimov, S. V. Pavlov: J. Alloys Comp. 242 (1996) 1361; [4] Y.-S. Kwon, K.B. Gerasimov, S.-K. Yoon: J. Alloys Comp. 346 (2002) 2708; [5] K. B. Gerasimov, I. G. Konstanchuck, S. A. Chizhik, J.-L. Bobbet: Int. J. Hydrogen Energy, 34 (2009) 1916.

Name: Concetta Giancola
Country: Italy
Year and place of birth: 1959, Campobasso, Italy
Current position and postal address: Associate Professor of Physical Chemistry, Department of Pharmacy, University of Naples Federico II, Via D.Montesano, 49—80131, Naples, Italy
Email: giancola@unina.it
Website: www.docenti.unina.it/concetta.giancola
ORCID: 0000-0002-0360-6062
Education and scientific degrees: Degree in Chemistry, 1983; Ph.D. in Chemical Sciences, 1992; Assistant Professor of Physical Chemistry, 1991; Associate Professor of Physical Chemistry, 2001
Workplaces: June-July 2001 and May 2003: Visiting Researcher, Department of Biochemistry and Molecular Biology, University College, London.; March 2007: Visiting Professor, Randall Division of Cell and Molecular Biophysics, King's College, London
Main fields of interest: Biophysical chemistry of biological macromolecules stability and their interactions with drugs or other bio-macromolecules, nanocapsules termodynamic stability
Relevant categories in thermal analyses: fields (pharmaceutical, biology, life); methods (DSC, ITC)
Professional activities: Board Member of Univ. Naples Federico II (2009–2012); Evaluation Board Member of Santobono-Pausilipon Hospital; Chair of the Interdivisional Group of Calorim. and Therm. Anal.(G.I.C.A.T.), formerly member of the Board of AICAT; Associate Editor of the J. Therm. Anal. Calorim.; Editor in Chief of the Int. Rev. Biophys. Chem.; Member of: Italian Chem. Soc. (S.C.I.), Division of Phys.Chem., Division of Chemistry of Biological Systems (D.C.S.B.); Scientific committees: IV Int. Symp. Med. Phys., Poland, 2009. XXX AICAT-GICAT, 2008. XXXII AICAT-GICAT, 2010. 10th MEDICTA 2011. Chair of Third Int. Meeting on G-Quadruplex and G-assembly, 2011. 11th MEDICTA 2013
Awards and acknowledgements: Invited speaker: XXXVI Congresso Nazionale di Chimica Fisica-Plenary lecture, 2008; Int. Conf. of Anticancer Res.—Keynote lecture, Greece, 2008; COST Action: European network on Self-assembled Guanosine Structures for Molecular Electronic Devices, Hungary, 2009; Randall Seminar Series, King's College London, U.K., 2009; IV Int. Symp. on Medical Phys., Poland, 2009—Plenary lecture; European Winter School on Phys. Org. Chem., Italy, 2010; Scuola Normale Superiore, Italy, 2010—Lesson-Seminar; XXXII AICAT-GICAT, Italy, 2010—Plenary lecture; Sigma-Tau, Pomezia (RM), Italy, 2010—Seminar
Publication record: papers (113), books (2 book chapters), citations (2000), h-index (25)
Equipments: Nano ITC Low Volume—TA Instruments, Nano DSC—TA Instruments
5 most important publications: [1] Fotticchia I, Giancola C, Petraccone L, Chem Commun (2013) 49 829488; [2] Giancola C, Pagano B, Top Curr Chem. (2013), 330, 211; [3] Petraccone L, Malafronte A, Amato J, Giancola C, J Phys Chem B. (2012) 116 (7) 2294; [4] Petraccone L, Spink C, Trent JO, Garbett NC, Mekmaysy CS, Giancola C, Chaires JB, J Am Chem Soc. (2011) 133 20951; [5] Cummaro A, Fotticchia I, Franceschin M, Giancola C, Petraccone L, Biochimie (2011) 931392, Top 25 Hottest Articles: Biochemistry, Genetics and Molecular Biology—Biochimie, July–September 2011.

Name: Paweł Gierycz
Country: Poland
Date and place of birth: 1953, Radom, Poland
Present position and address: professor; Faculty of Chemical and Process Engineering, Warsaw University of Technology, Waryńskiego 1, 00-645 Warsaw, Poland
Email: p.gierycz@ichip.pw.edu.pl
Website: http://www.ichip.pw.edu.pl/pl/gierycz
Education and scientific degrees: 1977—M.Sc., Ing., Faculty of Chemical and Process Engineering, Warsaw University of Technology, Poland; 1982—Ph.D., Institute of Physical Chemistry, Polish Academy of Sciences, Warsaw, Poland; 2001—D.Sc., Faculty of Process and Environmental Engineering, Łódź University of Technology, Poland; 2014—Professor, Faculty of Chemical and Process Engineering, Warsaw University of Technology, Poland
Workplaces: Institute of Physical Chemistry, Polish Academy of Sciences, Warsaw, Poland (1977–2014); Radom University of Technology, Faculty of Material Science and Footwear Technology, Radom, Poland (2004–2011); Faculty of Chemical and Process Engineering, Warsaw University of Technology, Warsaw, Poland (2011-present)
Main fields of interest: chemical thermodynamics and thermochemistry; chemical, process and environmental engineering; physical chemistry; information engineering (data bases, etc.)
Relevant categories in thermal analyses: fields (inorganic, materials, nano, minerals, complex, organic, pharmaceutical); methods (TG, DSC, calorimetry, microcalorimetry)
Awards and acknowledgments: Award in the Competition of "Young Master of Technology" (Warsaw 1977); The Commission of National Education Medal (Warsaw 2008); Awards for scientific achievements (Radom University of Technology: 2009, 2010)
Professional activities: Member of the board of Polish Society of Calorimetry and Thermal Analysis memorial of W.Świetosławski (2008–), member (1983–); Member of American Chemical Society (Division of Environmental Chemistry) (2006–); Member of ICTAC (2008–); Member of Editorial Board of Journal of Chemical and Engineering Data (2010–2012); Organizer of 35 int. conferences and 14 int. scientific exhibitions; Reviewer of national and international grants (for Polish Ministry of Science and Higher Education, Slovak Research and Development Agency, Portuguese Foundation for Science and Technology and scientific papers (15 int. journals); Director of the Information Processing Centre (1998–2004); Director of the Centre of Excellence TALES: "Thermodynamic Laboratory for Environmental Purposes", IPCh PAS (2002–2014)
Publication record: papers (96), books (7), chapters in books (25), patents (6), professional software packages (5), citations (333), h-index (12), sum of impact factors (109.268)
Equipments: TA Instruments DSC 910 and TG 951; Seteram TG-DSC 111; Microcal OMEGA ultrasensitive titration calorimeter
5 most important publications: [1] U. Domańska, A. Pobudkowska, A. Pelczarska, P. Gierycz, J. Phys. Chem. B, 113, 8941 (2009); [2] K. Kędra-Królik, M. Wszelaka-Rylik, P. Gierycz, J. Therm. Anal. Calorim., 101(2), 533 (2010); [3] P. Gierycz, "Simulation of $CaCO_3$ crystal growth in multiphase reaction." in 'Modern Aspects of Bulk Crystal and Thin Film Preparation', edited by N. Kolesnikov and E. Borisenko, In Tech, 555–578 (2012); [4] M. Wszelaka-Rylik, P. Gierycz, J. Therm. Anal. Calorim., 111 (3), 2029 (2013); [5] M. Kozbiał, P. Gierycz, J. Chem. Thermodyn. 72, 23 (2014).

Name: Ferdinando Giordano
Country: Italy
Date and place of birth: 1942, Mattuglie, Yugoslavia
Present position and address: Professor, Department of Pharmacy, University of Parma (Italy), now retired
Email: ferdinando.giordano@unipr.it
Education and scientific degrees: Maturitá classica, degree in Chemistry, Pavia (1967); degree in Pharmacy, Pavia (1972)
Workplaces: Lecturer in Pharmaceutical Chemistry Faculty of Pharmacy, University of Pavia (1968–1982); Senior Lecturer of Mathematics (1974–1977); Pharmaceutical Analysis (1977–1982); Faculty of Science and Faculty of Pharmacy, University of Pavia; (1982–1987); Associate Professor of Pharmaceutical Analytical Chemistry, Faculty of Pharmacy, University of Pavia (1987–1994); Associate Professor of Quantitative Analysis, Faculty of Pharmacy, University of Pavia; Full Professor, Chair of Industrial Pharmacy, Faculty of Pharmacy, University of Parma (1994–2007)
Main fields of interest: drug-drug and drug-excipient interactions in preformulation studies; thermal studies on pharmaceuticals; solid-state chemistry: polymorphism and pseudopolymorphism of drugs and excipients; preparation and characterization of drugcyclodextrin inclusion compounds; supercritical fluids applications in pharmaceutical technology; controlled drug release
Professional activities: Expertise for pharmaceutical industries
Publication record: papers (145)
5 most important publications: [1] A. Stassi, R. Bettini, A. Gazzaniga, F. Giordano, A. Schiraldi: Assessment of solubility of ketoprofen and vanillic acid in supercritical CO_2 under dynamical conditions, J. Chem. Eng. Data, 45 (2000) 161; [2] F. Giordano, A. Rossi, R. Bettini, A. Savioli, A. Gazzaniga, Cs. Novák: Thermal behavior of paracetamol-polymeric excipients mixtures, J. Therm. Anal. Cal., 68 (2002) 575; [3] C. Donini, D. N. Robinson, P. Colombo, F. Giordano, N. A. Peppas: Preparation of poly(methacrylic acid-g-poly(ethylene glycol)) nanospheres from methacrylic monomers for pharmaceutical applications, Int. J. Pharm., 245 (2002) 83; [4] A. Rossi, A. Savioli, V. Massarotti, D. Capsoni, M. Bini, R. Bettini, A. Gazzaniga, M. E. Sangalli, F. Giordano: Solid state characterization of paracetamol metastable polymorphs formed in binary mixtures with hydroxypropylmethylcellulose, Thermochim. Acta, 406 (2003) 55; [5] A. Foppoli, M. E. Sangalli, A. Maroni, A. Gazzaniga, M. R. Caira, F. Giordano: Polymorphism of NCX4016 an NO-releasing derivative of acetylsalicylic acid, J. Pharm. Sci., 93 (2004) 521.

Name: Maria Luisa Aleixo Gonçalves
Country: Brazil
Date and place of birth: 1953, Rio de Janeiro Brasil
Present position and address: Retired from Petrobras (Brazilian Oil Company); Thermal Analyis Adviser at Federal University of Rio de Janeiro, Chemistry Institute, Laboratory of Catalysis and Sustainable Energy (LACES), Rio de Janeiro, RJ, Brazil
Email: luisa.aleixo@infolink.com.br
Website: http://lattes.cnpq.br/1279245628759472
Education and scientific degrees: Graduate in Chemistry, Federal University of Rio de Janeiro, Rio de Janeiro, RJ, Brazil (1975), M.Sc, Federal University of Rio de Janeiro, Rio de Janeiro, RJ, Brazil (1980), D.Sc, University of São Paulo, São Paulo, SP, Brazil, (2002)
Workplaces: Research Center of Petrobras, Rio de Janeiro, RJ, Brazil (1979–2002), Fluminense Federal University, Niterói, RJ, Brazil (2002–2010); Federal University of Rio de Janeiro, Rio de Janeiro, RJ, Brazil (2011–)
Main fields of interest: Thermal Analysis applied to the petroleum industry
Relevant categories in thermal analyses: Fields: petroleum, catalysts, polymers, cements, ceramics. Methods: TG, DTA, DSC, TG-FTIR, TG-MS
Awards and acknowledgments: Jacques Abranavel Awards (1998)
Professional activities: Adviser of Thermal Analysis Laboratory at Fluminense Federal University (UFF-2002), Researcher of the National Council for Scientific and Technological Development (CNPq-2003–2007); Adviser of Thermal Analysis at Federal University of Rio de Janeiro (LACES-2011); Secretary of Brazilian Thermal analysis and Calorimetry Association (ABRATEC-1997–2002); Organizing member of the Brazilian Congress on Thermal Analysis and Calorimetry (1998 and 2000); Reviewer of: Fuel Processing Technology, Energy and Fuels, Journal of Thermal Analysis and Calorimetry, Fuel, Chemical Engineering Communications
Publication record: papers (32), books (2 book chapters), citations (79)
5 most important publications: [1] Goncalves, M. L. A. Petroleum Science and Technology, v. 17, pp. 1–3, 1999; [2] Goncalves, M. L. A.; Teixeira, M. A. G.; Pereira, R. C. L.; Matos, J. R.; Mercuri, L. P. Journal of Thermal Analysis and Calorimetry, v. 64, pp. 697–706, 2001; [3] Goncalves, M. L. A.; Mota, D. A. P.; Journal of Thermal Analysis and Calorimetry, v. 80, pp. 81–86, 2005; [4] Goncalves, M. L. A.; Ribeiro, D.A.; Mota, D. A. P.; Fuel (Guildford), v. 85, pp. 1151–1155, 2006; [5] Goncalves, M. L. A.; Mota, D. A. P.; Journal of Thermal Analysis and Calorimetry, v. 91, p. 341–316, 2008.

Name: Pierre Charles Gravelle
Country: France
Date and place of birth: 1931, Gap, Hautes Alpes, France
Present position and address: Director of Research (Honorary); 6 rue Commandant Faurax, 69006 Lyon, France
Email: gravelle.pm@numericable.fr
Education and scientific degree: Chemical Engineer, Ecole Supérieure de Chimie Industrielle de Lyon, ESCIL (1954); D.Sci., Lyon University (1963)
Workplaces: Institute of Research on Catalysis, IRC-CNRS Lyon-Villeurbanne (1957–1981); Energy and Material Resources Programme, Nat. Cter. for Sci. Resch, CNRS, Paris (1981–1996)
Main fields of interest: chemistry and physical chemistry of gas-solid interfaces; heterogeneous catalysis; adsorption calorimetry
Awards and acknowledgments: Prize of International Nickel Co.:1964; Frank G. Ciapetta Lectureship Award, American Chemical Society:1976; Lectureship Award, Japan Society for the Promotion of Science:1979; Honorary Membership, French Association of Calorimetry and Thermal Analysis, AFCAT:2003
Professional activities: CNRS Institute of Research on Catalysis: Deputy Director (1974–1982); Summer School on Calorimetry, Lyon (1971 and 1973): Co-organizer; French Association of Calorimetry and Thermal Analysis, AFCAT:President (1974–1979);ICTA, Affiliate Councillor (1977–1982); IUPAC Commission I-6, "Colloids and Surface Chemistry, including Catalysis" Associate (1978–1984), then Titular Member (1984–1989); Catalysis Division, French Chemical Society: President (1984–1989); NATO Special Programme on Systems of low Dimensionality, Chairman of the Committee (1985–1987); Scientific Council, French Institute of Petroleum, IFP: Member (1993–2001); Member of Editorial Advisory Boards: J. of Catalysis (1976–1982); J. of Chemical Thermodynamics (1976–1982); Thermal Analysis.Abstracts (1977–1982); CNRS Energy and Material.Resources. Programme: Assistant Director (1981–1996)
Publication record: papers (102), citations (1077), h-index (19)
List of the 5 most important publications: [1] Heat-flow microcalorimetry and its application to heterogeneous catalysis, P.C. Gravelle: Adv. in Catalysis and Rel. Subj., Academic Press 22 (1972) 191–263; [2] Calorimetry in adsorption and heterogeneous catalysis studies. P.C. Gravelle, Cat. Rev. Sci. Eng.16 (1977) 37–110; [3] J.J. Prinsloo, P.C. Gravelle: Volumetric and calorimetric study of the adsorption of hydrogen and carbon monoxide, at 296 K, on silica-supported nickel and nickel-copper catalysts. J. C. S. Faraday I, 76 (1980) 512 and 2221, and 78 (1982) 27; [4] P.C. Gravelle et al.: Recent Progress in Numerical Methods for the determination of Thermokinetics: results of a multinational programme on the comparison of the methods. J. Therm. Anal. 20 (1981) 47; [5] A. Auroux, J.C. Védrine, P.C. Gravelle: Characterization of small-pore zeolites by adsorption of ammonia,"Adsorption at the gas-solid and liquid-solid interface", J. Rouquerol and K.S.W. Sing, Eds, Elsevier, Amsterdam (1982) 305.

Name: Janusz Grębowicz
Country: USA
Date and place of birth: 1945, Skotninki, Poland
Present position and address: Associate Professor of Chemistry and Physics, University of Houston-Downtown, One Main Street, Houston TX, 77002.
Email: grebowiczj@uhd.edu
Website: http://www.uhd.edu/academic/colleges/sciences/ns/faculty/JGrebowicz.html
Education and scientific degrees: MS in Physics, Universtity of Lodz (1968), Ph.D. in Physical Chemistry, Center of Molecular and Macromolecular Studies, Polish Academy of Sciences, Lodz (1978), Postodctoral in Polymer Physics-Rensselear Polytechnic Institute, Troy, NY(1980–1985), Max Planck Institute, Mainz, Germany (1985–1987), University of Massachussetts, Amherst, MA (1987–1988).
Workplaces: SC Johnson Wax, Racine, WI (1988–1989), Shell Oil, Westhollow Technology Center, Houston TX (1989–2004), Univerrstity of Houston-Dwontown(2004–)
Main fields of interest: polymer physics, thermophysical properties of matrials; materials for energy: oil/gas shales, reposition of spent nuclear fuel, hybrid energy systems
Relevant categories in thermal analyses: fields (polymers, disoder in crystals, mesophases, oil/gas shales, geological materials); methods (DSC, opt. micr., light scattering, simultaneous TG/DTA/MS, TMA, Temperature range: −170 to 1,450 °C)
Awards and acknowledgments: Five Special Recognition Awards from Shell Technology Center, 2014/15 Fulbright Scholar at AGH University of Science and Technology, Krakow, Poland
Professional Activities: teaching college and university physics, thermophysical properties of materials, lecture and lab, organized advanced TA laboratory for research and instruction, co-inventor of the Summer School "The back-end of the nuclear fuel cycle"
Publication record: books (1 book, 1 book chapter), patents (1), citations (1000).
Equipments: Simultaneous TG/DSC/MS, TMA, Opy. Micr., DSC, TG, Laser
5 most important publications: [1] B. Wunderlich, M. Moeller, J. Grebowicz, H. Baur; Adv.Polymer Sci., 87, 1–121 (1988) (monograph); [2] B. Wunderlich, J. Grebowicz, Adv. Polymer Sci., 60-1, 1–59 (1984). 323; [3] J. Grebowicz, J. Therm. Analy. Calorim., 116 (3), 1481–1490 (2014); [4] J. Grebowicz, S.-F. Lau, B. Wunderlich; Polymer Sci., Polymer Symp. 71, 19–37, (1984); [5] B.J. Wang, C.Y. Li, J. Hanzlicek, S.Z.D. Cheng, P.H. Geil, J. Grebowicz, R.M. Ho, Polymer, 42 (16), 7171–7180 (2001).

Name: Jean Grenet
Country: France
Date and place of birth: 1944, Le Vésinet, Yvelines, France
Present position and address: Professor Emeritus, University of Rouen, Institute for Material Research, Avenue de l'Université, BP12, F 76801, Saint Etienne du Rouvray Cedex
Email: jean.grenet@univ-rouen.fr, jegrenet@gmail.com
Education and scientific degrees: M.Sc.1968, D. 3rd cycle 1973, D. in Physics 1983
Workplaces: University of Rouen (1970–1987), Ecole des Mines of Nancy (1987–1988)
Since 1989, Univ. Rouen, Material Science and presently: Institute for Material Reasearch, Rouen
Main fields of interest: slow relaxations processes in complex systems: physical ageing phenomena in glassy and vitreous materials (polymers and chalcogenides principally); instrumentation in thermal analysis; education in thermal methods
Relevant categories in thermal analyses: fields (polymers, biopolymers, composites, glasses); methods (DSC, TMDSA, specific heat, DMA and TMA, TSC and TSDC, dielectric spectroscopy, MFI, surface tension, mechanical testing machines)
Professional activities: Head of Laboratoire d'Etude et de Caractérisation des Amorphes et des Polymères (1996–2006). Head of Institute for Material Research (2006–2009). Chairman of French Association of Thermal Analysis and Calorimetry (AFCAT) (2006–2009). Organizer of French ATC annual meeting (2000, 2005, 2014). French councillor at ICTAC confederation (since 2009). Professor Emeritus University of Rouen since 2009
Publication record: papers (81), h-index (18)
5 most important publications: [1] Characterisation of structural relaxation phenomena in polymeric materials from thermal analysis investigations. Saiter A., Couderc H., Grenet J., Journal of Thermal Analysis and Calorimetry, 88: 483 (2007); [2] Study of Bisphenol A Polycarbonate relaxation kinetics at the glass transition temperature. Delbreilh L., Dargent E., Grenet J., Saiter J.M., Bernes A., Lacabanne C., European Polymer Journal, 43 : 249–254 (2007); [3] Glass transition temperature and value of the relaxation time at Tg in vitreous polymers. Saiter J.M., Grenet J., Dargent E., Saiter A., Delbreilh L., Macromol. Symp., 258, 1 : 152–161 (2007); [4] Average size of cooperative rearranging regions and fragility in a drawn Poly(Ethylene Terephtalate) at the glass transition, Saiter A., Dargent E., Saiter J.M., Grenet J. Journal of Non-Crystalline Solids, 354, 2-9 : 345–349 (2008); [5] Is the configurational entropic model able to predict the final equilibrium state reached by Se glasses after very long ageing durations? J. Grenet, E. Bouthegourd, A. Esposito; A. Saiter, J.M. Saiter; Philosophical Magazine, 93 (2), 2932–2946 (2013).

Name: Jean-Pierre Etienne Grolier
Country: France
Date and place of birth: 1936, Constantine, Algeria
Present position and address: Emeritus Professor. Thermodynamics. Institute of Chemistry of Clermont-Ferrand, Blaise Pascal University, 24 av.des Landais, 63177 Aubière, France
Email: j-pierre.grolier@univ-bpclermont.fr
Website: http://iccf.univ-bpclermont.fr, www.transitiometry.com
Education and scientific degrees: Ph.D., Physical Chemistry and Thermodynamics
Workplaces: Full Professor, Department of Chemistry Blaise Pascal University, Clermont-Ferrand, France (1983–2007); Research position, National Research Council of Canada, Ottawa (1972–1973); Associate Researcher, CNRS Microcalorimetry Research Center, Marseille (1974–1977); University of Delaware Department of Chemistry, Fulbright Professor, Newark (Del), USA (1979–1980; 1983); Visiting Researcher, Departments of Chemistry, Universities of Lethbridge (1982), Edmonton (1985), Canada; Head of the Chemistry Centre, Univ. Blaise Pascal, Clermont-Ferrand, France 2000–2007
Main fields of interest: thermodynamics of solutions and polymers; petroleum fluids, crude oils and biofuels; polymers under extreme conditions of T and p; high p and supercritical polymerisation; high pressure pVT- and calorimetric- experimental techniques; scanning transitiometry (patented), coupling with spectroscopic (UV/VIS/IR) techniques, TMDSC
Relevant categories in thermal analyses: fields (gases, solutions, natural (petroleum) and technological (automotive fuels) fluids, pharmaceutical and biology, polymers, nanoporous materials, repulsive clathrates); methods (high p, high T calorimetry; DSC, ITC, thermophysical properties, instrument development (patents via own company, BGR TEC))
Awards and acknowledgments: Calvet Prize (1985); Polish Association for Calorimetry and Thermal Analysis, Member of Honour (1991); Świętosławski Medal (1994); Kurnakov Medal (1994); H. Huffman Memorial Award (1997); Swiss Applied Chemical Thermodynamics Prize (1998); IUPAC, F.D. Rossini Award for excellence in thermodynamics (2004); Mège Prize, Académie des Sciences Belles-Lettres et Arts de Clermont-Ferrand, France (2005); AICAT-Setaram Award (2008); Order of Merit Commander Cross, Republic of Poland (1995); Officer, Academic Palms, France (2008)
Professional activities: ICTAC Council and Calorimetry Committee, Member (1988–1992); AFCAT, Member of Council, President (1989–1992); IUPAC Commission on Thermodynamics, Titular Member (1989–2001); International Association of Chemical Thermodynamics (IACT), President (2002–2008); Eurostar Science Vice-President (1999–); Board of Directors, American Calorimetry Conference, Member (1989–1991) and (2000–)
Publication record: papers (263), books (2), book chapters (12), patents (5), citations (2500), h-index (28), sum of impact factors (235.83)
Equipments: Mettler DSC (1) and Photo DSC (1), Setaram Titration Calorimeter (1), BGR TEC Transitiometers (2)
5 most important publications: [1] H. V. Kehiaian, J.-P. E. Grolier et al.: J Chim Phys 75 1031 (1978); [2] J.-P. E. Grolier et al.: Adv Polym Sci 238 137 (2011); [3] J.-P. E. Grolier et al.: J Chem Thermodyn 55 193 (2012); [4] M. A. Aquino-Olivos, J.-P. E. Grolier et al.: Energy and Fuels 27 1212 (2013); [5] Y. G. Grosu, V. Ievtushenko, V. A. Eroshenko, J.-M. Nedelec, J.-P. E. Grolier: Coll Surf A 441 549 (2014).

Name: Wei Guan
Country: P.R. China
Date and place of birth: 1968, Liaoning Province, P.R. China
Present position and address: Professor/vice director of school of environmental science, Liaoning University
No.66 Chongshan Middle Road, Huangu District, Shenyang, Liaoning, China, 110036
Email: guanweiy@sina.com
Education and scientific degrees: 2003–2006 The Institute of Salt Lakes, Chinese Academy of Science, Ph.D.; 1992–1995 Department of Chemistry, Liaoning University, MS.; 1986–1990 Department of Chemistry, Northeast Normal University, BS
Workplaces: College of Chemistry, Liaoning University; College of Environmental Science, Liaoning University
Main fields of interest: chemical thermodynamics and toxic properties of ionic liquids
Relevant categories in thermal analyses: fields (physical chemistry); methods (TG, DSC, thermomechanical analysis, specific heat, calorimetry)
Awards and acknowledgments: The first class of "Liaoning BaiQianWan Talents Program" (2013)
Professional activities: (1) member of Chemical Thermodynamics and Thermal Analysis Professional Committee of Chinese Chemical Society; (2) member of the First Ionic Liquids Professional Committee of Chemical Industry and Engineering Society of China
Publication record: papers (>100), books (4), patents (5), citations (60), h-index (10), sum of impact factors (100)
Equipments: TG209F1, Netzsch; DSC204F1 Phoenix®, Netzsch; SRC-100 solution-reaction isoperibol calorimeter, China.
5 most important publications: [1] W. Guan, et al.: J. Phys. Chem. B 115 (2011) 12915; [2] W. Guan, et al.: Ind. Eng. Chem. Res. 52 (2013) 9490; [3] W. Guan, et al.: J. Chem. Thermodynamics. 65 (2013) 91; [4] W. Guan, et al.: Thermochimica Acta, 437 (2005) 196; [5] W. Guan, et al.: J Therm Anal Calorim, 94(2) (2008) 507.

Name: Katarína Győryová
Country: Slovakia
Date and place of birth: 1946, Veľký Horeš, Czechoslovakia
Present position and address: Professor of Inorganic and Bioinorganic Chemistry, Department of Chemistry, Faculty of Science Šafarik University Košice, Moyzesova 11, 041 54 Košice Slovak Republic
Email: katarina.gyoryova@upjs.sk
Education and scientific degrees: M.Sc. Chemistry, University Košice Faculty of Science (1968); Dr.rer.nat.University Košice (1970); C.Sc. Ph.D. Hungarian Academy of Science (1981); Associate Professor University Bratislava (1990); D.Sc. Slovak Academy of Science Bratislava (1996); Professor Technical University Bratislava (1999)
Workplaces: Department of Biochemistry, Medical Faculty Šafarik University (1968–1971); Department of Inorganic Chemistry, Faculty of Science Šafarik University Košice (1971–).
Main fields of interest: the research activity since 1975 in the field of inorganic, bioinorganic, coordination chemistry; inorganic materials and substances (zeolites), specially their synthesis and characterization; thermal study using methods of thermal analysis (TG/DTG, DTA, ETA, DSC etc.); IR spectroscopy, mass and mössbauer spectroscopy and analysis of evolved gases; characterization of clay minerals during their interactions with gases and liquides
Professional activities: Member of Commitee for graduing of RNDr., Ph.D., D.Sc. degree in the field of Inorganic Chemistry; member of Working Group on Sustainable and Environmen-tal Prevention ICTAC; coordinator of several research projects in the frame of VEGA—Programme in Slovak Republic; principial investigator of the Project in the frame of bilateral agreements on scientific co-operation between the Slovak Republic and Czech Republic; reviewer for J. Thermal Anal.Cal., Chemical Papers, Thermochim.Acta, Spectrochim.Acta, Monatshefte f.Chemie, Ass.Ed. JTAC Bio-, Life science; member of the Slovak Chemical Society, American Chemical Society
Publication record: papers (195), books (8), patents (9), citations (575)
5 most important publications: [1] Győryová K.: Investigation of Thermal Behaviour of Solid Transition Metals Cyano Complexes with Organic Cations. Thermochim. Acta, 190 (1991)15–29; [2] Győryová K., Kovářová J., Andogová E., Zeleňák V., F. A. Nour El-Dien: Thermal, Spectral and Antimicrobial Properties of new Zinc(II) Isobutyrate Compounds. J. Therm. Anal. Calorim., 67(2002)119–128; [3] Erdelyiová A., Győryová K., Gyepes R., Halás L., Kovářová J. :Synthesis, Spectral, Thermal and Structural Study of Bis(2-bromobenzoato-O,O′)- bis(methyl-3-pyridylcarbamate-N)-zinc(II). Polyhedron 28(2009)131–137; [4] Findoráková L., Győryová K., Melník M., Koman M., Faten A.Nour El-Dien: Preparation, Thermal Decomposioin and Crystal Structure of Zinc(II) 2- Chlorobenzoate Complex with Nicotinamide. J. Coord. Chem., 63 (2010) 3348–3355; [5] Balek V., Malek Z., Ehrlicher U., Győryová K., Matushek G., Yariv S.:Emanation Thermal Analysis of TIXOTON (actived bentonite) treated with organic compounds. Appli. Clay Sci. 21 (2002) 295–302.

Name: Akbar Khodaparast Haghi
Country: I.R. Iran
Date and place of birth:1958, Tehran, Iran
Present position and address: Professor, Faculty of Engineering-University of Guilan, Rasht, Iran
Email: Haghi@Guilan.ac.ir
Education: Ph.D.: University of Franche-Comte (France), D.E.A.: University of Technology of Compiegne (France), M.Sc.: North Carolina A&T State University (USA), BSc: University of North Carolina (USA), G.C.E.: 'A" Level, Pure Mathematics(UK)
Workplaces: Member of Canadian Research and Development Center of Sciences and Cultures, Quebec, Canada; Senior Editor, Apple Academic Press (Canada), http://www.appleacademicpress.com/; Editor-In-Chief, IJCCE (USA), www.igi-global.com/ijcce; Editor-In-Chief, Polymers Research Journal (USA); Research Associate, Inst.for Engng.of Polymer Mat.(Poland); Editorial Board: Journal of Natural Fibers (Taylor & Francis); Editorial Board: Int. J. of Chemical Modeling (Nova Sci Pub)
Main fields of interest: nanotechnology, porous media, polymer
Publication record: papers (1000), books (115), patents (5).

Name: Tatsuko Hatakeyama
Country: Japan
Present position and address: President, Lignocel Research, 910-3558 Ytsumata, Fukui, Japan
Email: lignocel@mx3.fctv.ne.jp
Education and scientific degrees: Bachelor of Engineering; Tokyo University of Agriculture and Technology; Doctorate; Tokyo Institute of Technology
Workplaces: Scientist at Textile Research Institute, Ministry of International Trade and Industry, Senior Scientist at Research Institute of Polymers and Textiles, Ministry of International Trade and Industry, Senior Scientist at National Institute of Materials and Chemical Research, Ministry of International Trade and Industry, Professor Otsuma Women's University. From April 1997 to March 2005. Lignocel Research. From April 2005–
Main fields of interest: Water-biopolymer interaction
Relevant categories in thermal analyses: fields (polymers); methods (DSC, TG, TMA, DMA)
Awards and acknowledgments: Award from the Society of Fiber Science and Technology, Japan, (2006)
Professional activities: Editorial member of J. Fibre Science and Technology, Japan (1975–1981), Editorial member of J. Thermal Analysis and Calorimetry, Japan (1977–1979), Member of board of the Society of Thermal Analysis and Calorimetry, Japan, Editorial committee member of J. Fibre and Technology, Japan, (1987–1989). Editorial member of J. Thermal Analysis and Calorimetry, Japan, (1989–2000). Education committee member of International Confederation of Thermal Analysis and Calorimetry, International Standardization Organization (ISO), TC-61 Japanese committee member (1992–2003), Chief of Applied Thermal Analysis Committee, Japan (1992–1994), Chief editor of J. Fibre Science and Technology, Japan (1993–1995), Regional editor of J. Thermal Analysis and Calorimetry (1998–) Membership: Member of Chemical Society of Japan, Society of Polymer Science, Japan, Fibre Science and Technology, Japan, Japan Society of Thermal Analysis and Calorimetry
Publication record: papers (367), books (36), patents (3)
5 most important publications: [1] Hatakeyama T, and Quinn FX (1994) Thermal analysis, fundamentals and applications to polymer science. John Wiley and Sons, Chichester, UK; [2] Hatakeyama T, Liu Z (1999) Handbook of thermal analysis, John Wiley, Chichester, UK; [3] Hatakeyama T, Hatakeyama H (2004) Thermal properties of green polymers and biocomposites. Kluwer Academic, Dordrecht; [4] Hatakeyama H, Hatakeyama T (2010) Lignin structure, properties and application. Adv Polym. Sci. 232, 1–63; [5] Hatakeyama T, Tanaka M, Kishi A, Hatakeyama H (2012) Comparison of measurement techniques for identification of bound water restrained by polymers. Thermochim Acta 532 159–163.

Name: James Neilson Hay
Country: UK
Date and place of birth: 1935 Perth, Scotland
Present position and address: Professor in Polymer Chemistry, College of Engineering and Physical Sciences, School of Metallurgy and Materials, University of Birmingam, Edgbaston, Birmingham B152TT, UK
Email: j.n.hay@bham.ac.uk
Website: www.birmingham.ac.uk/schools/metallurgy-materials/people/
Education and scientific degrees: B.Sc. University of Glasgow 1957; Ph.D. University of Glasgow 1960; D.Sc. University of Birmingam 1972.; Chartered Chemist; FRSC.
Main fields of interest: crystallization kinetics; fracture mechanics; polymer degradation kinetics; physical ageing; polymer bonded magnets; characterisation of polyolefins; fuel cells
Relevant categories in thermal analyses: fields (polymers); methods (DSC; DTA; high temperature DSC; TG; minimat; thermomechanical analysis, instrument development; self referencing DSC; minimat; FTIR spectroscopy-TA)
Professional activities: Fellow Royal Society of Chemistry—thermal methods group—Macro Group—Polymer Physics Group.; Past editor of Thermochimica Acta; Past member of the Editorial Board of Polymer; Past UK ICTA Councillor
Publication record: papers (220), books (8 book chapters), patents (3)
Equipments: DSC; High Temperature DSC; Flash DSC: TA:TG;: TMA; FTIR Spectrometer TA; DMTA; DETA.
5 most important publications: [1] "Crystallization from the Melt" by J.N. Hay in "Flow induced Crystallization in Polymer Systems", Edited by Robert L. Miller, Published by Gordon and Breach Science Publishers, 1977; [2] The kinetics and mechanism of the thermal degradation of poly (methyl methacrylate) studied by Thermal Analysis-Fourier Transform IR spectroscopy, B. J. Holland, J. N. Hay, Polymer, 2001, 42, 4825; [3] Design and development of a self-referencing differential scanning calorimetry, B. J. Holland, J. R. Atkinson, J. N. Hay, Journal of Thermal Analysis and Calorimetry, 2002, 69, 371; [4] Structure Evolution in Melt Crystallized PEEK, M. J. Jenkins, J. N. Hay, N. J. Terrill, Polymer, 2003, 44, 6781; [5] The thermal analysis of poly (ethylene terephthalate) by FTIR Spectroscopy, Z. Chen, J. N. Hay, M. J. Jenkins, Thermochimica Acta, 2013, 552, 123.

Name: Klaus Heide
Country: Germany
Date and place of birth: 1938, Salzuflen, Germany
Present position and address: Prof. i.R., Grundweg 12, D-07749 Jena,Germany
Email: ckh@uni-jena.de
Website www.igw.uni-jena.de/mineralogie
Researcher ID: B-1858-2014
Education and scientific degrees: graduated in mineralogy (1961 diplom), Ph.D. (1964 Dr.rer.nat.), D.Sc. (1969 Dr.rer. nat.habil), University Professor 1993–2004
Workplaces: Mineralogisches Institut der Universität Jena (1962–1968), Otto-Schott-Institut Jena (1968–1992), Institut für Geowissenschaften Jena (1992–2004)
Main fields of interest: analysis of mineral phases, natural and artificial mixtures of phases, rocks and batches; kinetic non-isothermal processes; volatiles in natural and industrial processes; meteorites and impactites
Relevant categories in thermal analyses: fields (inorganic materials, minerals, glass, meteorites, archaeological artefacts); methods (TGA, instrument development, direct coupled-evolved-gas-analysis-system (DEGAS), mechanical spectroscopy)
Awards and acknowledgments: Netzsch-GEFTA Award (1984), Kurnakov medal (1985), asteroid 6506, discovered 1978 Mar.15 by S.J. Bus at Palomar was named "Klausheide = 1978 EN10" (2003)
Professional activities: 1968–1973 elected to the member of Committee on Standardization of the International Confederation for Thermal Analysis (ICTA). 1980–1985 Councillors-at-Large in the Council of ICTA; 1987 4th. ESTAC, Chairman of scientific program committee; 1996 Chairman and organiser of the 3rd International Conference on Natural Glasses in Jena Editor-in-Chief of Elsevier Journal "Chemie der Erde/Geochemistry" 1973–2013
Publication record: papers (210), books (1 book, 4 book chapters)
Equipments: Mettler TA 1 with quadrupol MS Balzers; Netzsch STA 425 with Quadrupol MS Balzers; Hot stage PolMicroscope Zeiss Jena; Derivatograph MOM
5 most important publications: [1] Heide, K.: Zum Mechanismus der Umbildungsvorgänge in Salzgesteinen. Chem. Erde 27 (1968) 353–368; [2] Heide, K.: "Kinetische Variationen"—Versuch einer Einordnung der dynamischen thermischen Analysenmethoden zur Beschreibung der Dynamik von Festkörper-Reaktionen in Anlehnung an Gedanken von Paul Klee, Thermochimica Acta 112 (1987) 13–29; [3] Heide K., Gerth K., Hartmann E., The detection of an inorganic hydrocarbon formation in silicate melts by means of a direct coupled-evolved-gas-analysis-system (DEGAS) Thermochim.Acta 354 (2000) 165–172; [4] Heide K., Woermann E., Ulmer G., Volatiles in pillows of the Mid-Ocean-Ridge-Basalt (MORB) and vitreous basaltic rims, Chem.Erde 68 (2008) 353–368; [5] Heide K. Gas release from minerals. Minerals as advanced materials II ed. By S. V. Krivivichev, Springer Verlag 2012, 25–36.

Name: John O. Hill
Country: Australia
Date and place of birth: 1942, UK
Present position and address: Emeritus Professor of Chemistry and Chemical Education, La Trobe Institute of Molecular Sciences (LIMS), La Trobe University, Melbourne, Victoria 3086, Australia.
Email: jce.hill@bigpond.com
Website: www.latrobe.edu.au/chemistry
Education and scientific degrees: BSc., Ph.D., D.Sc. (University of London); Grad. Dip. (Environmental Management) (La Trobe); Ph.D. (Chemical Education) (University of Melbourne)
Workplaces: La Trobe University, Melbourne, Australia; National University of Singapore.
Main fields of interest: thermochemistry (thermal analysis and calorimetry), environmental chemistry, sustainable chemistry, chemical education.
Awards and acknowledgments: B(Univ.) Honouris Causa, University of Surrey, UK
Publication record: number of papers (>200), books (1)
5 most important publications: [1] Development and Applications of a preparative scale Sample Controlled Thermogravimetric system. E. L. Charsley, J. J. Rooney, J. O. Hill and G. M. B. Parkes, P. A. Barnes, E. A. Dawson, J. Thermal Anal. Calorim. 72 1091–7 (2003); [2] Principles and Practices of Thermal Analysis and Calorimetry. John Hill, In: 'Characterisation of Materials' (2nd Edition) Kaufmann, E. N. (Editor), Volume 1, pp. 463–471, Wiley, Hoboken, NJ (2012); [3] A curriculum framework for education in Thermal Analysis. Ranjit K. Verma, John O. Hill, Lauri Niinisto, S. C. Mojumdar, David Kumar, J. Materials Education, 34 (3–4) 133–150 (2012); [4] A curriculum framework for education in Calorimetry. Ranjit K. Verma, John O. Hill, Lauri Niinisto, S. C. Mojumdar, David Kumar, J. Materials Education, 34 (5–6) 161–174 (2012); [5] Challenges for Chemical Education: Traversing the Chemical Sciences/Materials Science interface. John Hill, Ranjit K. Verma, David Kumar, J. Materials Education, 35 (1–2) 1–16 (2013).

Name: Klára Hódi
Country: Hungary
Date and place of birth: 1943, Kétegyháza, Hungary
Present position and address: H-6723 Szeged, Lugas u. 7/A, Hungary
Email: klara.hodi@pharm.u-szeged.hu
Website: http://www.pharm.u-szeged.hu/phtech/english/home
Researcher ID: A-6434-2008
Education and scientific degrees: Pharmacist diploma, Med. Univ. of Szeged (1968); Univ. doctorate, Medical Univ. of Szeged (SZOTE) (1973); Specialist pharmacist (pharmaceutical technology) (1975); CSc, Hugarian Academy of Sciences, (1983), Ph.D. degree, SZTE (1996); Habilitation, Univ. of Szeged (SZTE) (1996); D.Sc., Hugarian Academy of Sciences (2006), professor, SZTE, Dept. of Pharm. Technol. (1969–2013), professor emerita, SZTE (2013–)
Workplaces: OGV Emergé Gummi-factory, Szeged (1968–1969), Univ. of Szeged, Dept. of Pharmaceut. Technol. (1969–2013)
Main fields of interest: development and control of solid dosage forms (granules/pellets, tablets, coated solid dosage forms) with modern test methods (SEM, EXD, DSC, MTDSC, TG, TG-MS, dinamic force measurent, study of release kinetic, etc.)
Relevant categories in thermal analyses: fields (inorganic, materials, pharmaceutical, polymer); methods (TG-MS, DSC (normal and modulated program), kinetics, extremely high temperature (above 1,000 °C); specific heat, calorimetry)
Awards and acknowledgments: “Béla Spergely” prize (1998), “László Szebellédy” prize (2009), “László Batthyány Strattman prize (2010), “György Hintz” prize (2012), “Antal Végh” NIVO prize (2013), SZTE GYTK “Pro Facultate” prize (2013)
Professional activities: Hungarian Pharmaceutical Society; Industrial Board of Hungarian Pharmaceutical Society; Hungarian Microscopical Society; European Organisation for Quality; Hungarian Chemical Society
Publication record: papers (249), citations (355), h-index (13), sum of impact factors (178.4)
Equipments: Mettler DSC 821e with modulated program, Mettler TGA/DSC1 instrument with Pfeiffer mass spectrometer, Leica thermomicroscope
5 most important publications: [1] K. Nikowitz, K. Pintye-Hódi, G. Regdon jr.: Eur. J. Pharm. Sci. 48, 563–571 (2013); [2] G. Regdon Jr., D. Hegyesi, K. Pintye-Hódi: J. Therm. Anal. Calorim. 108, 347–352 (2012); [3] T. Hekmatara, G. Regdon jr., P. Sipos, I. Erős, K. Pintye-Hódi: J. Therm. Anal. Calorim. 86, 287–290 (2006); [4] J. Bajdik, G. Regdon jr., G. Lebák, O. Berkesi, K. Pintye-Hódi: Effect of stirring on film formation from Eudragit RS aqueous dispersion, Polym. Adv. Technol., 17, 814–817 (2006); [5] Károlyházy, L., Regdon, G. jr., Éliás, O., Beke, Gy., Tábi, T., Hódi, K., Erős, I., Mátyus, P.: J. Mol. Struct. (Theochem) 666–667, 667–680 (2003).

Name: Wolfgang Hohenauer
Country: Austria
Date and place of birth: 1958, Kufstein, Austria
Present position and address: AIT Austrian Institute of Technology GmbH, Head of the Accredited Laboratory for Thermo-Phsics and Thermo-Kinetics, University of Applied Sciences–Technikum Wien, Applied Physics and Material Science, Dr.-Ing. Wolfgang HOHENAUER, AIT Austrian Institute of Technology GmbH, Giefinggasse 2, 1210 Vienna, Austria
Email: wolfgang.hohenauer@ait.ac.at
Website: http://ait.ac.at
Education and scientific degrees: Phycicist, University Innsbruck (A) Mag.rer.nat. 1988, Rheinisch Westfälische Universität Aachen RWTH (D) Dr.-Ing. 1996.
Workplaces: Plansee AG (1989–1993), Forschungszentrum Jülich (1994–1996), AIT Austrian Institute of Technology GmbH (since 1997), University of Applied Sciences—Technikum Wien, Applied Physics and Material Sciences (since 1999)
Main fields of interest: thermo-physics, thermo-kinetics, uncertainty models
Relevant categories in thermal analyses: fields (measurement techniques, thermo-kinetic modelling); methods (dilatometry, DSC, flash methods, thermo-gravimetry, FTIR, mass spectroscopy)
Professional activities: Member of the managing committee of the GEFTA (Gesellschaft für Thermische Analyse), Member of the steering committee of the "Arbeitskreis Thermophysik in der GEFTA", Member of the International Organisation Committee of European Conference on Thermophysical Properties ECTP
Publication record: papers (>75), books (1), patents (3)
Equipments: Dilatometry, DSC, Flash methods, Transient Hot Bridge (THB), Thermo-Gravimetry, FTIR, Mass Spectroscopy
5 most important publications: [1] Wolfgang Hohenauer, Daniel Lager, Ijaz Ul Mohsin: Kinetic modelling of de-binding and sintering of pure Cu, Fe12Cu and W8Ni2Cu, High Temperatures–High Pressures, Vol.42 No.4 (2013) pp. 259–285; [2] Wolfgang Hohenauer, Daniel Lager, Ijaz Ul Mohsin: Improvement of thermo-physical data of pure Cu, Fe12Cu and W8Ni2Cu during and after sintering using kinetic modelling, High Temperatures–High Pressures, Vol.42 No.4 (2013) pp. 321–348; [3] Vozar, L.; Hohenauer, W: Uncertainty of Thermal Diffusivity Measurements Using the Laser Flash Method; Int. J. of Thermophysics 26(6) (2005) pp. 899–915; [4] Libor Vozar, Wolfgang Hohenauer: Flash method of measuring the thermal diffusivity A review; High Temperatures–High Pressures, Vol.35 No.36 (3) (2003) pp. 253–264; [5] Vozar L., Labudova G., Hohenauer W.: The Laser Flash Method with Repeated Pulses—Optimal Experimental Design Analysis; Int. J. of Thermophysics 23 (5) (2002) pp. 1157–1170.

Name: Pavel Holba
Country: Czech Republic
Date and place of birth: 1940, Prague, Czech Republic
Present position and address: senior researcher
New Technology Research Center, University of West Bohemia, Studentská 8, Plzeň
Email: holbap@gmail.com
Website: www.zcu.cz
Researcher ID/ORCID: O-1321-2013/0000-0002-3953-7679
Education and scientific degrees: Chemical Engineer, Institute of Chemical Technology (ICT) in Prague (1957–1962), C.Sc./Ph.D. (1986)
Workplaces: State Research Institute on Glass and Ceramic Industry (1963–4), Institute of Solid State Physics at Czech. Acad. Sci. (1964–1975), Institute of Inorganic Chemistry at Czech. Acad. Sci. (1976–1979), Design Inst. for Transport Structures and Other Infrastructures (1980–1989), Institute of Chief Architect of Capital Prague (1989–1990), Institute of Physics at Cz. Acad. Sci. (1990–1993), Capital Prague (vicemayor 1993–1994), Institute of Municipal Informatics (director 1995–2006), Institute of Chemical Technology, Prague (external lecturer 2009–2013), University of West Bohemia in Pilsen (2012–)
Main fields of interest: phenomenological thermodynamics, thermal analysis, phase equilibria, crystal defects equilibria, nonstoichiometry, computer science, history, politics
Relevant categories in thermal analyses: **fields** (inorganic, materials, ceramics, theory); **methods** [DTA, DSC, dilatometry, TG, high temperatures (>1,000 °C)]
Awards and acknowledgments: Hanuš Award (2012) of Czech Chemical Society
Professional activities: member of TA Group of Czech Chemical Society, lectures
Publication record: papers ($\sim$ 50), books (5 book chapters), citations (275), h-index (9)
5 most important publications: [1] Holba, P., Šesták, J.: Kinetics with regards to the equilibrium of processes studied at increasing temperatures, Z. phys. Chem. N.F. 80 (1972) 1-20; [2] Holba, P., Nevřiva, M., Šesták, J.: Analysis of DTA curves and calculation of kinetic data using computer technique, Thermochimica Acta 23 (1978) 223–231; [3] Holba, P. (1992): Thermodynamics of Partially Open Systems, Czech. J. Physics 42 (6) (1992) 549–575; [4] Holba, P., Šesták, J., Sedmidubský, D. (2013): Heat transfer and phase transition in DTA experiment, Chapter 5 in book: "Thermal analysis of micro-, nano and non-crystalline materials: Transformation, crystallization, kinetics and thermodynamics", Ed. J. Šesták and P. Šimon, Springer 2013, pp. 99–134 [ISSN 1571-3105 ISBN 978-90-481-3149-5 ISBN 978-90-481-3150-1; [5] Holba, P., Sedmidubský, D. (2013): Heat capacity equations for nonstoichiometric solids, Journal of Thermal Analysis and Calorimetry 113 (1) (2013) 239–245.

Name: Bob A. Howell
Country: USA
Date and Place of birth: 1942, Jefferson, North Carolina, USA
Present position and address: Professor, Center for Applications in Polymer Science/Department of Chemistry, Central Michigan University, Mt. Pleasant, MI 48859-0001 USA
Email: bob.a.howell@cmich.edu
Education and Scientific Degrees: B.A. Chemistry, Berea College, 1964; Ph.D., Physical Organic Chemistry, Ohio University, 1971; Postdoctoral Associate, Organometallics, Iowa State University, 1971–1974; Industrial Internship, Polymer Science, Dow Chemical Company, 1985
Workplace: Center for Applications in Polymer Science, Department of Chemistry, Central Michigan University, Mt. Pleasant, MI, 1976–
Main fields of interest: thermogravimetry; polymer degradation; flame retardancy; polymer additives from renewable biosources; vinylidene chloride barrier polymers; polymer-supported organoplatinum antitumor agents
Awards and acknowledgments: NATAS Outstanding Achievement (Mettler-Toledo) Award, 2012; NATAS Distinguished Service Award, 2012; NATAS Fellow, 2002; American Chemical Society (ACS) Fellow, 2011; E. Ann Nalley ACS Service Award, 2013
Professional Activities: NATAS Executive Board, 2001—(Secretary, President-Elect, President, Executive Councilor, Education Councilor, Meetings Councilor, Conference Chair, Awards Chair); Councilor and former President, Midland Section—ACS; ASC Committee on Professional Training; ACS Committee on Patents and Related Matters (former Chair); ACS Committee on Nomenclature, Terminology and Symbols; ACS Organic Examination Committee; ACS Polymer Education Committee; symposium organizer (several) for both NATAS and ACS; various other
Publication record: paper (250), books (7 books, 14 book chapters), patents (4)
5 most important publications: [1] B.A. Howell, "Utilization of Thermogravimetry in the Study of Reaction Mechanism", J. Therm. Anal. Calorim., 2008, 93, 27–34; [2] B.A. Howell, "The Utilization of TG/GC/MS in the Establishment of the Mechanism of Poly(styrene) Degradation", J. Therm. Anal. Calorim., 2007, 89, 393–398; [3] B.A. Howell, "Kinetics of the Thermal Dehydrochlorination of Vinylidene Chloride Barrier Polymers", J. Therm. Anal. Calorim., 2006, 83, 53–55; [4] B.A. Howell and D. Fan, "Poly(amidoamine) Dendrimer-Supported Organplatinum Antitumor Agents", Proc. R. Soc. A., 2010, 466, 1515–1526; [5] B.A. Howell, "Degradation of Vinylidene Chloride/Phenyl Acetylene (VDC/PA) Copolymers. Effect of Internal Unsaturation on Poly(vinylidene chloride) Stability", J. Polym. Sci,. Part A, Polym. Chem., 1987, 25, 1681–1695.

Name: Günther W. H. Höhne
Country: Germany
Date and place of birth: 1937, Berlin, Germany
Present position and address: retired from University of Ulm
Email: gwh.hoehne@t-online.de
Education and scientific degrees: Chemistry, Physics and Mathemathics at the Technical University of Berlin (1955–1968); Dr. rer. nat. (1970); Privatdozent (Adj. Prof.) after habilitation in Experimental Physics (1997)
Workplaces: University of Ulm: Head of Section Calorimetry (1970–1999); after retirement: Visiting Professor at the Technical University of Eindhoven (Netherlands) (1999–2008)
Main fields of interest: calorimetry of all kind (theory and applications) in particular temperature modulated DSC and high pressure DSC
Relevant categories in thermal analyses: fields (materials, polymer, life); methods (DSC, high pressure DSC (500 MPa), specific heat, calorimetry, microcalorimetry, instrument theory)
Awards and acknowledgments: Science award of the German Society of Thermal Analysis GEFTA (2002)
Professional activities: Founder and leader of the German Calorimetry Conference "Ulmer Kalorimetrietage" (1975–1994)
Publication record: papers (110), books (3)
5 most important publications: [1] W. Hemminger and G.W.H. Höhne: "Calorimetry, Fundamentals and Practice" Verlag Chemie, Weinheim (1984); [2] G.W.H. Höhne, W. Hemminger and H.J. Flammersheim: "Differential Scanning Calorimetry An Introduction for Practitioneers" 2nd enlarged Ed. Springer Verlag Berlin (2003); [3] S.M. Sarge, G.W.H. Höhne and W. Hemminger: "Calorimetry—Fundamentals, Instrumentation and Applications" Wiley-VCH Verlag GmbH Weinheim (2014); [4] G.W.H. Höhne, S. Rastogi and B. Wunderlich: "High pressure Differential Scanning Calorimetry of Poly(4-methyl-pentene-1)" Polymer 41 8869–8878 (2000); [5] G.W.H. Höhne: "Temperature modulated differential scanning calorimetry (TMDSC) in the region of phase transitions. Part1: Theoretical considerations" Thermochim. Acta 86 93–99 (1999).

Name: Rongzu Hu
Country: China
Date and place of birth: 1938, Wu Xi, P.R. China
Present position and address: Retired from the Xi'an Modern Chemistry Research Institute; Prof. Rongzu Hu, Zhang Ba east road 83#, Xi'an, Shaanxi, 710065, P.R. China
Email: hurongzu88@163.com
Education and scientific degrees: Zhejiang University (1958–1963); Xian Modern Chemistry Research Institute (1963–2013); Lecturer (1978), Associate Professor (1983), Professor (1987)
Workplaces: Xi'an Modern Chemistry Research Institute (1963–2013); Doctoral Co-Advisor and Advisor (1990–)
Main fields of interest: study on thermochemistry and thermal analysis of energetic materials
Relevant categories in thermal analyses: fields (energetic materials (EMs)); methods (DSC, kinetics, microcalorimetry)
Awards and acknowledgments: The CTTA outstanding contribution award of the Chemical Thermodynamics and Thermal Analysis (CTTA) Commission of the Chinese Chemical Society (CCS) in 2012; The Award of Kwang-Hua Science and Technology Foundation in 1993; 12 Science and technology prizes (one being the 1st rank, four being the 2nd rank, seven being the 3rd rank) issued by China Ordinance Industry Ministry and Shaanxi Provincial Government in 1993–2012; The Government Special Allowance Prize from the state Council, China in 1993;The contribution Prize of Shaanxi Provincial Academic Degree and Post-graduate Student Education by Shaanxi Provincial Academic Degree Committee in 2010;The National Science Conference Award from The State Council, China in 1978
Professional activities: Councilor of International Committee of Thermal Analysis and Calorimetry (1998–2000); Vice-Director of the Committee of Thermodynamics and Thermal Analysis of the CCS (1996–2002); Member of the Editorial Board of the Chinese Journal of EMs and a Regional Editor of JTA and JTAC (1993–2000); Part-time Professor of Sichuan University from 1995 to 1998, Northwest University since 2000, Qufu Normal University since 2000 and Institute of Chemical Materials, China Academy of Engineering Physics since 2000
Publication record: papers (350), books (4)
Equipments: Calvet microcalorimeter, type BT215, RD496-2000, C08 and C-500, CDR-1 type DSC
5 most important publications: [1] Hu Rongzu, Chen Xuelin, Chu Shijin, Li Nan, Qin Jiao, Yan Zhe. J. Therm. Anal, 42(2-3) (1994) 505–520; [2] Hu Rongzu, Meng Zihui, Kang Bing. Thermochim. Acta, 275(2) (1996) 159–172; [3] Hu Rongzu, Chen Sanping, Gao Shengli, Zhao Fengqi, Luo Yang, Gao Hongxu, Shi Qizhen, Zhao Hongan, Yao Pu, Li Jing. J. Hazard. Mater. 117(2-3) (2005) 103–110; [4] Hu Rongzu, Zhao Feng-qi, Gao Hong-xu, Ma Hai-xia, Zhang Hai, Xu Kang-zhen, Zhao Hong-an, Yao Er-gang. Acta Phys.-Chim. Sin. 29(10) (2013) 2071–2078; [5] Hu Rongzu, Zhao Fengqi, Gao Hongxu, Song Jirong. Calorimetry Fundamentals and Application. Beijing: Science Press, 2011.

Name: Wenbing Hu
Country: China
Date and place of birth: 1966, Zhejiang, China
Email: wbhu@nju.edu.cn
Website: http://hysz.nju.edu.cn/hwb/
Education and scientific degrees: Ph.D.(1995), Fudan University
Workplaces: School of Chemistry and Chemical Engineering, Nanjing University (2004–)
Main fields of interest: polymer crystallization, DSC, molecular simulations
Relevant categories in thermal analyses: fields (polymer); methods (TMDSC, chip-calorimeter)
Professional activities: Co-organizer for China-France Bilateral Workshop on Polymer Crystallization at Nanjing, China, June 6-9, 2006. Co-organizer for European Discussion Meeting on Polymer Crystallization at Waldau, Germany, Oct. 3-6, 2007. Co-organizer for International Discussion Meeting on Polymer Crystallization at Shanghai, China, Aug. 12–15, 2009. Local organizer for KITPC program: Growth of Hierarchical Functional Materials in Complex Fluids at Beijing, China, July 5- Aug. 5, 2011, including International Conference on the Hierarchical Structures in Complex Fluids July 4-8, Summer School on "Growth of Hierarchical Functional Materials in Complex Fluids" July 10–30, and International Discussion Meeting on Polymer Crystallization Aug. 1-5. Editorial Board of "Chinese Journal of Polymer Science". Boards of Polymer Chemistry Committee and of Computer Chemistry Committee, Chinese Chemical Society. Vice-chairman of Thermal Analysis Committee, Jiangsu Analytical and Testing Society
Publication record: papers (80), books (5)
Equipments: Flash DSC1, TGA/DSC
5 most important publications: [1] Hu, W.-B; Frenkel, D. "Polymer crystallization driven by anisotropic interactions" Advances in Polymer Science 191, 1–35 (2005). Review; [2] Hu, W.-B. "Intramolecular crystal nucleation", Lecture Notes in Physics: Progress in Understanding of Polymer Crystallization, Ed. G. Strobl and G. Reiter, Springer-Verlag, Vol. 714, 47–63 (2007). Review; [3] Hu, W.-B; Cai, T. "Regime transitions of polymer crystal growth rates: molecular simulations and interpretation beyond Lauritzen-Hoffman model" Macromolecules 41, 2049–2061 (2008). Article; [4] Hu, W.-B. Principle of Polymer Crystallization (in Chinese), Chemical Techology Publisher, Beijing, 2013. Book; [5] Hu, W.-B. Polymer Physics: A Molecular Approach, Springer-Verlag, Vienna, 2013. Book.

Name: Nicolae Hurduc
Country: Romania
Date and place of birth: 1956, Iasi, Romania
Present position and address: Professor, Piata Unirii 3, sc. B, et.4, ap, 22, 70056-Iasi, Romania
Email: nhurduc@ch.tuiasi.ro
Website: http://omicron.ch.tuiasi.ro/~inor/nicolae_hurduc/index_en.html
Researcher ID: 7005440544
Education and scientific degrees: Chemical Engineer—"Gheorghe Asachi" Technical University of Iasi (1975–1980); Ph.D. (1993); Professor at "Gheorghe Asachi" Technical University of Iasi (from 2000)
Workplaces: Synthetic Fibre Company of Iasi (1980–1984); Institute of Macromolecular Chemistry "P. Poni" Iasi (1984–1986); "Gheorghe Asachi" Technical University of Iasi (1986-present)
Main fields of interest: supramolecular chemistry, liquid crystals, molecular modelling
Relevant categories in thermal analyses: fields (polymers, biology); methods (DSC, TG, DTA, MS-TG)
Awards and acknowledgments: Prize of the Romania Academy (1994)
Professional activities: Member in Organizing Committee of the 1st Central and Eastern European Conference on Thermal Analysis and Calorimetry CEEC-TAC1, Craiova, Romania (2011); Member in Scientific Committee of the 2nd Central and Eastern European Conference on Thermal Analysis and Calorimetry CEEC-TAC2, Vilnius, Lithuania (2013); Member of the Editorial Board—Environmental Engineering and Management Journal; Associated Professor at Paris Nord University (2000–2001, 2005)
Publication record: papers (125), books (8), patents (6), citations (760), h-index (16), sum of impact factors (168)
Equipments: Derivatograph, Mettler Toledo TGA/SDTA 851e, DSC 1 Mettler Toledo
5 most important publications: [1] Raicu Luca, I. Moleavin, N. Hurduc, M. Hamell, L. Rocha, Applied Surface Science, 290 (3) 172 (2014); [2] G. Lisa, C. Păiuş, A. Raicu, N. Hurduc, High Performance Polymers, 24 (6) 530 (2012); [3] E. Taran, B. Donose, Ko Higashitani, A. D. Asandei, D. Scutaru, N. Hurduc, European Polymer Journal, 42 (1), 119 (2006); [4] A. Creanga, N. Hurduc, G. Pokol, C. Novák, S. Alazaroaie, Natalia Hurduc, Journal Thermal Analysis and Calorimetry, 70, 877 (2002); [5] N. Hurduc, M. Prajinaru, B. Donose, D. Pavel, N. Hurduc, Polymer Degradation and Stability, 72 (3), 441 (2001).

Name: John M. Hutchinson
Country: Spain
Date and place of birth: 1948, Bexhill-on-Sea, Sussex, UK
Present position and address: University professor. Universitat Politècnica de Catalunya, ETSEIAT, Carrer Colom 11, 08222 Terrassa, Spain
Email: hutchinson@mmt.upc.edu
Website: www.upc.edu
Education and scientific degrees: BA, Oxford University, Engineering Science (1967−1970); MA, DPhil, Oxford University (1974)
Workplaces: Post-doc: Centre de Recherches sur les Macromolécules, Strasbourg, France (1974−1975); Post-doc: Cranfield University, UK (1975−1978); Lecturer: Department of Metallurgy and Materials Science, University of Nottingham, UK (1978−1985); Senior Lecturer, Reader: Department of Engineering, University of Aberdeen, Scotland (1985−2004); Investigador Ramón y Cajal, Professor Agregat: Departament de Màquines i Motors Tèrmics, Universitat Politècnica de Catalunya (2004−)
Main fields of interest: glass transition of polymers; physical aging; temperature modulated differential scanning calorimetry (TMDSC); thermal analysis of epoxy-based layered silicate nanocomposites
Relevant categories in thermal analyses: fields (materials, nano, polymer, glass); methods (DSC, thermomechanical analysis, kinetics)
Publication record: papers (>100), chapters in books (5), citations (>3,700), h-index (30)
Equipments: DSC, TMDSC (Alternating DSC), TOPEM (stochastic TMDSC), UV-DSC, TGA, DSC/TGA, DMA, DEA, TMA
5 most important publications: [1] A. J. Kovacs, J. J. Aklonis, J. M. Hutchinson, A. R. Ramos. Isobaric volume and enthalpy recovery of glasses. II. A transparent multiparameter theory. J. Polym. Sci., Polym. Phys. Ed., 17 (1979) 1097–1162; [2] J. M. Hutchinson. Physical aging of polymers. Prog. Polym. Sci., 20 (1995) 703–760; [3] J. M. Hutchinson, M. Ruddy. Thermal cycling of glasses. II. Experimental evaluation of the structure (or nonlinearity) parameter x. J. Polym. Sci. Part B, Polym. Phys., 26 (1988) 2341–2366; [4] J. M. Hutchinson. Characterising the glass transition and relaxation kinetics by conventional and temperature-modulated differential scanning calorimetry. Thermochim. Acta, 324 (1998) 165–174; [5] F. Román, S. Montserrat, J. M. Hutchinson. On the effect of montmorillonite in the curing reaction of epoxy nanocomposites. J. Thermal Anal. Calorim., 87 (2007) 113–118.

Name: Cornelia Iditoiu
Country: Romania
Date and place of birth: 1938, Bucharest, Romania
Present position and address: Retired from the Aurel Vlaicu University of Arad, 300021-TIMISOARA, C.D. Loga nr.17, Romania
Email: corneliaiditoiu@yahoo.com
Education and scientific degrees: Chemical Engineer, "Traian Vuia" Polytechnic Institute of Timisoara (1956–1961); Ph.D., Polytechnic University of Bucharest (1975–1980); Dr.eng. (1981); Univ.Professor (1993)
Workplaces: "Aurel Vlaicu "University of Arad (1990–2010); "Traian Vuia" Polytechnic Institute of Timisoara (1977–1990); Vest University of Timisoara (1961–1977)
Main fields of interest: physical chemistry, catalysis in heterogeneous processes, thermal analysis and kinetics
Awards and acknowledgments: Outstanding Professional Award of National Minister of Education (1987); Aurel Vlaicu Medal fromUniversity of Arad (2003); "Petru Poni "Medal and Honorary Diploma of Romanian Chemical Society (2004); "Virtute Et Sapientia" Honorary Diploma for Scientific Research (2008) University of Bucharest
Professional activities: Fulbright scholar in USA, Rensselaer Polytechnic Institute (1974–1975); Founder member of FIATPM (Food Engineering and Enviromental, Protection) Faculty of Aurel Vlaicu University from Arad, Dean, Department Director, Chemical Research Center founder and director (2003–2010), Editor of The "Scientific and Technical Buletin" of Aurel Vlaicu—University from Arad (1991–2010)
Publication record: papers (103), books (4), citations (114)
5 most important publications: [1] Maria-Raluca Szabo, Dorina Chambre, Cornelia Iditoiu, TG/DTG/DTA for the oxidation behaviour characterization of vegetable and animal fats, J.T. A.C., 110, 2012, p 281–285; [2] D. Chambre, M. R. Szabo, C. Popescu, C. Iditoiu, Heterogeneous soft acid catalysis of sucrose hydrolysis, J.T.A.C., 94(2), 2008, p 417–420; [3] D. Chambre, C. Iditoiu, E. Segal, Non-isotermal dehydration kinetics of acrylic ion-exchange resins, J.T.A.C., 88 (3), 2007, pp. 673–679; [4] D. Chambre, C. Iditoiu, E. Segal, Attilo Cesaro, The study of non-isotermal degradation of acrylic ion-exchange resins, J.T.A.C., 82 (3), 2005, pp. 803–811; [5] C. Iditoiu, E. Segal, D. Chambre, Kinetics of non-isothermal behaviour of synthetic cationites with lou acidity, ESTAC-7, J.T.A.C., 55(1), 1999, 407–417.

Name: Emília Illeková
Country: Slovakia
Date and place of birth: 1949, Praha, Czechoslovakia
Present position and address: Emeritus chief research scientist, Institute of Physics, Slovak Acadmy of Sciences, Dubravska cesta 9, 845 11 Bratislava, Slovakia
Email: emilia.illekova@savba.sk
Website: www.fu.sav.sk/rq/illekova
Education and scientific degrees: Applied and experimental physics, Comenius University in Bratislava (1972), RNDr., Ph.D., DrSc.
Workplaces: Institute of Physics, Slovak Academy of Sciences, Bratislava (1972–2013); Emeritus Scientist (2014–)
Main fields of interest: kinetics of transformations in solids (structural relaxation and crystallization of glasses, solidification of confined liquids and their melting); thermophysical and thermodynamic properties of non-equilibrium solids; material research in glasses, especially metallic glassy ribbons; mechanically alloyed powders; bulk metallic, chalcogenide and oxide glasses, composites consisting of confined liquid medium and solid matrix
Relevant categories in thermal analyses: fields (glasses, metallic glasses, nano, confined liquids, non-equilibrium solids); methods (DTA, DSC, TG, thermomagnetometry, kinetics, specific heat)
Awards and acknowledgments: 1st Prize, Young Phys. Competition, Inst. Phys. SAS (1984); D. Ilkovič Silver medal of Merit in Physical and Chemical Sciences (2000), Slovak Physical Society Medal (2009)
Professional activities: Member of the Org. Committee of RQ9, Int. Conf. on Rapid Quenched and Metastable Materials (1996); Member of the Committee of the Slovak Physical Society (1981–3, 2000–3); External lecturer at the Faculty of Materials Science and Technology, Slovak Technical University (1989–2001)
Publication record: papers (158), book chapters (6), citations (388)
Equipments: DSC7, DSC8500, DTA7 (PerkinElmer), SDT Q600 (TA Instruments)
5 most important publications: [1] E. Illeková, B. Aba, F.A. Kuhnast: Measurements of Accurate Specific Heats of Metallic Glasses by Differential Scanning Calorimetry (DSC): I. Analysis of Theoretical Principles and Accuracies of Suggested Procedures, Thermochim. Acta 195 (1992) 195–209; [2] E. Illeková, I. Maťko, P. Duhaj, F.A. Kuhnast: The complex characteristics of crystallization of the Fe75Si15B10 glassy ribbon, J. Mater. Sci. 32 (1997) 4645–4654; [3] E. Illeková: FINEMET-type nanocrystallization kinetics, Thermochim. Acta 387 (2002) 47–56; [4] E. Illeková, P. Švec, M. Miglierini: Thermokinetic analysis of the multistep crystallization of a NANOPERM-type ribbon, J. Non-Cryst. Solids, 353 (2007) 3342–3347; [5] E. Illeková: Kintics of structiral relaxation in glasses, as Chapter 8, in Thermal Analysis of Micro, nano-, and Non-Crystalline Materials, J. Šesták, P. Šimon, Eds., Springer, 2013, pp. 175–194, in Transformation, Crystallization, Kineitcs and Thermodynamics Series: Hot Topics in Thermal Analysis and Calorimetry, Vol. 9 (2013) XXI, 484 p., ISBN 978-90-481-3150-1.

Name: Ewa Ingier-Stocka
Country: Poland
Date and place of birth: 1946, Jelenia Góra, Poland
Present position and address: Retired from the Wroclaw University of Technology, Faculty of Chemistry; D.Sc. Ewa Ingier-Stocka, M. Bacciarellego 48/8 str., 51-649 Wroclaw, Poland
Email: ewa.ingier-stocka@pwr.edu.pl
Education and scientific degrees: Chemical Engineer, Technical University of Wroclaw (1964–1969); M.Sc. (1969); Ph.D. (1977); D.Sc. (2008)
Workplaces: Chemical Metallurgy Group, Faculty of Chemistry, Wroclaw University of Technology (1972–2012)
Main fields of interest: kinetics of the thermal decomposition of complex compounds; study of lanthanide halide–alkali metal halide binary systems (determination of phase diagrams, electrical conductivity and thermodynamic functions)
Relevant categories in thermal analyses: fields (inorganic, complex); methods (TG, DTA, EGA, DSC, kinetics, specific heat, calorimetry, microcalorimetry)
Publication record: papers (66), book chapters (3), patents (2), citations (92), sum of impact factors (63)
Equipments: Derivatograph C (MOM, Budapest), Setaram DSC 121 differential scanning calorimeter, Calvet-type high-temperature microcalorimeter
5 most important publications: [1] E. Ingier-Stocka: J. Thermal. Anal., 40 (1993)1357; [2] E. Ingier-Stocka: J. Thermal.Anal., 50 (1997) 603; [3] E. Ingier-Stocka, M. Maciejewski: Thermochim. Acta, 354 (2000) 45; [4] E. Ingier-Stocka, M. Maciejewski: Thermochim. Acta, 432 (2005) 56; [5] L. Rycerz, E. Ingier-Stocka, S. Gadzuric, M. Gaune-Escard: J. Thermal.Anal. Calorim., 101 (2010) 455.

Name: Massao Ionashiro
Country: Brazil
Date and place of birth: 1945, Paraguaçu Paulista, Brazil
Present position and address: Collaborative Professor in Departamento de Química Analítica Intituto de Química —Universidade Estadual Paulista "Júlio de Mesquita Filho"—UNESP 55 Prof. Francisco Degni Street Quitandinha—Araraquara, São Paulo State Zip code: 14800-060
Email: massaoi@iq.unesp.br
Website: http://www.iq.unesp.br/
Researcher ID/ORCID: D-7644-2013/0000-0003-2980-5841
Education and scientific degrees: Degree in Chemistry, Faculty of Philosophy, Sciences and Letters of Araraquara (1969); Ph.D. in Inorganic Chemistry, Institute of Chemistry, University of São Paulo (1976); "Privatdozent" in Analytical Chemistry, Institute of Chemistry of Araraquara—UNESP (1989). Became Professor at the Institute of Chemistry of Araraquara-UNESP in 1995. Retired in 1998 from Universidade Estadual Paulista, Instituto de Química de Araraquara and he is currently associate at the Department of Analytical Chemistry IQ—Araraquara—UNESP. Emeritus Professor at UNESP (insignia received in 2008).
Workplaces: UNESP—Intituto de Química—Departamento de Química Analítica—Laboratório de Análise Térmica Ivo Giolito (LATIG)
Main fields of interest: analytical chemistry with emphasis on thermal analysis, aiming the synthesis, characterization and thermal properties and spectroscopic study of compounds of various metal ions such as lanthanides and transition metals with organic and inorganic ligands
Relevant categories in thermal analyses: fields (inorganic, materials, complexes, organic, pharmaceuticals); methods (TG, DTA, EGA and DSC)
Awards and acknowledgments: Honorary member since 2006 (The Thermoanalytical Group of the Hungarian Chemical Society awards Honorary Membership)
Publication record: papers (160), books (1), citations (597), h-index (13)
Equipments: TG-DTA, model SDT 2960, TA instruments; DSC, model Q10, TA instruments; TG-DSC (Mettler Toledo) coupled to a FTIR spectrophotometer (Nicolet iS10): TG-DSC/FTIR
5 most important publications: [1] Caires, F.J.; Lima, L.S.; Gomes, D.J.C.; Gigante, A. C.; Treu-Filho, O.; Ionashiro, M. Thermal and spectroscopic studies of solid oxamate of light trivalent lanthanides. J Therm Anal Calorim 111, 349–355, 2013; [2] Gomes, D.J.C.; Caires, F.J.; Lima, L.S.; Gigante, A.C.; Ionashiro, M. Thermal behaviour of α-hydroxyisobutyric acid, sodium α-hydroxyisobutyrate and its compounds with some bivalent transition metal ions. Thermochim Acta (Print), v. 564, pp. 7-12, 2013; [3] Siqueira, A. B.; Bannach, G.; Rodrigues, E.C.; Carvalho, C.T.; Ionashiro, M. Solid-state 2-methoxybenzoates of light trivalent lanthanides. Synthesis, characterization and thermal behaviour. J Therm Anal Calorim 91, 897–902, 2008; [4] Bannach, G.; Siqueira, A.B.; Ionashiro, E.Y. Rodrigues, E.C.; Ionashiro, M. Solid-State Compounds of 2-Chlorobenzylidenepyruvate with some bivalent metal ions: Synthesis, characterization and thermal behaviour. J Therm Anal Calorim 90, 873–879, 2007; [5] Schnitzler, E.; Melios, C.B.; Ionashiro, M. Solid-state compounds of 4-methoxybenzylidenepyruvates and cinnamylidenepyruvates with thorium (IV) preparation and thermal studies. J Therm Anal Calorim, 70, 581–592, 2002.

Name: Zeljko K. Jacimovic
Country: Montenegro
Date of birth: 1966
Present position and address: Full prof, Faculty of Metallurgy and Techolology, University of Montenegro, Dzordza Vasingtona bb, Podgorica, Montenegro
Email: zeljkoj@ac.me
Website: www.mtf.ac.me
Researcher ID/ORCID: C-6674-2014/0000-0001-6803-3835
Education and scientific degrees: Chemist, Faculty of Natural Sciences—Chemistry, Ph.D. in Chemistry, Full professor.
Workplaces: University of Montenegro, Podgorica, Montenegro
Main fields of interest: inorganic chemistry, coordination chemistry and crystallography
Relevant categories in thermal analyses: fields (inorganic, complex compounds); methods (TG, DTA and DSC)
Professional activities: Chemical Society of Montenegro—president, Chairman of14th European Meeting on Environmental Chemistry, 1st and 2nd Symposium of Chemistry and Environment (Organised by Chemical Society of South-Eastern European County)
Publication record: papers (48), books (2), citations (276), h-index (10)
5 most important publications: [1] K. Mészáros-Szécsényi, V. M. Leovac, Ž. K. Jaćimović, V. I. Česljevic, A. Kovács, G. Pokol: "Transition metal complexes with pyrazole-based ligands. Part 12. Characterization and Thermal decomposition of CuCl2 complexes with di- and trisubstituted pyrazoles", J. Thermal. Anal. Cal., (2001) 66, 573; [2] K. Mészáros Szécsényi, V. M. Leovac, Ž. K. Jaćimović, G. Pokol: "Transition metal complexes with pyrazole-based ligands. Part 15. Cobalt(III)-, nickel(II)-, and copper(II)- complexes with 3,5-dimethyl-1- thiocarboxamidepyrazole", J. Thermal. Anal. Cal., (2003) 74, 943; [3] K. Mészáros Szécsényi, V. M. Leovac, A. Kovács, G. Pokol, Ž. K. Jaćimović: "Transition metal complexes with pyrazole-based ligands. Part 21. Thermal decomposition of copper and cobalt halide complexes with 3,5-Dimethyl-1-thiocarboxamidepyrazole", J. Thermal. Anal. Cal, (2006) 85, 289; [4] Ž. K. Jaćimović, V. M. Leovac, G. Giester, Z. D. Tomić, and K. Mészáros-Szécsényi: "Transition metal complexes with pyrazole-based ligands. Part 56. Structural and thermal characterization of Fe(III) and Characterization and Fe(II) complexes with tridentate ONO pyridoxal semicarbazone ligand", J. Thermal. Anal. Cal., (2007) 90, 549; [5] K. Mészáros Szécsényi, V. M. Leovac, R. Petkovic, Ž. K. Jaćimović and G. Pokol: "Transition metal complexes with pyrazole-based ligands. Part XXV. Deaquation of isostructural cobalt(II) and NI(II) complexes with 3,5-dimethylpyrazole-1-carboxamidine", J. Thermal. Anal. Cal., (2007) 90, 899.

Name: Michael Jaffe
Country: USA
Date and place of birth: 1942, New York City, USA
Present position and address: Professor of Biomedical Engineering, New Jersey Institute of Technology, 111 Lock Street, Newark, NJ 07103
Email/Website: jaffe@njit.edu/biomedical.njit.edu/people/faculty.php
Education and scientific degrees: Cornell University, B.A. Chemistry, 1963, Ph.D. Rennselaer Polytechnic Institute, Chemistry, 1967
Workplaces: New Jersey Institute of Technology 1999-present, Celanese, Hoechst Celanese 1967–1998
Main fields of interest: polymer science, materials science, biotechnology, biomaterials, high performance materials
Relevant categories in thermal analyses: fields (materials characterization); methods (DSC, TGA, DMA, TMA)
Awards and acknowledgments: NATAS Fellow (1991), Fellow of PMSE, Fellow of ACS
Professional activities: ACS, NATAS, Fiber Society, APS
Publication record: papers (71), books (18), patents (22)
5 most important publications: [1] Feng, X., East, A. J., Hammond, W., Ophir, Z., Zhang, Y., and Jaffe, M., "Thermal analysis characterization of isosoribide containing thermosets", Journal of Thermal Analysis and Calorimetry (2012), 109(30, 1267–1275; [2] M. Jaffe, Thermal Analysis of Fibers, E. Turi, Ed., Academic Press (1981); [3] M. Jaffe, J. Menczel, W. Bessey, Thermal Characterization of Polymeric Materials, 2nd Ed., Vol. 1, (E. A. Turi, Ed.), Chapter 7: "Fibers" and Chapter 8: "Films." Academic Press, San Diego, CA. (1997); [4] J. Menczel, M. Jaffe, W. Bessey, Thermal Characterization of Polymeric Materials, 2nd Ed., Vol. 2, (E. A. Turi, Ed.), Chapter 8: "Films." Academic Press, San Diego, CA (1997); [5] Eichhorn, S., Hearle, j., Jaffe, M., and Kikutani, T.eds. (2009), Handbook of textile fibre structure, vol. 1&2, Woodhead.

Name: Emma Jakab
Country: Hungary
Date and place of birth: 1954, Felpéc, Hungary
Present position and address: senior research fellow, Reseach Center for Natural Sciences, 1117 Budapest, Magyar tudósok krt. 2.
Email: jakab.emma@ttk.mta.hu
Education and scientific degrees: 1978: M.Sc. in chemistry, Loránd Eötvös University; 1983: Dr. Univ. in chemical technology, Loránd Eötvös University; 1984–1985: Postdoctoral fellow, Biomaterials Profiling Center, Univ. Utah, Salt Lake City, USA; 1996: C. Sc. in chemistry.
Workplaces: 1978-present: Institute of Materials and Environmental Chemistry, Research Center for Natural Sciences, Hung. Acad. Sci. and its predecessors. 1993–1994: Visiting scientist, Center for Microanalysis and Reaction Chemistry, Univ. Utah, Salt Lake City, USA
Main fields of interest: thermal decomposition of wood, lignin, biomass materials and synthetic polymers using pyrolysis-GC-MS and TG-MS; the effect of inorganic and organic additives on the decomposition mechanisms of natural and synthetic polymers; the influence of various physical and chemical pretreatments on the composition and thermal properties of biomass.
Relevant categories in thermal analyses: fields (polymer, biomass); methods (TG, EGA (MS))
Professional activities: Member of Working Group for Thermal Analysis of Hung. Acad. Sci., Member of Working Group for Natural Polymers of Hung. Acad. Sci. Conference secretary of the 17th Int. Symp. on Anal. Appl. Pyrolysis, 2006, Budapest, Hungary
Publication record: papers (117), book chapter (1), citations (2040), h-index (26), sum of impact factors (142.9)
Equipments: TG-MS, analytical pyrolyzer-GC-MS
5 most important publications: [1] Z. Sebestyén, Z. May, K. Réczey, E. Jakab, The effect of alkaline pretreatment on the thermal decomposition of hemp, J. Therm. Anal. Calor., 105(3) (2011) 1061–1069; [2] E. Jakab, E. Mészáros, J. Borsa, Effect of Slight Chemical Modification on the Pyrolysis Behavior of Cellulose Fibers, J. Anal. Appl. Pyrolysis, 87 (2010) 117–123; [3] E. Jakab, Md. A. Uddin, T. Bhaskar, Y. Sakata, Thermal decomposition of flame-retarded high impact polystyrene, J. Anal. Appl. Pyrolysis, 68–69 (2003) 83–99; [4] E. Jakab, G. Várhegyi, O. Faix, Thermal decomposition of polypropylene in the presence of wood-derived materials, J. Anal. Appl. Pyrolysis, 56 (2000) 273–285; [5] E. Jakab, O. Faix, F. Till: Thermal decomposition of milled wood lignins studied by thermogravimetry/mass spectrometry, J. Anal. Appl. Pyrolysis, 40–41, 171–186 (1997).

Name: Bojan Ž. Janković
Country: Serbia
Date and place of birth: 1974, Smederevska Palanka, Serbia
Present position and address: Assistant Professor, Faculty of Physical Chemistry, University of Belgrade, Serbia; Studentski Trg 12-16, P.O. Box 137, 11001 Belgrade, Serbia
Email: bojanjan@ffh.bg.ac.rs; bojanjankovi78@gmail.com
Website: www.ffh.bg.ac.rs
Researcher ID: 11238857600
Education and scientific degrees: 2000. Graduated at Faculty of Physical Chemistry, University of Belgrade, Serbia; 2005. M.Sc. at Faculty of Physical Chemistry, University of Belgrade, Serbia; 2012. Ph.D. at Faculty of Physical Chemistry, University of Belgrade, Serbia
Workplaces: Faculty of Physical Chemistry, University of Belgrade, Serbia
Main fields of interest: nanotechnology, biomaterials, polymer degradation, extraction of rare earth elements from the mineral matrix by thermal procedures, thermal stability of pharmaceuticals and "shelf-life" analysis, food chemistry, applied energy and fuels
Relevant categories in thermal analyses: fields (inorganic, materials, nano, minerals, organic, pharmaceutical, polymer, food); methods (TG, DTA, DSC, kinetics; specific heat, calorimetry)
Awards and acknowledgments: The American Association of Pharmaceutical Scientists: Award for Top 12 Reviewers and Pavle Savić Award for the scientific contribution
Professional activities: Participant in National basic research Project funded by the Ministry of Science and Technology of Republic of Serbia; Participant in International project FR 1713/13-1: "Kinetics and mechanism of thermal zinc-ferrite decomposition" funded by the Deutsche Forschungsgemeinschaft DFG in Bonn (Germany); A member of the Serbian Chemical Society (SCS) and a member of the Society of Physical Chemists of Serbia (SPCS)
Publication record: papers (68), citation index (295), h-index (10), sum of impact factors (81.874)
Equipments: TA Instruments SDT 2960 for simultaneous TGA-DTA analyses, and Netzsch STA 409 CD Thermal Analyzer for simultaneous TGA-DTA analyses.
5 most important publications: [1] B. Janković, B. Adnađević, J. Jovanović, Thermochimica Acta 452 (2007) 106–115; [2] B. Janković, S. Mentus, Journal of Thermal Analysis and Calorimetry 94 (2008) 395–403; [3] B. Janković, Chemical Engineering Journal 162 (2010) 331–340; [4] B. Janković, Bioresource Technology 102 (2011) 9763–9771; [5] B. Janković, Carbohydrate Polymers 95 (2013) 621–629.

Name: Mohamed Jemal
Country: Tunisia
Date and place of birth: 1945, Sfax, Tunisia
Email: jemal@planet.tn
Education and scientific degrees: Ph.D.
Workplaces: Tunis El Manar University, Faculty of Science, Chemistry Department, Tunisia
Main fields of interest: education and research in chemical thermodynamics and kinetics
Relevant categories in thermal analyses: fields (inorganic materials, phosphates, nitrates, glasses); methods (TG, DTA and DSC (theory and applications), calorimetry, microcalorimetry)
Awards and acknowledgments: Tunisian Chemical Society (TCS) Award (2008), Chemists Arab Union (CAU) Award (2010)
Professional activities: Past President of TCS, Past President of CAU, Editor of Inorganic section of the Journal of TCS, Organizer of 24th IVMT Conference (1991), Co-organizer of 23rd JEEP (1997), Organizer of 32nd JCAT (2001) and of 41st JCAT (2010)
Publication record: papers (85), book chapters (2)
Equipments: Thermobalance (up to 1,000 °C), locally made calorimetres, Reaction Calorimeter, Microcalorimeter (up to 300 °C)
5 most important publications: [1] M. Jemal, Thermodynamics and kinetics of the reactions of apatite phosphates with acid solutions, in Application of Thermodynamics to Biological and Materials Science Tadashi Mizutani Editor, INTECH open access, 2011; [2] M. Jemal, Purification of water by zone melting. A phase diagramm interpretation, J. Chem. Educ., 2004, 81(7), 952; [3] M. Jemal and I. Khattech, Simultaneous thermogravimetry and gas chromatography during decomposition of carbonate apatites, Thermochimica Acta, 1989, 152, 65; [4] H. Gmati-Ben Khaled, I. Khattech and M. Jemal, Standard enthalpy of formation of disodium hydrogen phosphate and sodium diphosphate, J. Chem. Thermodynamics, 2011, 43(4), 521–526; [5] M. Fertani-Gmati and M. Jemal, Thermochemistry and kinetics of silica dissolution in NaOH aqueous solution, Thermochimica Acta, 2011, 513(1-2), 43–48.

Name: Karol Jesenák
Country: Slovakia
Date and place of birth: 1951, Bratislava, Slovakia
Present position and address: Professor, Department of Inorganic Chemistry, Faculty of Natural Sciences, Comenius University, Mlynská Dolina CH2, 842 15 Bratislava, Slovakia
Email:jesenak@fns.uniba.sk
Website: http://www.fns.uniba.sk/index.php?id=jesenak
Education and scientific degrees: Chemical engineer, Technical University, Bratislava (1969–1973), Ph.D. 1988, Docent 2008, Assoc. Professor 2008, Professor 2014
Workplaces: Department of Inorganic Chemistry, Faculty of Natural Sciences, Comenius University, Bratislava, Slovakia (1974–2014)
Main fields of interest: materials chemistry, particle size analysis, sol-gel methods
Relevant categories in thermal analyses: fields (inorganic nanomaterials, carbon nanotubes, inorganic aerogels, sorbents, clay minerals); methods (TG, DTA, DSC)
Awards and acknowledgments: Faculty of Natural Sciences Comenius University (2011), Slovak Chemical Society (2011), Slovak Mineralogical Society (2013)
Publication record: papers (280), books (7)
Equipments: Differential Scanning Calorimeter (PerkinElmer), Thermogravimetric Analyser (PerkinElmer)
5 most important publications: [1] Jesenák, K., Kubinec, R., Kuchta, Ľ., Fajnor, V.: Calcination of SiO2-aerogel in inert atmosphere, Journal of Thermal Analysis and Calorimetry, 67, No. 1 (2002), 207; [2] Jesenák, K., Hlavatý, V.: Estimation of sorption parameters of solids by means of derivatograph Q. Journal of Thermal Analysis and Calorimetry. 67, No. 1 (2002), 93; [3] Pastorková, K., Jesenák, K., Kadlečíková, M., Breza, J., Kolmačka, M., Čaplovičová, M., Lazišťan, F., Michalka, M.: The growth of multi-walled carbon nanotubes on natural clay minerals (kaolinite, nontronite and sepiolite), Applied Surface Science, 258, No. 7 (2012), 2661; [4] Jesenák, K.: Polymer-clay nanocomposites, Chem. Papers, 101, (2007) 657. review; [5] Jesenák, K.: Particle size analysis, Bratislava, Comenius University (2008) ISBN 978-80-223-2464-9.

Name: Małgorzata Jóźwiak
Country: Poland
Date and place of birth: 1954, Warsaw, Poland
Present position and address: professor, Department of Physical Chemistry, University of Lodz, Pomorska 165, 90-236 Lodz, Poland
Email: mjozwiak@uni.lodz.pl
Website: http://www.chemia.uni.lodz.pl/kchf/eng.htm
Education and scientific degrees: master (1979) Department of Organic Chemistry, Faculty of Mathematic, Physic and Chemistry, University of Lodz, Ph.D. (1989) Department of Physical Chemistry, Faculty of Chemistry, University of Warsaw, D.Sc. (Doctor habilitus) (2005) Department of Physical Chemistry, Faculty of Physic and Chemistry, University of Lodz
Workplaces: Department of Physical Chemistry, University of Lodz
Main fields of interest: thermochemical behaviour, water-organic solvent mixtures, solvation process, preferential solvation, hydrophobic hydration, cyclic ethers, complexes of crown ether with cation, thermodynamic function of complex formation, density and viscosity of solutions
Relevant categories in thermal analyses: fields (solutions, cyclic ethers, complexes); methods (calorimetry)
Professional activities: 1994–2000 secretary of the Polish Society of Calorimetry and Thermal Analysis; 2006- member of the board of the Polish Society of Calorimetry and Thermal Analysis; 2007- treasurer of the Polish Society of Calorimetry and Thermal Analysis; Member of the scientific committee of several Conferences Calorimetry and Thermal Analysis; Member of the organizing committee of several Conferences Calorimetry and Thermal Analysis
Publication record: papers (63), citations (470), h-index (13), sum of impact factors (105,677)
Equipments: isoperibol type calorimeter, DSC 111
5 most important publications: [1] H. Piekarski, M. Jóźwiak, Thermochemistry of interactions of Na + with benzo-15-crown-5 ether in acetonitrile-water mixtures at 298.15 K, J. Thermal Anal., 48 (1997) 1283–1291; [2] G. Della Gatta, M. Jóźwiak, B. Brunetti, L. Abate, Enthalpies and entropies of fusion and of sublimation at the temperature 298.15 K of thiourea and seven N-alkylthioureas, J. Chem. Thermodyn., 32 (2000) 979–997; [3] M. Jóźwiak, The effect of properties of water-organic solvent mixtures on the solvation enthalpy of of 12-crown-4, 15-crown-5, 18-crown-6 and benzo-15-crown-5 ethers at 298.15 K, Thermochim. Acta, 417 (2004) 31–41; [4] M. Jóźwiak, Thermochemical behaviour of crown ethers in the mixtures of water with organic solvents. Part VIII. Hydrophobic hydration and preferential solvation of 1,4-dioxane in {(1-x)amide + H_2O} at T = 298.15 K, J. Chem. Thermodyn., 39 (2007) 433–437; [5] M. Jóźwiak, The effect of carbonyl carbon atom replacement in acetone molecule (ACN) by sulphur atom (DMSO) Part III. Effect of base-acid properties of the mixtures of water with acetone or dimethylsulfoxide on the solution enthalpy of cyclic ethers in mixed solvent, J. Therm. Anal. Calorim. 101 (2010) 1039–1045.

Name: Lawrence Judovits
Country: USA
Date and place of birth: 1955, Bronx, USA
Present position and address: Senior Research Scientist, Arkema Inc., 900 First Avenue, King of Prussia, PA 19406
Email: larry.judovits@hotmail.com
Education and scientific degrees: Rensselear Polytechnic Institute, Ph.D.
Workplaces: Arkema Inc./ATOFINA Chemicals/Elf Atochem NA/Atochem NA/Pennwalt
Main fields of interest: thermal analsis of polymers, ferroic polymers
Relevant categories in thermal analyses: fields (polymer); methods (DSC, TGA, TMA, DMA, modulated techniques)
Awards and acknowledgments: Union Carbide Teaching Incentive Award, 1982; 2001 NATAS Fellow; ASTM Award of Merit and Fellow, 2003; 2005 NATAS Service Award; Certificate of Recognition for 25 years of Service at Arkema (2011)
Professional activities: North American Thermal Analysis Society (NATAS): Short Course Chair for 2011 and 2012, Technical Program chair for 2010 and 2008, Vice-Conference chair for 2006, Conference chair for 2005, Executive Councilor for 2003, President for 2002, Vice-President (president-elect) for 2001, Secretary 1999–2000, Education Secretary 1997–1998. ASTM E37 Committee on Thermal Measurements: Chairman 2006–2011; co-edited 2 STPs. JTAC: Appointed to the Editorial Advisory Board 2003
Publication record: papers (>26), patents (1)
Equipments: DSC, TGA, TMA, DMA, TGA/sDTA, PDSC, calorimeter
5 most important publications: [1] J. D. Menczel, L. Judovits, R. B. Prime, H. E. Bair, M. Reading, and S. Swier in Differential Scanning Calorimetry in Thermal Analysis of Polymers, J. D Menczel and R. B. Prime eds., Wiley, 2009; [2] L. Judovits, Nucleating Agents for Increased Melting Point for PVDF, 34th NATAS Conference Proceedings, (2006) on CD; [3] Polyvinylidene fluoride having an increased melting point, US patent 8263695; [4] ASTM STP 1402 on Characterization by Dynamic and Modulated Thermal Analysis Analytical Techniques A. T. Riga and L. Judovits, eds. (2001); [5] L. Judovits, J. Menzcel, and A.-G. Leray, Molecular Weight Effects on the Reorganization of Poly(vinylidene Fluoride), Polyamide 12, and Poly(phenylene Sulfide), Journal of Thermal Analysis and Calorimetry, 54 (1998) 605.

Name: Andrei G. Kabo
Country: Belarus
Date and place of birth:1962, Samara, Russia
Present position and address: Chief of the Development Department, Belaquilon ODO, Minsk, Belarus; A.G. Kabo, Krasnoarmeiskaya str., 19-8, Minsk, 220030 Belarus
Email: a.kabo@mail.ru
Education and scientific degrees: Radiophysicist, Belarusian State University (1979–1984); Ph.D., Department of Physical Chemistry of the Belarusian State University (2010)
Main fields of interest: development of new devices for chemical laboratories; technique of chemical experiments
Relevant categories in thermal analyses: fields (materials, nano, organic, polymer, ionic liquids); methods (DSC, kinetics, specific heat, combustion calorimetry, calorimetry, microcalorimetry, instrument development)
Publication record: h-index (9)
5 most important publications: [1] Shevelyova M.P. [et al.]: Physicochemical properties of imidazolium-based ionic nanofluids: Density, heat capacity, and enthalpy of formation, J. Phys. Chem. C, 2013, 117 (9), pp 4782–4790; [2] Kabo G.J. [et al.]: Experimental determination of enthalpy of 1-butyl-3-methylimidazolium iodide synthesis and prediction of enthalpies of formation for imidazolium ionic liquids, J. Chem. Thermodyn., Vol. 42, Issue 10, 2010, pp 1292–1297; [3] Paulechka Y.U. [et al.]: Heat capacity of ionic liquids: Experimental determination and correlations with molar volume, J. Chem. Eng. Data, 2010, 55 (8), pp 2719–2724; [4] Paulechka Y.U. [et al.]: Calorimetric determination of the enthalpy of 1-Butyl-3-methylimidazolium bromide synthesis: A key quantity in thermodynamics of ionic liquids, J. Phys. Chem. B, 2009, 113 (44), pp 14742–14746; [5] Kabo G.J [et al.]: Thermodynamic properties of 1-butyl-3-methylimidazolium hexafluorophosphate in the condensed state, J. Chem. Eng. Data, 2004, 49 (3), pp 453–461.

Name: Erwin Kaisersberger
Country: Germany
Date and place of birth: 1942, Adlmühle, Germany
Present position and address: Consultant for Thermal Analysis, Bahnhofstr. 26, 95168 Marktleuthen, Germany
Email: e.kaisersberger@t-online.de
Education and scientific degrees: Physicist, diploma at L. M. University Munich (1973)
Workplaces: GSF Munich Neuherberg (1971–1973), NETZSCH-Gerätebau GmbH (1973–2007), Consultant (2007–)
Main fields of interest: application of thermal analysis in general, interpretation of results
Relevant categories in thermal analyses: fields (inorganic and organic materials, polymers, pharmaceuticals, food, glass and ceramics); methods (DSC, TG, DTA, EGA, simultaneous and combined methods)
Awards and acknowledgments: NETZSCH-GEFTA Award (2008)
Professional activities: Founder member of GEFTA (Society for Thermal Analysis in German speaking countries), application laboratory manager at NETZSCH-Gerätebau (1979–1989), regional sales manager Europe of NETZSCH-Gerätebau (1992–2002), sales and application support NETZSCH-Gerätebau (2002–2007), independent consultant for Thermal Analysis (2007–)
Publication record: papers (61), books (1)
Equipments: Netzsch STA 429, Balzers QMS
5 most important publications: [1] E. Kaisersberger, W.-D. Emmerich, C. Urso, C. Politis, W. Krauss: Preparation and characterization of high temperature superconductors by thermal analysis, high temperature superconductors, (Ed.) P. Vicenzini (1991) 755; [2] E. Kaisersberger, E. Post: Practical aspects for the coupling of gas analytical methods to thermal analysis instruments, Thermochim. Acta, 295 (1997) 73; [3] E. Kaisersberger: Techniques for a high temperature coupling of quadrupole mass spectrometers and a thermoanalyzer, Int. J. Mass Spectrometry Ion Physics, 46 (1983) 155; [4] E. Kaisersberger, W.-D. Emmerich: New TA-MS coupling system with increased sensitivity for low volatile materials, Thermochim. Acta, 85 (1985) 275; [5] Editor of Special Issue: Coupling thermal analysis and gas analysis methods, Thermochim. Acta, Vol. 295. 1–2 (1997) pp. 1–186.

Name: Aivaras Kareiva
Country: Lithuania
Date and place of birth: 1960, Lithuania
Present position and address: Professor; Department of Inorganic Chemistry, Vilnius University, Naugarduko 24, LT-03225 Vilnius, Lithuania
Email: aivaras.kareiva@chf.vu.lt
Education and scientific degrees: Master's (1983). Ph.D. (1989), Vilnius University, Lithuania.
Workplaces: Vilnius University, Lithuania
Main fields of interest: inorganic, solid state, materials and analytical chemistry
Relevant categories in thermal analyses: fields (inorganic, materials); methods (TG, DTA, EGA, DSC)
Awards and acknowledgments: Vilnius University Rector's Research Award (2003, 2007, 2011). Lithuanian Republic Research Award. Lithuanian Academy of Sciences Juozas Matulis Award. The Honorary International Chair Professor at National Taipei University of Technology, Taiwan
Professional activities: Dean of Faculty of Chemistry of Vilnius University. Deputy Editor of Lithuanian Journal "Chemija" (Chemistry). Member of Lithuanian Academy of Sciences.
Publication record: papers (230), books (8), patents (4), h-index (20)
Equipments: TG-DTA, DSC
5 most important publications: [1] A. Katelnikovas, P. Vitta, P. Pobedinskas, G. Tamulaitis, A. Zukauskas, J.-E. Jørgensen and A. Kareiva. Photoluminescence in sol-gel derived YAG:Ce phosphors. J. Cryst. Growth, 304 (2007) 361–368; [2] A. Katelnikovas, J. Barkauskas, F. Ivanauskas, A. Beganskiene and A. Kareiva. Aqueous sol-gel synthesis route for the preparation of YAG: Evaluation of sol-gel process by mathematical regression model. J. Sol-Gel Sci. Techn., 41 (2007) 193–201.; [3] I. Bogdanoviciene, A. Beganskiene, K. Tõnsuaadu, J. Glaser, H.-J. Meyer, A. Kareiva. Calcium hydroxyapatite, (Ca10(PO4)6 (OH)2, HA) ceramics prepared by aqueous sol-gel processing. Mater. Res. Bull., 41 (2006) 1754–1762; [4] M. Veith, S. Mathur, A. Kareiva, M. Jilavi, M. Zimmer and V. Huch. Low temperature synthesis of nanocrystalline Y3Al5O12 (YAG) and Ce-doped Y3Al5O12 via different sol-gel methods. J. Mater. Chem., 9 (1999) 3069–3079; [5] A. Kareiva, C. J. Harlan, D. B. MacQueen, R. Cook and A. R. Barron. Carboxylate substituted alumoxanes as processable precursors to transition metal-aluminum and lanthanide-aluminum mixed metal oxides: atomic scale mixing via a new transmetalation reaction. Chem. Mater. 8 (1996) 2331–2340.

Name: Etienne Karmazsin
Country: France
Date and place of birth: 1942, Ungvàr, Hungary
Present position and address: Chimie Physique et Electronique de Lyon; 17 Montée Saint Sébastien—69002 Lyon.
Email: etienne.karmazsin@cpe.fr
Workplaces: University of Lyon 1, Asian Institute of Technology Bangkok, Ecole Normale Supérieure de Lyon, Ecole Supérieure de Chimie Physique et Electronique de Lyon
Main fields of interest: development of a specific calorimeter for plaster hydration, a calorimeter-conductimeter, development of a simultaneous DTA-magnetometer, of a microwave dilatometer, thermal analysis techniques with microwaves, development of a high temperature thermo-conductimeter, development of a surface treatment device with microwave plasma; economic intelligence
Awards and acknowledgments: Medal of The Russian Academy of Sciences for works in the field of materials for aeronautics. Selected to propose candidates for Nobel prize of Chemistry
Professional activities: ICTAC Councilor (1992), member of the AFCAT council (1988) Editor of AFCAT Bulletin (1990), Member of The Editorial Board of the j. of Therm. Anal. Cal. Estac organization Comity member (1991), ICTAC representant at the SFC (French Chemical Society) and SFT (French Thermical Society) (1992), (President of ARESAD (Association pour la Reflection sur les Etude Scientifiques Apliquées à la Défense) (1987), Member of EDF "Club Rayonnement", Directed 14 PHD, A big number of international collaborations: in Europe, Asia and Africa. 40 sucessful industrial projects (EADS, CNET, Saint Gobain)
Publication record: papers (96), patents (12)
5 most important publications: [1] E. Karmazsin, Simultaneous thermomagnetic and dilatometric measurements in the study of the Fe2O3 metastable transformation, Thermal Analysis 1983—28/2, pp. 279–284; [2] E. Karmazsin., Use of microwaves in thermal analysis: in DTA and thermodilatometry, Journal of Thermal Analysis, 30 (1985) 43–47; [3] E. Karmazsin., Développement d'un dispositif original de traitement de surface par plasmas micro-ondes. Applications aux polymères, Le Vide, les Couches Minces, 253 (1990) 268–273; [4] E. Karmazsin., Face à la mondialisation: l'entreprise métallurgique en 2020, Colloque VILS Académie des Sciences de Moscou, Moscou 13–14 juin 2006; [5] E. Karmazsin. Le décryptage du traitement médiatique en période de crise, 27 mars 2003 au Sénat, Palais du Luxembourg, Paris.

Name: Maarit Karppinen
Country: Finland
Date and place of birth: 1959, Helsinki, Finland
Present position and address: Professor, Deputy Head of Department, Department of Chemistry, School of Chemical Technology, Aalto University; P.O. Box 16100, FI-00076 Aalto Finland
Email: maarit.karppinen@aalto.fi
Website: http://chemistry.aalto.fi/en/, http://www.academia-net.org/alias/Profil/Prof-Maarit-Karppinen/1186882
Researcher ID: G-8035-2012
Education and scientific degrees: Helsinki University of Technology, Department of Chemical Technology, Finland M.Sc. (Tech) 1987, Lic.Tech. 1990, Ph.D. (Tech) 1993
Workplaces: Aalto University, Department of Chemistry, Finland, Professor in Inorganic Chemistry 2006–; Academy Professor 2009–2013; Head of Department 2008–2013; Deputy Head of Department 2014–2016; Tokyo Institute of Technology, Materials and Structures Laboratory, Japan; Associate Professor (Regular Chair) 2001–2006; Adjunct Professor 1996–2000, 2006–2008; Visiting Associate Professor 1995–1996
Main fields of interest: functional oxide materials, Inorganic-organic hybrid materials
Relevant categories in thermal analyses: fields (inorganic materials); methods (TG, DSC)
Awards and acknowledgments: European Research Council (ERC) Advanced Grant 2013
Publication record: papers (340), books (10), patents (8), h-index (31)
Equipments: Three TG apparatuses, two DSC apparatuses, one TG/DTA apparatus
5 most important publications: [1] M. Karppinen, L. Niinistö and H. Yamauchi, Studies on the oxygen stoichiometry in superconducting cuprates by thermoanalytical methods, Journal of Thermal Analysis 48, 1123–1141 (1997); [2] M. Karppinen, H. Yamauchi, S. Otani, T. Fujita, T. Motohashi, Y.-H. Huang, M. Valkeapää and H. Fjellvåg, Oxygen nonstoichiometry in $YBaCo_4O_{7+\delta}$: Large low-temperature oxygen absorption/desorption capability, Chemistry of Materials 18, 490–494 (2006); [3] M. Lappalainen and M. Karppinen, Techniques of differential scanning calorimetry for quantification of low contents of amorphous phases, Journal of Thermal Analysis and Calorimetry 102, 171–180 (2010); [4] S. Räsänen and M. Karppinen, Thermogravimetric study of water-based $LiFePO_4$ composite electrode powders, Thermochimica Acta 547, 126–129 (2012); [5] O. Parkkima, H. Yamauchi and M. Karppinen, Oxygen storage capacity and phase stability of variously substituted $YBaCo_4O_{7+\delta}$, Chemistry of Materials 25, 599–604 (2013).

Name: Mimi Y. Keating
Country: USA
Date and place of birth: 1944, Chengdu, China
Present position and postal address: Retired Research Fellow from Dupont Company. Currently persue art projects. 1709 Gunning Drive, Wilmington DE 19803, USA
Email: mimiy.keating@gmail.com
Website: www.mkeatingstudio.com
Education and scientific degrees: BS in Chemistry, National Taiwan University (1966); Ph.D. in Physical Organic Chemistry, Pennsylvania State University, (1970)
Workplaces: DuPont Company (1985–2011)
Main fields of interest: She spent her entire career in thermal analysis, first in the Thermal lab of Polymer Products, continuing in Corporate Center of Analytical Sciences when it became part of Central Research and Development. She is credited with running a diversified Thermal Lab with high efficiency and accuracy in dealing with almost all DuPont products. Techniques that she introduced include DSC isothermal crystallization rate, DMA accelerated creep, Laser flash thermal conductivity in through-plane and in-plane directions, Thermal fractionation, Nano-DSC for biomaterials, and Offgassing by direct TGA-MS.
Awards and acknowledgments: NATAS Outstanding Service Award (1997), NATAS Fellow (2004)
Professional activities: She was active in professional society NATAS from 1986, and later in ICTAC. She served in many Councilor roles and as the President of NATAS in 2006. She initiated a two-year CRDF Project (RCO-889, 2000–2001) and was the US Project Dirctor in collaboration with Professor V. A. Bershtein, Principal Investigator-Project Director of Ioffe Physico-Technical Instititute of the Russian Academy of Sciences. Her expert analyses and her testimony at the European Patent Office contributed to successfully overcoming Dow Chemical's appeal DuPont's opposition to a Packaging and Industrial Polymer's Devolatilization Extruder case in Munich in 2004. Guest Editor for the Proceedings of the 11th ICTAC, 1996. Published the reviewed papers in J. Therm. Anal. Calorim., 49(1-3), 1997, pp. 1-1729. Published a millennium special polymer issue honoring Dr. Edith Turi, J. Therm. Anal. Calorim., 59(1-2), 2000, pp. 1-609.
Publication record: papers (47 journal papers, 500 DuPont internal analytical technical reports)
5 most important publications: [1] Polymorphism in polyamide of Dytek-A and dodecanedioic acid J. Thermal Anal. Cal., 56(3) (1999) 1133–1140; [2] Dynamics in poly (oxymethylene) and poly(oxymethylene-co-oxyethylene) as studied via a combined creep rate spectroscopy/differential scanning calorimetry approach. J. Macrom. Sci., Physics (2002), B41(4-6), 797–831; [3] Structural and morphological inhomogeneity of short-chain branched polyethylenes in multiple-step crystallization. J. Macrom. Sci., Physics (2000), B39 (3), 317–331; [4] Glass transition, crystallinity, resin stiffness, and branch distribution in metallocene and Ziegler-Natta ethylene 1-olefins. J. Macrom. Sci., Physics (1999), B38(4), 379–401; [5] Thermal fractionation of ethylene polymers in packaging applications. Thermochim. Acta (1996), 284(1), 47–56.

Name: Tamás Kemény
Country: Hungary
Date and place of birth: 1946, Budapest, Hungary
Present position and address: external co-worker, Wigner Research Center HAS, 1525 Budapest P.O.Box 49, Hungary
Email: kemeny.tamas@wigner.mta.hu
Website: http://www.szfki.hu/~kemeny
Education and scientific degrees: Dipl. Phys., Roland Eötvös University, Budapest, 1969
Candidate of Physical Sciences (including Ph.D.) 1985; Doctor of the Hungarian Academy of Sciences 2001
Workplace: Wigner Research Center of the Hungarian Academy of Sciences, Institute of Solid State Physics and Optics
Main fields of interest: crystallization of metallic and semiconducting amorphous alloys and its relation to other materials properties (e.g. chemical and topological short range order of non-crystalline materials); evaluation of kinetical parameters from non-isothermal experiments; nanocrystalline materials produced by thermal crystallization of melt-spun amorphous ribbons
Relevant categories in thermal analyses: fields (non-isothermal calorimetry); methods (DSC)
Awards and acknowledgments: Janossy Award (1983); Zoltan Gyulai Award of the Eotvos Society (1984); Academic Award of the Hungarian Academy of Sciences (1991), Physics Award of the Hungarian Academy of Science (2005); Officer's cross of the Order of Merit of Hungary (2014)
Professional activities: Member of the Solid State Physics' Committee of the Hungarian Academy of Science
Publication record: papers (130), citations (1400), h-index (18), sum of impact factors (175)
Equipments: PerkinElmer DSC2-based home developed instrumentation
5 most important publications: [1] T. Kemény, I. Vincze, B. Fogarassy, S. Arajs: Structure and crystallization of Fe-B metallic glasses, Phys. Rev. B20 (1979) 476; [2] L. Gránásy, T. Kemény: The non-existence of a general correction term in continuous heating experiments, Thermochim. Acta, 42 (1980) 289; [3] T. Kemény, J. Sestak: Comparison of crystallization kinetics determined by isothermal and non-isothermal experiments, Thermochim. Acta, 110 (1987) 113; [4] T. Kemény, D. Kaptás, L. F. Kiss, J. Balogh, I. Vincze, S. Szabó, D. L. Beke: Structure and magnetic properties of nanocrystalline soft ferromagnets, Hyperfine Interactions, 130 (2000) 181; [5] G. K. Panova, N. A. Chernoplekov, A. A. Shikov, T. Kemény, L. F. Kiss; Vibrational and electronic characteristics of Zr70Pd30, Zr80Pt20 icosahedral quasicrystals and their amorphous counterparts; Phys. Solid State; 49 (2007) 1617.

Name: Michael R. Kessler
Country: USA
Date and place of birth: 1974, California, USA
Present position and address: Professor and Director, School of Mechanical and Materials Engineering, Washington State University, Pullman, WA, 99164-2920
Email: MichaelR.Kessler@wsu.edu
Website: http://www.mme.wsu.edu/
Researcher ID/ORCID: C-3153-2008/0000-0001-8436-3447
Education and scientific degrees: University of Illinois at Urbana-Champaign; Ph.D., Theoretical and Applied Mechanics; 2002, University of Illinois at Urbana-Champaign; M.S., Theoretical and Applied Mechanics; 1998 LeTourneau University; B.S., Engineering (Mechanical), Math Minor, Summa Cum Laude; 1996
Workplaces: 2013–, School of Mechanical and Materials Engineering, Washington State Univ., Pullman, WA; 2005–2013, Depart. of Materials Science and Eng., Iowa State Univ., Ames, IA; 2002–2005, Depart. of Mechanical Eng., Univ. of Tulsa, Tulsa, OK
Main fields of interest: mechanics, processing, and characterization of polymer matrix composites and nanocomposites, and the evaluation of these materials using experimental mechanics and thermal analysis
Awards and acknowledgments: Fellow of the North American Thermal Analysis Society, 2013.; National Science Foundation's CAREER Award, 2010.; Elsevier Young Composites Researcher Award from the American Society for Composites, 2009.; Air Force Office of Scientific Research Young Investigator Award, 2008.; Army Research Office Young Investigator Award, 2004
Professional activities: Program Chair (2011–2012), Division Chair (2012–2013), Materials Division, American Society of Engineering Education—a division of over 800 materials engineering educators. Academic Liaison for the North American Thermal Analysis Society (NATAS), 2006
Publication record: papers (106), books (1), book chapters (7), patents (6), citations (3740), h-index (25)
5 most important publications: [1] S. A. Madbouly, Y. Xia, M. R. Kessler: Rheological Behavior of Environmentally-Friendly Castor Oil-Based Waterborne Dispersions, Macromolecules, 2013, 46 (11), 4606–4616; [2] Y. Li, P. Badrinarayana, M. R. Kessler: Liquid Crystalline Epoxy Resin Based on Biphenyl Mesogen: Thermal Characterization, Polymer, 2013, 54(12), 3017–3025; [3] S. A. Madbouly, Y. Xia, M. R. Kessler: Rheokinetics of Ring Opening Metathesis Polymerization of Bio-Based Castor Oil Thermosets, Macromolecules, 2012, 45(19), 7729–7739; [4] P. Badrinarayanan, K. B. Dowdy, M. R. Kessler: A Comparison of Crystallization Behavior for Melt and Cold Crystallized Poly (L- lactide) Revealed Using Rapid Scanning Rate Calorimetry, Polymer. 2010; 51, 4611–4618; [5] S. R. White, N. R. Sottos, P. H. Geubelle, J. S. Moore, M. R. Kessler, S. R. Sriram, E. N. Brown, S. Viswanathan: Autonomic Healing of Polymer Composites, Nature. 2001; 409, 794–797.

Name: Detlef Klimm
Country: Germany
Date and place of birth: 1957, Halle/Saale, Germany
Present position: Leibniz Institute for Crystal Growth, Dept. Head "Simulation and Characterization"
Email: detlef.klimm@ikz-berlin.de
Website: www.ikz-berlin.de
Researcher ID/ORCID: F-2627-2011/0000-0002-7961-9337
Education and scientific degrees: Diplom-Kristallograph (University Leipzig, Germany), Dr. rer. nat. habil. (Universität Leipzig, Germany), Privatdozent (Humboldt-Universität zu Berlin)
Workplaces: Universität Leipzig, Göttingen, Tübingen (Germany), Lomonossov-University Moscow (Russia), Leibniz-Institut für Kristallzüchtung (Berlin, Germany)
Main fields of interest: phase diagrams, crystal growth
Relevant categories in thermal analyses: fields (inorganic materials, mainly oxides and halides); methods (DTA, DSC, TG, evolved gas analysis (extremely high temperature))
Awards and acknowledgments: Alexander von Humboldt Scholarship (1990)
Professional activities: member: IUCr Commission on Crystal Growth and Characterization of Materials
Publication record: papers (109), books (1), patents (3), h-index (18)
Equipments: NETZSCH: STA 409 (2,000 °C) with Balzers QMS, STA 429 (2,400 °C), STA 449 (1,650 °C)
5 most important publications: [1] G. Brauer et al., Identification of Zn-vacancy-hydrogen complexes in ZnO single crystals: A challenge to positron annihilation spectroscopy, Phys. Rev. B 79 (2009) 115212; [2] Y. Tomm et al., Czochralski grown Ga_2O_3 crystals, J. Crystal Growth 220 (2000) 510–514; [3] D. Klimm et al., The growth of ZnO crystals from the melt, J. Crystal Growth 310 (2008) 3009–3013; [4] D. Klimm et al., Growth of Cr:$LiCaAlF_6$ and Cr:$LiSrAlF_6$ by the Czochralski method, J. Crystal Growth 210 (2000) 683–693; [5] Phase Equilibria, D. Klimm, in "Nishinaga, Rudolph, Kuech—Handbook of Crystal Growth, 2nd Edition", to be published ca. 2014.

Name: Nobuyoshi Koga
Country: Japan
Date and place of birth: 1963, Fukuoka, Japan
Present position and address: Professor, Chemistry Laboratory, Department of Science Education, Graduate School of Education, Hiroshima University, 1-1-1 Kagamiyama, Higashi-Hiroshima, 739-8524 Japan.
Email: nkoga@hiroshima-u.ac.jp
Education and scientific degrees: Bachelor, Hiroshima University (1986); M.Ed., Hiroshima University (1988); Ph. D.-CSc., Institute of Chemical Technology at Pardubice, Czech (1991)
Workplaces: Faculty of School Education at Hiroshima University (1991–2000), Graduate School of Education at Hiroshima University (2000–)
Main fields of interest: kinetics of solid-state reactions; reactivity of solids, development of thermoanalytical techniques; chemistry education
Relevant categories in thermal analyses: fields (inorganic, materials, nano, minerals, glass, ceramics); methods (TG, DTA, EGA, DSC, thermomechanical analysis, kinetics, instrument development, microscopy)
Awards and acknowledgments: ICTAC Young Scientist Award (ICTAC:1996), The Jaroslav Heyrovsky Honorary Medal for Merit in Chemical Sciences (Czech Academy of Sciences: 2013), Pametni medaili Univerzity Pardubice (Univ. Pardubice: 2013), President Award of Hiroshima University (Hiroshima Univ.: 2013), The Award of The Japan Society of Thermal Analysis and Calorimetry (JSCTA: 2014)
Professional activities: Thermochimica Acta (Editor: 2013–)
Publication record: papers (150), books (15), citations (>1900), h-index (24)
Equipments: TG-DTA (×4), TG (×3), TG/DTA-MS, TG-EGA, TG-EGD, Humidity controlled TG, Vaccum TG, DSC (×3), DTA, TMA (×2), TPD
5 most important publications: [1] Koga N. Ozawa's Kinetic Method for Analyzing Thermoanalytical Curves. J. Therm. Anal. Calorim. 2013; 113(3):1527–41; [2] Koga N, Yamada S, Kimura T. Thermal Decomposition of Silver Carbonate: Phenomenology and Physicogeometrical Kinetics. J. Phys. Chem. C. 2013; 117(1):326–36; [3] Koga N, Tanaka H. A Physico-Geometric Approach to the Kinetics of Solid-State Reactions as Exemplified by the Thermal Dehydration and Decomposition of Inorganic Solids. Thermochim. Acta. 2002; 388(1-2):41–61; [4] Koga N, Šesták J. Crystal Nucleation and Growth in Lithium Diborate Glass by Thermal Analysis. J. Am. Ceram. Soc. 2000; 83(7):1753–60; [5] Koga N, Criado JM. The Influence of Mass Transfer Phenomena on the Kinetic Analysis for the Thermal Decomposition of Calcium Carbonate by Constant Rate Thermal Analysis (CRTA) under Vacuum. Int. J. Chem. Kinet. 1998; 30(10):737–44.

Name: Romana Cerc Korošec
Country: Slovenia
Date and place of birth: 1968, Ljubljana, Slovenia
Present position and address: Assistant Professor, Faculty of Chemistry and Chemical Technology, University of Ljubljana, Aškerčeva 5, SI-1000 Ljubljana, Slovenia
Email: romana.cerc-korosec@fkkt.uni-lj.si
Website: www.romanacerc-korosec.si
Education and scientific degrees: B.Sc. in Chemistry (1994), Ph.D. (2001)
Workplaces: Faculty of Chemistry and Chemical Technology (1994–)
Main fields of interest: TG and DSC analysis of thin films
Relevant categories in thermal analyses: fields (thin films, materials, nano-materials, complex compounds, wood, polymers); methods (TG, DTA—temperatures up to 1500 °C, EGA, DSC, HP DSC, DSC-microscopy, DSC-chemiluminescence)
Professional activities: Member of ICTAC, Member of Slovenian Chemical Society
Publication record: papers (35), book chapters (2), citations (329), h-index (11), sum of impact factors (78,3)
Equipments: Mettler Toledo TG/DSC 1, coupled with MS Thermostar and with FTIR Spectrometer Nicolet; Mettler Toledo DSC 1, coupled with Optical Microscope Gloor Instruments; Mettler Toledo High Pressure DSC827^{e} (from room temperature to 600 °C, maximal pressure 100 bars; DSC is coupled with Camera PCO SensiCam).
5 most important publications: [1] R. Cerc Korošec and P. Bukovec: Optimisation of the thermal-treatment of chemically prepared electrochromic nickel oxide thin films, their electrochromic properties and structural investigations. In: P. R. Somani (Ed.). Chromic materials, phenomena and their technological applications (Multifunctional materials and devices), Pune, Applied Science Innovations Private Limited, 2010, pp. 241–282; [2] R. Cerc Korošec and P. Bukovec: Thermal methods of analysis as a tool for quantitative composition determination of "water-in-oil" emulsions. In: T. J. Janssen (Ed.). Explosive materials: classification, composition, and properties. Hauppauge, New York, Nova Science Publishers,. 2011, pp. 125–144; [3] R. Cerc Korošec and P. Bukovec: Acta chim. slov. 53 (2006) 136. review; [4] B. Genorio, K. Pirnat, R. Cerc Korošec, R. Dominko and M. Gaberšček: Angew. Chem. 49 (2010) 7222; [5] R. Cerc Korošec, B. Lavrič, G. Rep, F. Pohleven, P. Bukovec: J. Therm. Anal.Cal. 98 (2009) 189.

Name: Mohamed Fathy Kotkata
Country: Egypt
Date and place of birth: 1943, Cairo, Egypt
Present position and address: Prof. Materials Science, Head of the Semiconductors Technology Lab., Faculty of Science, Ain Shams University, Cairo-11566, Egypt.
Email: mfk.asu@gmail.com
Education and scientific degrees: B.Sc. (Applied Physics, 1966), M.Sc. (1970) and Ph.D. (1974), Faculty of Science, Ain Shams University, Cairo; D.Sc. (1997); Professor (1985)
Workplace: Semiconductors Technology Lab., Faculty of Science, Ain Shams University
Main fields of interest: studying of the physical properties, characteristics, fundamental transports and structural peculiarities for different nano-scale, non-crystalline and crystalline semiconductors issued in technological applications
Professional activities: Head of the Permanent Committee of Physics, Supreme Councils of Egyptian Universities, 2013–; Member of the Permanent Committee of Physics (2008–2012 and 2001–2004), Supreme Councils of Egyptian Universities.; Chairman of the Physics Department (2001–2004), Faculty of Science, Ain Shams University, Cairo, Egypt.; Regional Editor of "J. Therm. Anal. Cal." for Middle East Countries (1992–2007)
Publication record: papers (113), citations (1021); h-index (16); i10 index (28)
List of the 5 most important publications: [1] M.F. Kotkata et al.: Acta Phys. Hung., 54 (1983) 303–312. Review; [2] M.F. Kotkata: J. Mater. Sci., 27 (1992) 4858–4870 and 4847–4857. Review; [3] M.F. Kotkata et al.: Thin Solid Films, 240 (1994) 143–146; [4] M.F. Kotkata et al.: Physica E, 41 (2009) 640–645 and 1457–1465; [5] M.F. Kotkata: J. Non-Cryst. Solids, 358 (2012) 420–426 and 3342–3347. Review.

Name: Zdeněk Kožíšek
Country: Czech Republic
Date and place of birth: 1955, Chomutov, Czech Republic
Present position and address: Deputy head of the Department of Optical Materials, Institute of Physics, Academy of Sciences of the Czech Republic, Cukrovarnicka 10, 16200 Praha 6
Email: kozisek@fzu.cz
Website: http://www.fzu.cz/ ~ kozisek/
Education and scientific degrees: 1974–1979: Faculty of Mathematics and Physics, Charles University, Praha; 1979: appointed at the Institute of Physics ASCR, Praha; 1986: submitted and defended Thesis—title CSc. (equivalent Ph.D.)
Workplaces: Institute of Physics, Academy of Sciences of the Czech Republic, Cukrovarnicka 10, 16200 Praha 6 (1974–); in 1999: Hiroshima University, Japan (visiting professor)
Main fields of interest: nucleation and growth theory
Relevant categories in thermal analyses: fields (glass, polymer, nano); methods (nucleation and growth kinetics, numerical simulation)
Professional activities: Advisory committee: ICCGE-17, Warsav, Poland, 2014; chairman of the Czechoslovak Association for Crystal growth (2012–); Conference chair of DMSRE22 (2012, Lednice, Czech Republic) and DMSRE24 (2014, Lednice, Czech Republic)
Publication record: papers (63), book chapters (4), citations (532), h-index (10)
5 most important publications: [1] Z. Kožíšek, Crystal nucleation kinetics in confined systems, CrystEngComm 15 (2013) 2269–2274.; [2] Z. Kožíšek, P. Demo, A. Sveshnikov, Kinetics of crystal nucleation in closed systems (Chapter 9) J. Šesták, P. Šimon (Eds.): Thermal analysis of micro, nano- and non-crystalline materials: Transformation, crystallization, kinetics and thermodynamics, vol. 9, Elsevier (2013) pp. 195–208, ISBN: 978-90-481-3149-5; [3] Z. Kožíšek, M. Hikosaka, K. Okada, P. Demo, Nucleation on active centers in confined volumes, J. Chem. Phys. 136 (2012) 164506 (10 pages.); [4] M. Rodová, M. Liška, K. Nitsch, Z. Kožíšek, Solidification of molten zinc chloride: Experimental and theoretical studies, J. Therm. Anal. Calorim. 91 (2008) 181–185; [5] Z. Kožíšek, P. Demo, K. Sato, Nucleation on active sites: evolution of size distribution, J. Cryst. Growth 209 (2000) 198–202.

Name: Mustafa Verşan Kök
Country: Turkey
Date and place of birth: 1960, Turkey
Present position and address: Prof.Dr. Dept. of Petroleum and Natural Gas Eng.; Middle East Tech. Univ; Universiteler Mah. Dumlupinar Bulv. No:1; 06800 Cankaya/Ankara
Email: kok@metu.edu.tr
Website: http://www.metu.edu.tr/~kok
ORCID: 0000-0002-1180-5862
Education and scientific degrees: Post-Doc: CNRS-INSA (Instiitute National Des Sciences Appliquees, Laboratoire de Thermochimie Minerale) Lyon, France, 1993–1994., Ph.D. : Middle East Technical University, Dept. of Petroleum Eng., 1990.; 1996–2001: Associate Professor; 2001–: Prof. Dr.
Workplaces: Dept. of Petroleum and Natural Gas Eng. Middle East Tech. University (1983–2001 and 2006–); The Scientific and Technological Research Council of Turkey (TUBITAK) (2001–2006); National Contact Point of Turkey in FP-6 Under the priority of Energy and Environment; SHELL (1986–1987)
Main fields of interest: fossil fuel characterization, enhanced oil recovery
Relevant categories in thermal analyses: fields (fossil fules (coal, lignite, oil shale, crude oil, biomass) cement and clay); methods (TG, EGA, DSC, kinetics, wax deposition)
Awards and acknowledgments: 2 Science Awards; 3 Research and Achievement Awards; 2 Thesis of the Year Awards; 15 Outstanding Researcher Awards
Professional activities: Associate Editor (PSE, JTAC and IJOGCT); Member of Editorial Board (Energy and Fuels, JTAC, Oil Shale, PST); Reviewer (in 42 journals)
Publication record: Papers (174), book chapters (3), citations (2154), h-factor (28), i-factor (64)
Equipments: Q Series TA (Differential scanning Calorimetry and Thermogravimetric Analysis with EGA)
5 most important publications: [1] Thermal Analysis and Kinetics of Biomass Samples, Fuel Processing Technology, 106, 2013, 739–743; [2] Simultaneous Thermogravimetry-Calorimetric Study on the Combustion of Coal Samples: Effect of Heating Rate, Energy Conversion and Management, 53-1, 2012, 40–44; [3] Thermo-oxidative Reactions of Crude Oils" JTAC, 105-8, 2011, 411–414; [4] Temperature Controlled Combustion and Kinetics of Different Rank Coals" JTAC, 79, 2005, 175–180; [5] Crude Oils: Characterization of Waxes Precipitated on Cooling by DSC and Thermomicroscopy" Fuel, 74-6, 1995, 810–817.

Name: János Kristóf
Country: Hungary
Date and place of birth: 1950, Gödöllő, Hungary
Present position and address: Full professor, Head, Department of Analytical Chemistry, University of Pannonia, H-8200 Veszprém, Egyetem 10.
Email: kristof@almos.vein.hu
Education and scientific degrees: Chemical engineer (1974), CSc (1985), D.Sc. (1995), Dr.Habil (1995)
Workplaces: University of Pannonia, Department of Analytical Chemistry (1974–)
Main fields of interest: inorganic, materials, nano, minerals, complex, polymer
Relevant categories in thermal analyses: fields (surfaces, thin films, nanohybrids); methods (TG-DTA(DSC)-MS, FTIR, Raman, XRD)
Awards and acknowledgments: Knight's cross of the order of merit of the Republic of Hungary; Albert Szent-Györgyi Award
Professional activities: ICTAC National councilor; Co-editor (Journal of Thermal Analysis and Calorimetry; ISRN Spectroscopy); Board member (Hungarian Clay Minerals Society; Hungarian Spectrochemical Society); Member (Analytical and Environmental Chemical Committee of the Hungarian Academy of Sciences (HAS); Member (Working Committee for Thermal analysis of the HAS); Co-chairman (Working Committee for Spectrochemistry of the HAS)
Publication record: papers (170), books (10), patents (2), citations (1500), h-index (26), sum of impact factors (230)
Equipment: Netzsch TG 209/Balzers QMS; Netzsch STA 409 CD; Bruker Eqinox 55 FTIR; Bruker Raman RFS 100/S; Bruker Senterra Raman microscope; Bruker Vertex 70 FTIR microscope; Derivatograph PC
5 most important publications: [1] R.L. Frost, J. Kristóf, E. Horváth, J.T. Kloprogge: J. Phys. Chem. A, 103 (1999) 9654–9660; [2] J. Kristóf, R.L. Frost, W.N. Martens, E. Horváth: Langmuir, 18 (2002) 1244–1249; [3] E. Horváth, J. Kristóf, R.L. Frost, V. Vágvölgyi, T. Cseh: J. Colloid Interf. Sci., 71 (2003) 707–714; [4] J. Kristóf, T. Szilágyi, E. Horváth, A. De Battisti, R.L. Frost, Á. Rédey: Thermochim. Acta, 413 (2004) 93–99; [5] É. Makó, J. Kristóf, E. Horváth, V. Vágvölgyi: J. Colloid Interf. Sci., 330 (2009) 367–373.

Name: Jiří Kučerík
Country: Germany
Date and place of birth: 1975, Třebíč, Czech Republic
Present position and address: researcher and teacher at University Koblenz-Landau, campus Landau, Institute for Environmental Sciences, Fortstr. 7, Landau, 768 29, Germany
Email: kucerik@email.cz
Education and scientific degrees: Chemical Engineer (1998); Ph.D. (2003); Assoc. prof. (2009) all at Brno University of Technology, Czech Republic; Assoc. prof. (2011) at University Koblenz-Landau.
Workplaces: researcher/teacher at Brno University of Technology (2003–2010); researcher/teacher at University Koblenz-Landau, (2010–); internship at University of Naples, Italy (2002); visiting scientist at University of Palermo (2013); (2004–) visiting teacher at universities in Naples and Palermo (Italy), Poitiers (France), Turku (Finland), Konya (Turkey)
Main fields of interest: development of thermo-analytical methods for determination of soil properties, nature of water in soils and soil aging, colloidal properties of dissolved organic matter, thermal analysis of native and modified polysaccharides
Relevant categories in thermal analyses: fields (materials, organic, pharmaceutical, (bio) polymers, soils, humic substances); methods (TG, DTA, DSC, kinetics, calorimetry)
Professional activities: member of Czech Chemical Society, International Humic Substances Society, German Soil Science Society, organizing committee of conferences CEEC TAC1 and CEEC TA2, editorial board of Global Journal of Analytical Chemistry (2010–2011)
Publication record: papers (>50), book chapters (2), citations (>200)
Equipment: DSC Q1000 with RSC90, TA Instruments, STA 449 F3 Jupiter with MS detector, Netzsch
5 most important publications: [1] J. Kučerík, J. Schwarz, A. Jaeger, M. Bertmer, G.E. Schaumann. J. Therm. Anal. Calorim. (2014) doi:10.1007/s10973-014-3971-4; [2] J. Kučerík, C. Siewert. J. Therm. Anal. Calorim. 116 (2014) 563–570; [3] C. Siewert, M.S. Demyan, J. Kučerík. J. Therm. Anal. Calorim. 110 (2012) 413-419; [4] Y. Kunhi Mouvenchery, J. Kučerík, D. Diehl, G.E.Schaumann. Rev. Environ. Sci. Biotech. 11 (2012) 41–54; [5] J. Kučerík, A.Průšová, A. Rotaru, K. Flimel, J. Janeček, P. Conte. Thermochim. Acta. 523 (2011) 245–249.

Name: Queenie S.M. Kwok
Country: Canada
Date and place of birth: 1974, Hong Kong
Present position and address: Explosives Hazards Scientist; Canadian Explosives Research Laboratory, Natural Resources Canada, Canmet Bells Corners Complex Building 12, 1 Haanel Drive, Ottawa, Ontario, K1A 1M1, Canada
Email: queenie.kwok@nrcan.gc.ca
Website: www.nrcan.gc.ca/explosives
Education and scientific degrees: Bachelor of Engineering (Chemical), McGill University (1994–1998); Master of Science (Chemistry), Carleton University (2000–2001).
Workplaces: Canadian Explosives Research Laboratory (1998–)
Main fields of interest: hazard characterization for improving the safety and security of energetic materials during manufacture, transportation and use
Relevant categories in thermal analyses: fields (energetic materials); methods (TG, DTA, EGA, DSC, kinetics, calorimetry, heat flow calorimetry, accelerating rate calorimetry, isothermal nanocalorimetry)
Awards and acknowledgments: NATAS PerkinElmer Student Award (2001); CTAS Student Award (2001); Frank Carver Bursary of the International Pyrotechnic Society (2001); ASTM E37 Awards of Appreciation (2004, 2007); NATAS Executives Appreciation Award (2011); NRCan MMS Crystal Award for Integration (2013)
Professional activities: CTAS Executive Committee (2002–2010) and President (2006–2008); Member of the ASTM Committee E37 Thermal Measurements (2003–); Annual NATAS Conference as Session Chair (2005, 2006, 2008, 2011) and Conference Chair (2015); NATAS Publication Councilor (2008–)
Publication record: papers (21), citations (278), h-index (10)
Equipments: DSC, TG, Simultaneous TG-DSC-FTIR-MS, C600 heat flow calorimetry, accelerating rate calorimetry, adiabatic Dewar calorimetry, isothermal nanocalorimetry
5 most important publications: [1] Q.S.M. Kwok, R.C. Fouchard, A.-M. Turcotte, P.D. Lightfoot, R. Bowes and D.E.G. Jones, Propellants, Explosives, Pyrotechnics, 27 (2002) 229; [2] D.E.G. Jones, Q.S.M. Kwok, M. Vachon, C. Badeen and W. Ridley, Propellants, Explosives, Pyrotechnics, 30 (2005) 140; [3] R. Turcotte, M. Vachon, Q.S.M. Kwok, R. Wang, and D.E.G. Jones, Thermochim. Acta, 433 (2005) 105; [4] C.M. Badeen, Q.S.M. Kwok, M.C.R. Vachon, R. Turcotte, and D.E.G. Jones, J. Therm. Anal. Calorim., 81 (2005) 225; [5] Q.S.M. Kwok, B. Acheson, R. Turcotte, A. Janès, and G. Marlair, J. Therm. Anal. Calorim., 111 (2013) 507.

Name: Maria Lalia-Kantouri
Country: Greece
Date and place of birth: 1949, Thessaloniki, Greece
Present position and address: Professor of Inorganic Chemistry, Laboratory of Inorganic Chemistry, Department of General and Inorganic Chemistry, Faculty of Chemistry, Aristotle University of Thessaloniki, GR-54124 Thessaloniki, Greece
Email: lalia@chem.auth.gr
Website: http://www.chem.auth.gr/
Education and scientific degrees: Chemist, Aristotle University of Thessaloniki BSc.(1972); Ph.D.(1983), Postgraduated studies a) on Thermal Analysis 1977 and 1979 at the Institute of Gesteinshuttenkunde, Aachen, Germany and b) on Coordination Chemistry 1990–1991 at Imperial College of London, UK
Main fields of interest: synthesis and structural characterization of coordination compounds with transition metals and lanthanides incorporating ligands containing nitrogen, oxygen and sulfur donor atoms (methods used: FT-IR, UV-Vis, `H-NMR, 13C-NMR, ESR, CV, X-ray)
Relevant categories in thermal analyses: fields (inorganic, materials, minerals, complex, organic, pharmaceutical, biology, life, ceramics); methods (TG/DTG-DTA, QIA, EGA, DSC, MS, kinetics, coupled TG-DTA-MS)
Awards and acknowledgments: She has been awarded the Prize for her outstanding and long-time contribution to the Hellenic Society of Thermal Analysis (HSTA) in 2010
Professional activities: (A) conference organizing: THERMA 2002 (1st National Conference in Therm. Anal.) Thessaloniki, MEDICTA 2005 in Thessaloniki, Greece and Co-chairperson in THERMA 2004, Ioannina Greece; (B) Vice-President of HSTA (Hellenic Society of Thermal Analysis-foundation year 1995) 1995–2002 and President 2003–2009. (C) Councilor of HSTA in ICTAC 2003–2005 and 2005–2009–2012. (D) Member of the Editorial Board-Associate Editor in J.Therm.Anal.Calor., according to the subject area: "Materials science, Inorganics and Geoscience" since 2005 and member of the Editorial Board of the Global Journal of Inorganic Chemistry since 2010. (E) Coordinator of a Training-Summer School in Thermal analysis (Erasmus-IP program), July 2014
Publication record: papers (70), books (7), patents (1)
Equipments: SETARAM Model Setsys 1200
5 most important publications: [1] M. Lalia-Kantouri, G.E. Manoussakis, J. Thermal Anal., 29 (1984) 1151; [2] M. Lalia-Kantouri, G.A. Katsoulos, C.C. Hadjikostas, P. Kokorotsikos, J.Thermal Anal., 35 (1989) 2411; [3] C. D. Papadopoulos, A. G. Hatzidimitriou, G. P. Voutsas, M. Lalia-Kantouri, Polyhedron, 26 (2007) 1077; [4] M. Lalia-Kantouri, L. Tzavellas, D. Paschalidis, J. Thermal Anal. Calorim., 91 (2008) 93; [5] M. Lalia-Kantouri, M. Gdaniec, T. Choli-Papadopoulou, A. Badounas, C. D.Papadopoulos, A. Czapik, G. D. Geromichalos, D. Sahpazidou, F. Tsitouroudi, J. Inorg Biochem 117 (2012) 25.

Name: Isaak L. Lapides
Country: Israel
Date and place of birth: 1936, Odessa, Ukraine
Present position and address: Professor Emeritus for Physical Chemistry, Instutute of Chemistry, the Hebrew University of Jerusalem, 91904 Israel
Email: Isaak8@hotmail.com, Isaak_L@yahoo.com
Education and scientific degrees: State University in Petrozavodsk (USSR) (1958), Ph.D. , 1968, D.Sci. 1989.
Workplaces: Institute of Chemistry, the Hebrew University of Jerusalem
Main fields of interest: themal analysis of inorganic materials, mechanochemistry, Intercalation by clay minerals, order-disorder problem
Relevant categories in thermal analyses: fields (materials, nano particles, minerals, organic); methods (IR-spectroscopy, powder XRD, TG, DTA, EGA, DSC, thermo-XRD, thermo-FTIR, thermo-UV/visible)
Professional activities: Editor of Annual Book of Institute of Geochemistry Siberian Branch of Academy USSR (1969–1976), Editor of Book "Cation ordering in the mineral structures" (Nauka, Novosibirsk, 1977); conference organizer ("Cation ordering in minerals", 1976, Irkutsk, USSR)
Publication record: papers (> 200), books (3), patents (1)
Equipments: Fillips XRD, Bruker FTIR (Tensor 27), UV/VIS Spectrometer PerkinElmer (Lambda 20).
List of the 5 most important publications: [1] I.Lapides, V.Kovalenko, P.Koval, Micas from rare metal granitoids (Chemistry and thermal properties) (Nauka, Novosibirsk, 1977); [2] I.Lapides, T.Valetov, Cation ordering in amphiboles. (Nauka, Moscow, 1986); [3] I.Lapides, Evaluation of kinetic parameters from single TG curve based on the similarity theory and process symmetry., J. Therm. Anal. Calorim., 1997, 50, N1-2, 269–277; [4] I.Lapides, L.Heller-Kallai, Novel features of smectite settling, Colloid. Polym. Sci., 2002, 280, 554–561; [5] Lapides I, Borisover M, Yariv S. Thermal-analysis of hexadecyltrimethylammonium-montmorillonites. Part 1: thermogravimetry, carbon and hydrogen analysis and thermo-IRspectroscopy-analysis. J. Therm. Anal. Calorim., 2011;105:921–9.

Name: Mika Lastusaari
Country: Finland
Date and place of birth: 1971, Turku, Finland
Present position and address: Adjunct professor, University of Turku, Department of Chemistry, FI-20014 Turku, Finland.
Email: mika.lastusaari@utu.fi
Website: https://update.utu.fi/en/units/sci/units/chemistry/research/mcca/Pages/home.aspx
Education and scientific degrees: Ph.D. (Chemistry), 2001
Workplaces: University of Turku, Department of Chemistry, since 1996
Main fields of interest: luminescent solid state materials, magnetic materials, minerals, chemistry education and learning
Relevant categories in thermal analyses: fields (inorganic, materials, nano, minerals, glass, ceramics); methods (TG, DTA, DSC, extremely high temperature (above 1,000 °C), thermoluminescence, high-temperature X-ray powder diffraction)
Professional activities: Member of FinTAC (Finnish Thermal Analysis and Calorimetry Association) board since 2011; organizer of the ESTAC-11 (11th European Symposium on Thermal Analysis and Calorimetry) conference in 2014; guest co-editor of the JTAC (Journal of Thermal Analysis and Calorimetry) special issue on ESTAC-11
Publication record: papers (105), h-index (20)
Equipments: TA Instruments Q600, TA Instruments SDT2960, MikroLab Thermoluminescent Materials Laboratory Reader RA'04, Huber high-temperature X-ray powder diffraction setup
5 most important publications: [1] Brito, H.F., Hölsä, J., Jungner, H., Laamanen, T., Lastusaari, M., Malkamäki, M., and Rodrigues, L.C.V., Persistent Luminescence Fading in $Sr_2MgSi_2O_7$:Eu^{2+},R^{3+} Materials: a Thermoluminescence Study, Opt. Mater. Express 2 (2012) 287–293; [2] Hölsä, J., Laamanen, T., Lastusaari, M., Malkamäki, M., Welter, E., and Zajac, D.A., Valence and Environment of Rare Earth Ions in $CaAl_2O_4$:Eu^{2+},R^{3+} Persistent Luminescence Materials, Spectrochim. Acta B 65 (2010) 301–305; [3] Antal, T., Harju, E., Pihlgren, L., Lastusaari, M., Tyystjärvi, T., Hölsä, J., Tyystjärvi, E., Use of Near-Infrared Radiation for Oxygenic Photosynthesis via Photon Up-Conversion, Int. J. Hydrogen Energ. 37 (2012) 8859–8863; [4] Lastusaari, M., Laamanen, T., Malkamäki, M., Eskola, K.O., Kotlov, A., Carlson, S., Welter, E., Brito, H.F., Bettinelli, M., Jungner, H., and Hölsä, J., The Bologna Stone: History's First Persistent Luminescence Material, Eur. J. Mineral. 24 (2012) 885–890; [5] Carlson, S., Hölsä, J., Laamanen, T., Lastusaari, M., Malkamäki, M., Niittykoski, J., and Valtonen, R., X-Ray Absorption Study of Rare Earth Ions in $Sr_2MgSi_2O_7$:Eu^{2+},R^{3+} Persistent Luminescence Materials, Opt. Mater. 31 (2009) 1877–1879.

Name: Giuseppe Lazzara
Country: Italy
Date and place of birth: 1979, Palermo, Italy
Present position and address: Researcher in Physical-Chemistry at the Department of Physics and Chemistry, University of Palermo, viale delle Scienze pad 17. Palermo, Italy
Email: giuseppe.lazzara@unipa.it
Researcher ID/ORCID: O-3356-2013/0000-0003-1953-5817
Education and scientific degrees: 2008—Ph.D. degree in Chemical Science at the University of Palermo
Workplaces: (2004–2005), Universität Bayreuth Physikalische Chemie I Bayreuth (Germany); (2006), "Stranski Laboratorium für Physikalische und Theoretische Chemie" Technische Universität of Berlin (Germany); (2007), "Stranski Laboratorium für Physikalische und Theoretische Chemie" Technische Universität of Berlin (Germany); (2009–2010), Chemistry Department, Lund University (Sweden).
Main fields of interest: physico-chemical characterization of nanomaterials and composites; sustainable materials and applications in cultural heritage
Relevant categories in thermal analyses: fields (materials, nano, polymer, clay, cyclodextrin, soft matter, wood, cultural heritage); methods (TG, DSC, specific heat, microcalorimetry, DMA)
Awards and acknowledgments: 2005—IACT Doctoral Award; 2006—Alberto Lucci Award; 2007—W. F. Giauque Award; 2012—"Grant Programme for Young Scientists"
Professional activities: Principal investigator of FIRB Futuro in Ricerca 2012; 2007–2013, member of the scientific committee of the Italian Group of Calorimetry and Thermal Analysis; 2012, Co-chairman of the Winter School "Renewable Energy Systems and Green Nanotechnologies for a Clean Environment", Drobeta, Romania; at present, member of the scientific committee of the Italian Association of Calorimetry and Thermal Analysis
Publication record: papers (49), citations (490), h-index (14)
Equipments: TGA Q5000 IR (TA Instruments), nano-ITC200 (MicroCal), micro-DSC III (SETARAM), DSC 2920 CE (TA Instrument), DMA Q800 (TA Instruments)
5 most important publications: [1] Lazzara, G.; Milioto, S. J. Phys. Chem. B 2008, 112, 11887–11895; [2] Donato, D.; Lazzara, G.; Milioto, S. J. Therm. Anal. Calorim. 2010, 101, 1085–1091; [3] Lazzara, G.; Prevost, S.; Gradzielski, M. Soft Matter 2011, 7, 6082–6091; [4] Cavallaro, G.; Donato, D. I.; Lazzara, G.; Milioto, S. J. Phys. Chem. C 2011, 115, 20491–20498; [5] Cavallaro, G.; Lazzara, G.; Milioto, S. J. Phys. Chem. C 2012, 116, 21932–21938.

Name: Pierre Le Parlouër
Country: France
Date and place of birth: 1951, Paimpol, France
Present position and address: Retired, Scientific Director (SETARAM Instrumentation), 48 Chemin de Crépieux, 69300 Caluire, France
Email: leparlouer@orange.fr
Education and scientific degrees: Chemical Engineer, Chemical Engineer School of Lyon (ESCIL), France (1974); Ph.D., University of Saint Etienne, France (1976)
Workplaces: University of Saint Etienne, France (1974–1976); Setaram, Application Manager, R§R Manager, General Manager (1976–1998); Therm@l Consulting, Owner and Consultant (1998–2008); Setaram, Scientific Director (2009–2014)
Main fields of interest: thermal analysis and calorimetry instrumentation; development of new thermal and calorimetric techniques and applications; training and promotion of thermal analysis and calorimetry; expertise, consulting and organisation of conferences and seminars
Relevant categories in thermal analyses: fields (inorganic, organic, polymers, pharmaceuticals, food, bio products, oil and gas, biomass, thermal storage materials, hydrogen storage, CO_2 capture, minerals, ceramics, catalysts); methods (DTA, TGA and combined techniques (DTA, DSC, EGA), TMA, DSC, microDSC, Calvet calorimetry, high pressure DSC and calorimetry, drop calorimetry)
Awards and acknowledgments: STK (Swiss Society of Thermal Analysis and Calorimetry) Award (2012)
Professional activities: ICTAC treasurer (2012–), ICTAC french convenor (1999–2008), Member of the scientific board of LNE (French National Metrological Institute) (2013–), Past president of the French association of Thermal Analysis and Calorimetry AFCAT (2002–2006), Past ISO convenor for thermal analysis standards (ISO/TC61) (1993–2003), Member of the Editorial Advisory Board of Journal of Thermal Analysis and Calorimetry
Publication record: papers (38), books (2), patents (2), citations (144)
5 most important publications: [1] P. Le Parlouër, C. Dalmazzone, B. Herzhaft, C. Mathonat, Characterization of gas hydrates using a new high pressure MicroDSC, J. Therm. Anal. Calorim., 78 (2004) 165–172; [2] P. Le Parlouër, I. Chan, Calorimetry, dilatometry at high temperature: recent developments in instrumentation and applications, Am. Ceram. Soc. Bull. 80 (2001) 31; [3] P. Le Parlouër, B. Benoist, Integrated circuit thermal analysis: a new technique for polymer characterization, J. Therm. Anal. Calorim., 59 (2000) 351; [4] P. Le Parlouër, Selection of an optimum calorimetry method for measuring specific heat of ceramics at high temperature, Thermochimica Acta 192 (1991) 55; [5] P. Le Parlouër, Simultaneous TG-DSC: a new technique for thermal analysis, Thermochimica Acta 121 (1987) 307.

Name: Vesa-Pekka Lehto
Country: Finland
Date and place of birth: 1967, Karjala Tl, Finland
Present position and address: Professor, Department of Applied Physics, University of Eastern Finland, P.O.B 1627, 70211 Kuopio, Finland
Email: vesa-pekka.lehto@uef.fi
Website: www.uef.fi/sovfys/vesa-pekka-lehto
Education and scientific degrees: Ph.D. in Physics (1999), Docent in Industrial physics (2002), Docent in Physical pharmacy (2005), Professor in Industrial physics (2008), Professor in Materials physics (2013).
Workplaces: University of Turku, Finland (1991–2002), Focus Inhalation Ltd (2002–2003), University of Turku, Finland (2003–2008), University of Kuopio, Finland (2008–2010), University of Eastern Finland (2011–).
Main fields of interest: utilization of inorganic mesoporous materials in intelligent drug delivery
Relevant categories in thermal analyses: fields (inorganic, materials, nano, pharmaceutical); methods (TG, DSC, kinetics, microcalorimetry, instrument development, XRPD)
Awards and acknowledgments: CRS/Eurand Award Grand Prize, 2007
Professional activities: Chairman of FinTAC (www.fintac.org/), chairman and organizer of ESCTAC-11 (www.estac11.fi/)
Publication record: papers (141), book chapters (2), patents (4), citations (2253), h-index (24)
Equipments: DSC, TGA, XRPD, isothermal calorimetry
5 most important publications: [1] Xu, W., Riikonen, J. and Lehto, V-P., Mesoporous systems for poorly soluble drugs. Int. J. Pharm. 453:181 (2013).; [2] Riikonen, J, Salonen, J. and Lehto V-P., Utilising thermoporometry to obtain new insights into nanostructured materials—Part 1. J. Therm. Anal. Cal. 105:811 (2011).; [3] Riikonen, J., Salonen, J., Kemell, M., Kumar, N., Murzin, D., Ritala, M. and Lehto, V-P., A novel method of quantifying u-shaped pores in SBA-15. J. Phys. Chem. C, 113:20349 (2009); [4] Bimbo, L., Sarparanta, M., Santos, H., Airaksinen, A., Mäkilä, E., Laaksonen, T., Peltonen, L., Lehto, V-P., Hirvonen, J. and Salonen, J., Biocompatibility of thermally hydrocarbonized porous silicon nanoparticles and their biodistribution in rats. ACS Nano 4(6):3023 (2010); [5] Salonen, J., Laitinen., L. Kaukonen, A. M., Tuura, J., Björkqvist, M., Heikkilä, T., Vähä-Heikkilä, K., Hirvonen, J. and Lehto, V-P., Mesoporous silicon particles for oral drug delivery: Loading and release of five model drugs. J. Contr. Release, 108:362–374 (2005).

Name: Maria Inês Gonçalves Leles
Country: Brazil
Date and place of birth: 1963, Ceará, Brazil
Present position and address: Professor, Instituto de Química, Universidade Federal de Goiás, C.P. 131—Goiânia, GO
Email: leles@ufg.br
Website: http://lames.quimica.ufg.br/pages/39547
Education and scientific degrees: B.Sc. in Chemistry—UFC—Universidade Federal do Ceará (1983–1988); Postgraduate (master degree) USP—Universidade de São Paulo (1989–1991); Postgraduate (Doctor degree) UNESP—Universidade Estadual Pailista (1991–1995); Post-doc—University of Huddersfield (2014).
Workplaces: UFG—Universidade federal de Goiás
Main fields of interest: development methodology in food, pharmaceuticals, biodiesel and other
Relevant categories in thermal analyses: fields (pharmaceutical, oil, biodiesel, and foods); methods (TG, DSC)
Professional activities: Reviewer of the Journal Acta Scientiarum Technology, the Journal Ciência e Agrotecnologia and the Brazilian Journal of Thermal Analysis
Publication record: papers (22), citations (74), h-index (5)
Equipments: TGA/SDTA 851, DSC 822 (Mettler Toledo)
5 most important publications: [1] Pascoal, A. M.; Di-Medeiros, M. C. B.; Batista, K. A.; Leles, M. I.G.; Liao, L.M.; Fernandes, K. F. Carbohydrate Polymers, v. 98, pp. 1304–1310, 2013; [2] Antoniosi Filho, N. R.; Menezes R.S.; Leles, M. I. G.; Soares, A. T.; Franco, Pedro I B M; Antoniosi Filho, N. R.; Sant´Anna C.L.; Vieira, A. A. H. Química Nova, v. 36, pp. 10–15, 2013; [3] Carvalho, K.P.; Rocha, T. C.; Leles, M. I. G. Brazilian Journal of Thermal Analysis, v. 1, pp. 79–83, 2012; [4] Filho, R. O. C.; Franco, P. I. B. M.; Conceicao, E. C.; Leles, M. I. G. Journal of Thermal Analysis and Calorimetry, v. 47, pp. 1–5, 2009; [5] Garcia, C. C.; Franco, P. I. B. E. M.; Zuppa, T. O.; Antoniosi Filho, N. R.; Leles, M. I. G. Journal of Thermal Analysis and Calorimetry, v. 87, pp. 645–648, 2007.

Name: Markku Leskelä
Country: Finland
Date and place of birth: 1950, Kokkola, Finland
Present position and address: Department of Chemistry, PO Box 55, FI-00014 University of Helsinki, Finland
Email: markku.leskela@helsinki.fi
Website: http://www.helsinki.fi/kemia/epaorgaaninen/personnel/leskela.htm
Education and scientific degrees: M.Sc. 1974, Ph.D. 1980, Helsinki University of Technology
Workplaces: 1974–1979, 1981–1982 Helsinki University of Technology, 1979–1981, 1982–1986, University of Turku, 1983, University of Utrecht, 1986–1990 University of Turku, 1987–1988 University of Florida, 1990– University of Helsinki
Main fields of interest: chemical deposition of thin films, volatile metal complexes, homogeneous catalysis
Relevant categories in thermal analyses: TG, DTA, DSC
Awards and acknowledgments: Magnus Ehrnrooth Foundation Award in Chemistry 2002, ISI Highly cited scientist (materials science) 2004, A.I Virtanen award 2011, American Vacuum Society ALD award 2012, Honorary award of Finnish Academy of Sciences and Letters 2014, Honorary member of Finnish Chemical Society 2014
Professional activities: Finnish Chemical Society, member of the board 1991–2011; Nordic Society for Thermal Analysis and Calorimetry, member of the board 1982–2000; International Confederation for Thermal Analysis and Calorimetry, member of the council 1985–92, Federation of Finnish Learned Societies, member of the board 1994–1999; 2008–; Delegation of the Finnish Academies of Sciences and Letters 2008–; International Union of Pure and Applied Chemistry, Inorganic Chemistry, titular member 2006–2015, Member of Board of University of Helsinki (2010–2017), Board member of several foundations
Publication record: papers (635), reviews (62), books (16), patents (44), citations (17100), h-index (61)
Equipments: Mettler TGA850, PerkinElmer DSC-2
5 most important publications: [1] J. Valo and M. Leskelä: Thermoanalytical Methods in Studies of High Temperature Superconductors. Handbook of Thermal Analysis and Calorimetry. Vol. 2, M.E. Brown and P.K. Gallagher (eds.), Elsevier, Amsterdam 2003, pp. 817–879; [2] T. Hatanpää, M. Vehkamäki, M. Ritala and M. Leskelä: Study of Bismuth Alkoxides as Possible Precursors for ALD. Dalton Trans. 39 (2010) 3219–3226; [3] T. Hatanpää, K. Kukli, M. Ritala and M. Leskelä: Crystal Structures and Thermal properties of Some Rare-earth Alkoxides with Tertiary Alcohols. Possible Precursors for Atomic Layer Deposition of Rare Earth Oxides. J. Therm. Anal. Calorim. 105 (2011) 61–71; [4] M. Mäntymäki, M. Ritala and M. Leskelä: Double Metal Alkoxides of Lithium: Synthesis, Structure and Applications in Materials Chemistry. Coord. Chem. Rev. 256 (2012) 854–877; [5] T. Hatanpää, M. Ritala and M. Leskelä: Precursors as Enablers of ALD Technology: Contributions from University of Helsinki. Coord. Chem. Rev. 257 (2013) 3297–3322.

Name: Irina A. Letyanina
Country: Russia
Date and place of birth: 1986, Nizhny Novgorod, Russia
Present position and address: researcher, Institute of Chemistry, Saint Petersburg State University, Universitetskiy prospect 26, Peterhof, 198504, Saint-Petersburg, Russia
Email: irina-letyanina@mail.ru; irene.letyanina@gmail.com; ezzaria@rambler.ru
Researcher ID/ORCID: N-1623-2013/0000-0001-6799-2583
Education and scientific degrees: Bachelor of Chemistry, Chemical Department of Lobachevsky State University of Nizhny Novgorod (2003–2007), Master of Chemistry (2009), Ph.D. (2012)
Workplaces: Lobachevsky State University of Nizhny Novgorod (2007–2013), Saint Petersburg State University (2013–)
Main fields of interest: thermodynamic properties of individual compounds, particularly metalorganic compounds, within wide temperature range (from T → 0); thermodynamic properties of splitting systems with chemical interaction
Relevant categories in thermal analyses: fields (organic, metalorganic, polymer, inorganic); methods (TG, DSC, adiabatic vacuum calorimetry, specific heat, Calvet calorimetry, microcalorimetry)
Awards and acknowledgments: Scholarship of the President of the Russian Federation (2011–2012); State Scholarship of the Government of the Russian Federation (2010–2011); Scholarships of the Government of Nizhny Novgorod Region (2010–2011, 2011–2012); Honorable mention prize in the XVIII International Conference on Chemical Thermodynamics in Russia RCCT-2011
Professional activities: tmember of the Organizing Committee of the VIII All-Russian conference "Mendeleev-2014"
Publication record: papers (19), citations (45), h-index (6), sum of impact factors (29.148)
Equipments: adiabatic vacuum calorimeter BCT-3 (Termis), DSC 204 F1 Phoenix (Netzsch), TG 209 F1 Iris (Netzsch), Calvet calorimeter C80 (Setaram), μDSC3 evo (Setaram)
5 most important publications: [1] Smirnova N.N., Letyanina I.A., Larina V.N., Markin A. V., Sharutin V.V., Senchurin V.S. Thermodynamic properties of pentaphenylantimony Ph5Sb over the range from T → 0 K to 400 K, J. Chem. Thermodynamics. 2009. 41. 1. 46–50; [2] Smirnova N.N., Letyanina I.A., Zakharova Yu.A., Pimerzin A.A., Vishnevskaya E.E. Thermodynamic properties of neopentylbenzene over the range from T → (0 to 350) K, J. Chem. Thermodynamics. 2012. 48. 118–122; [3] Letyanina I.A., Smirnova N.N., Markin A.V., Ruchenin V.A., Larina V.N., Sharutin V.V., Molokova O.V. Thermodynamics of tetraphenylantimony acetophenoneoxymate, J. Therm. Anal. Calorim. 2011. 103. 1. 355–363; [4] Markin A.V., Letyanina I.A., Smirnova N.N., Sharutin V.V., Molokova O.V. Standard thermochemical characteristics of formation of triphenylantimony bis(acetophenoneoxymate), J. Therm. Anal. Calorim. 2013. 111. 2. 1499–1502; [5] Letyanina I.A., Markin A.V., Smirnova N.N., Sologubov S.S., Sharutin V.V. Heat capacity and thermodynamic functions of triphenylantimony bis(1-adamantanecarboxilate) over the temperature range from (0 to 520) K, J. Chem. Eng. Data. 2013. 58. 3087–3095.

Name: György Liptay
Country: Hungary
Date and place of birth: 1932, Budapest, Hungary
Present position and address: retired Professor, Technical University of Budapest, Institute for Inorganic and Analytical Chemistry, H–1521, Budapest, Szt. Gellért tér 4. Hungary
Email: liptay.g@mail.bme.hu
Education and scientific degrees: Chemical engineer, Technical University of Budapest (1954); Ph.D. (1963); C. Sc. (1971); D.Sc. (1994)
Workplaces: Technical University of Budapest Institute for General Chemistry (1954–1966); Institute for Applied Chemistry (1967–1973); Institute for Inorganic and Analytical Chemistry (1974–)
Main fields of interest: analytical precipitates; thermoanalytical and structural experimental of complex compounds; development of simultaneous thermoanalytical methods and their extension in practical application
Relevant categories in thermal analyses: fields (inorganic, complex, pharmaceutical, biology); methods (TG, DTA, (simultanous methods))
Awards and acknowledgments: Károly Than Award: 1985, MTESZ Medal: 2002, Rudolf Fabinyi Medal: 2007, Paulik Brothers Award: 2007, ICTAC Distinguished Service Award: 2008, Náray-Szabó I. Award: 2013
Professional activities: IV. ICTA Conference (Budapest) (1974), Executive Committee Chairman; ESTAC-7 Conference in Balatonfüred (1998), Chairman; ESTAC Committee (1994–1998), Depute Chairman; EUROSTAR (1995–2001), national representative; Hungarian Academy of Sciences Thermoanalytical Committee: President (1992–2010); Hungarian Chemical Society Vice President (2003–); Hung. Chem. Soc. Analytical Division: President (1995–2003); Journal of Thermal Analysis: Member of Editorial Advisory Board (1969–) and Consulting Editor; Thermal Analysis Abstracts: Regional Editor of Eastern Europe (1972–1991); Chemie Analityzne: Member of Advisory Board (2001–2009)
Publication record: papers (260), books (6), patents (2), citations (1700), h-index (20)
Equipments: Derivatograph (MOM)
5 most important publications: [1] L. Erdey, S. Gál, G. Liptay: Thermoanalytical properties of analytical-grade reagents, Talanta 11 (1964) 913; [2] G. Liptay, E. Papp-Molnár, K. Burger: The thermal analysis of metal complexes I. Relationship between the thermal stability and the equilibrium stability of complexes, J. Inorg. Nucl. Chem., 31 (1969) 247; [3] M. Berényi, G. Liptay: The use of thermal analysis in medical science with special reference to nephrolits, J. Thermal Anal., 3 (1971) 437; [4] G. Liptay, G. Kenessey, L Bihátsi, T. Wadsten, J. Mink: Pyridine type complexes of transition- metal halides I, J. Thermal Anal., 38 (1992) 899. XIV, J. Therm. Anal Cal., 68 (2002) 81.; [5] G. Kenessey, G. Liptay: Solid primary aliphatic amine complexes of transition metal halides I, J. Thermal Anal., 39 (1993) 332. III. J. Thermal Anal., 50 (1997) 167.

Name: Marek Liška
Country: Czech Republic
Date and place of birth: 1951, Bratislava, Czechoslovakia
Present position: A. Dubček University of Trenčín, Professor, Head of the VILA Glass Center
Email: marek.liska@tnuni.sk
Website: www.tnuni.sk; www.erbbk.sk
Education and scientific degrees: Ing.—1974, Slovak Technical University, Bratislava; Ph.D.—1978, Slovak Technical University, Bratislava; D.Sc.—1996, Slovak Academy of Sciences, Bratislava, Assoc. Prof.—1997, Slovak Technical University, Bratislava (Physical Chemistry and Chemical Physics); Prof.—2002, Slovak Technical University, Bratislava (Physical Chemistry)
Workplaces: Faculty of Chemical Technology, Slovak Technical University, Bratislava (1974–1983); Institute of Inorganic Chemistry, Slovak Academy of Sciences, Bratislava (1984–1997); Alexander Dubček University of Trenčín, Trenčín (1997–)
Main fields of interest: physical chemistry, glass science and technology
Relevant categories in thermal analyses: fields (inorganic, glass, ceramics, materials); methods (DTA, DSC, TMA, crystallization kinetics)
Awards and acknowledgments: Vitrum pro Futurum award of the Euroregion Bílé-Biele Karpaty
Professional activities: Member of the Editorial board of: International Journal of Applied Glass Science, European Journal of Glass Science of Technology: Part A—Glass Technology, Part B—Physics and chemistry of Glasses, Ceramics-Silikáty, Sklář a keramik. Member of the Technical Committee TC03 of the International Commission on Glass. Member of the Steering Committee of the Czech Glass Society. Chairman of the Board of Region Biele Karpaty, Vice-chairman of the Board of the Euroregion Bílé-Biele Karpaty. Thermodilatometric and calorimetric study of relaxation phenomena in glasses. Structure-property dependences interpretation based on the thermodynamic models
Publication record: papers (215), books (2), patents (2), citations (220)
5 most important publications: [1] P. Pelikán, M. Čeppan, M. Liška: Computational Methods in Molecular Spectroscopy. CRC Press, Boca Raton 1994, 341 p.; [2] P. Lichvár, M. Liška, P. Šajgalík, P. Bezdička: Crystallisation, Thermal Expansion and Density of Si-Al-Y-O Glasses for Ceramics. Key Engineering Materials 175–176, 125–130 (1999); [3] M. Liška, J. Antalík, I. Štubňa: Enthalpic relaxation of 25 Na2O.xTiO2.(75-x)SiO_2 glasses. J. Therm. Anal. Calorimetry 55, 155–164 (1999); [4] M. Liška, J. Antalík: Enthalpy Relaxation in Glasses-Regression Analysis of Integral DSC Data. J. Therm. Anal. Calorimetry 67, 213–222 (2002); [5] O. Gedeon, M. Liška. Rings in covalent glass and an evaluation of configurational entropy associated with rings. J. Non-Cryst. Solids 360, 41–48 (2013).

Name: Zhen-Hai Liu
Country: P.R. China
Date and place of birth: 1936, Liaoning province, China
Present position and address: Research professor, Changchun Institute of Applied Chemistry, Chinese Academy of Sciences, Changchun, People str.5625, Postcode 130022, China
Email: liuzh@ciac.ac.cn
Education and scientific degrees: Graduated from Chemical Department, Jilin University (1959)
Workplaces: Changchun Institute of Applied Chemistry, Chinese Academy of Sciences (1959–)
Main fields of interest: thermal analysis of macromolecules; standardization expression of scientific writing.
Publication record: papers (> 100), books (18)
5 most important publications: [1] Handbook of thermal analysis, Wiley, Chichester (1998) (in English); [2] Thermal analysis, Chemical Industrial Press, Beijing (1982, 1985) (in Chinese); [3] Calorimetric measurements of polymers, Chemical Industrial Press, Beijing (2002) (in Chinese); [4] Instrument and application of thermal analysis. Chemical Industrial Press, Beijing (2011); [5] How to write scientific papers in Chinese and English (2nd edition), High Education Press, Beijing (2012) (in Chinese).

Name: Vladimir Logvinenko
Country: Russia
Date and place of birth: 1937, Tomsk, Russia
Present position and address: Principal research scientist, Lab of the synthesis of cluster compounds and materials; Nikolaev Institute of Inorganic Chemistry, Siberian Branch of Russian Academy of Sciences, Professor, Chair of Analytical Chemistry, Novosibirsk State University; Ac. Lavrentyev Ave. 3, Novosibirsk-90, 630090, Russia.
Email: val@niic.nsc.ru
Education and scientific degrees: Physicochemical engineer, Leningrad Technological Institute (1960); Ph.D. (1969); Ass. Prof. (1972); D.Sc. (1992); Professor (2008)
Workplaces: Nikolaev Institute of Inorganic Chemistry of Russian Academy of Sciences (1960–); Novosibirsk State University (1971–1989; 2006–); Novosibirsk Technologvical Institute (1994–2013); Siberian University of Consumer`s Cooperation (2001–2011)
Main fields of interest: thermal analysis; solid-phase reactions; non-isothermal kinetics; coordination chemistry; supramolecular chemistry
Relevant categories in thermal analyses: fields (inorganic, materials, nano, organic, coordination compounds, cluster compounds, supramolecular compounds); methods (TG, DTA, EGA, DSC, thermomechanical analysis, kinetics)
Professional activities: Head of a Laboratory of Thermal methods of investigation, Nikolaev Institute of Inorganic Chemistry (1989–1997); member of Honorary board of Journal of Thermal Analysis and Calorimetry (1989–); member of Honorary Committees: the 10th International Seminar on Thermal Analysis and Calorimetry to the memory of Prof. St. Bretsznaider (Poland, 2011) and the 2nd Central and Eastern European Conference on TA&C (Lithuania, 2013)
Publication record: papers (225), books (2), patents (1), citations (428), h-index (11)
Equipments: Netzsch thermal analyser TG 209 F1; Derivatograph-C
5 most important publications: [1] V. Logvinenko, F. Paulik, J. Paulik. Quasi-equilibrium thermogravimetry in the up-to-date inorganic chemistry. Novosibirsk, Nauka; 1982; [2] V. Logvinenko, V. Fedorov, Yu. Mironov, V. Drebushchak. Kinetic and thermodynamic stability of cluster compounds under heating. J Therm Anal Calorim. 2007. V. 88. pp. 687–692; [3] V. Logvinenko, D. Dybtsev, V. Fedin, V. Drebishchak, M. Yutkin. The stability of inclusion compounds under heating. J Therm Anal Calorim. 2010. V.100. pp. 183–189; [4] V. Logvinenko. Stability of supramolecular compounds under heating. J Therm Anal Calorim. 2010. V.101. pp. 577–583; [5] V. A. Logvinenko, V. P. Yutkin, M. S. Zavakhina, V. P. Fedin. Porous metal–organic frame-works (MOFs) as matrices for inclusion compounds. Kinetic stability under heating. J Therm Anal Calorim. 2012. V. 109. pp. 555–560.

Name: Dénes Márton Lőrinczy
Country: Hungary
Date and place of birth: 1945, Pécs, Hungary.
Present position and address: Professor of Biophysical Department University of Pécs Faculty of Medicine, H-7624 Pécs Szigeti str. 12. Hungary.
Email: denes.lorinczy@aok.pte.hu
Education and scientific degrees: Graduated (MS) physicist, University of Lóránd Eötvös Budapest (1963–1968). CSc (Ph.D.) 1980, Dr. habil 2002, D.Sc. 2006.
Workplaces: Biophysical Department of Medical University of Pécs (1968–2000), Biophysical Department University of Pécs Faculty of Medicine (2000–)
Main fields of interest: biopysics, biochemistry, life sciences, food
Relevant categories in thermal analyses: fields (muscle proteins, food physics R&D, application of thermal analysis in medical sciences); methods (DSC, TA/TG, microcalorimetry, instrument development)
Awards and acknowledgments: Pro Facultatae Medicinae silver medal, University of Pécs Faculty of Medicine (2005). Eugene Ernst award of Hungarian Biophysical Society (2007)
Professional activities: President of the International Society of Biological Calorimetry (2006–2008), President of the Thermoanalytical Technical Commission of the Hungarian Chemical Society (2011–). 1990–1993: sponsor and instrument exhibition manager of the 11th International Biophysics Congress (Budapest, Hungary). 1993–1994: sponsor and instrument exhibition manager of the 1st International Foodphysics Congress (Budapest, Hungary). 2003–2004: President of the 6th International Foodphysics and Dairy Congress (Pécs, Hungary). 2008: President of the 15th International Conference of Biological Calorimetry (Pécs, Hungary). 2009: member of the International Scientific Board of 2nd International Biophysics Congress (Diyarbakir, Turkey). 2011: member of the International Scientific Board of 1st Conference of Central and Eastern European Committee for Thermal Analysis and Calorimetry (CEEC-TAC1, Craiova, Romania). 2012: member of Organising Committee of 27th European Cytoskeletal Forum Meeting (Pécs, Hungary). 2013: member of the International Scientific Board of 4th Joint Czech-Hungarian-Polish-Slovak Thermoanalytical Conference (Pardubice, Czech Republik).
Publication record: papers (166), books (6), citations (1084), sum of impact factors (224.3)
Equipments: SETARAM Micro DSC-II, Seiko SSC 5200 TA/TG
5 most important publications: [1] Visegrády, B., Lőrinczy, D., Hild, G., Somogyi, B., Nyitrai, M.: The effect of phalloidin and jaspaklinolide on the flexibility and thermal stability of actin fila-ments. FEBS Letters 2004. 565. 163–166; [2] Lőrinczy, D.: Effect of nucleotides and environmental factors on the intermediate states of ATP hydrolysis cycle in skeletal muscle fibres. In: The Nature of Biological Systems as Revealed by Thermal Methods (Ed.: D. Lőrinczy) Kluwer Academic Publisher 2004. 159–186; [3] Visegrády, B., Lőrinczy, D., Hild, G., Somogyi, B., Nyitrai, M.: A simple model for the cooperative stabilisation of actin filaments by phalloidin and jasplakinolide. FEBS Letters 2005. 579. 6-10; [4] Bugyi, B., Papp, G., Hild, G., Lőrinczy, D., Nevalainen, E. M., Lappalainen, P., Somogyi, B., Nyitrai, M.: Formins regulate actin filament flexibility through long-range allosteric interactions. J. Biol. Chem. 2006. 281. 10727–10736; [5] Katics, L., Lőrinczy, D.: Strength Training. Biomechanics, exercises and methods. Akadémiai Kiadó, 2012. ISBN 978 963 05 8999 4.

Name: Werner Ludwig
Country: Germany
Date and place of birth: 1933, Jena, Germany
Present position and address: Retired, D-7749 Jena, Eugen Diederichs Str. 4, Germany
Education and scientific degrees: Diplome chemist, Friedrich Schiller University of Jena (1959), Dr.rer.nat. (1964)
Workplaces: Friedrich Schiller University, Institute for Inorganic Chemistry (1960–1969); Dept. of Chemistry WB Inorg. Solid State Chemistry (1969–1990); Institute of Inorg. and Analytical Chemistry (1990–1995)
Main fields of interest: thermoanalytical methods in solid-state chemistry of vitreous systems; thermal properties of coordination compounds; thermal analysis in environmental chemistry
Awards and acknowledgments: Kurnakov medal ANCCR Moskow (1985)
Professional activities: Chairman of the Thermal Analysis Group by the Chemical Society of the GDR (1974–1990); excecutice chairman of the ESTAC 4 in Jena (1987); chairman ESTAC Committee (1987–1991); secretary of ESTAC Committee (1991–1995); Journal of Thermal Analysis: member of Editorial Advisory Board (1976–2000); vice president of the Society for Thermal Analysis GEFTA (1995–1997)
Publication record: papers (41)
5 most important publications: [1] W. Ludwig: Studies on the thermal characterization of vitreous chalcogenides, J. Thermal. Anal., 8 (1975) 73; [2] B. Voigt, W. Ludwig: Eine neue Methode zur Abschätzung der Glasdbildungstendenz von Schmelzen durch DTA am Beispiel des Systems Pb-Ge-Se, J. Thermal Anal., 25 (1982) 579; [3] W. Ludwig, M. Döring, R. Fischer, A. Friedrich, W. Seidel, E. Uhlig, D. Walther: Some applications of thermal analysis to coordination chemistry, J. Thermal Anal., 38 (1992) 231; [4] M. Döring, W. Ludwig, H. Görls, J. Wuckelt: Thermal latent coordination compounds I/II, J. Thermal Anal., 42/50 (1994/1997) 443; [5] W. Ludwig, J. E. Einax: Thermische Analyse an Flußsedimenten der Saale, J. Therm. Anal. Calorim., 52 (1998) 65.

Name: Boris V. L'vov
Country: Russia
Date and place of birth: 1931, Leningrad, Russia
Present position and address: Professor, Department of Physical Chemistry, St Petersburg State Polytechnic University, Politekhnicheskaya ul 29, St Petersburg 195251, Russian Federation
Email: borislvov@rambler.ru
Researcher ID: 7004807320
Education and scientific degrees: Chemist-Investigator, Leningrad State University, (1954); C.Sc. (1962); D.Sc (1973); Professor (1977)
Workplaces: State Institute of Applied Chemistry (1955–1975); St Petersburg State Polytechnic University (1975–)
Main fields of interest: electrothermal atomic absorption spectrometry; mechanisms and kinetics of heterogeneous reactions; thermal analysis; heterogeneous catalysis
Relevant categories in thermal analyses: fields (solid-state decomposition reactions and heterogeneous catalysis); methods (thermochemical approach including the congruent dissociative vaporization (CDV) mechanism of decomposition)
Awards and acknowledgments: Talanta Gold Medal (1974); Doctor of Science Honorius causa of the University Strathclyde (Scotland) (1988); James Waters Symposium Award (1991) Pittsburg Conference; Bunsen-Kirchhoff Prize of the German Chemical Society (1993); Marcus Marcy Medal of the Czech Spectroscopic Society (1995); Gold Medal of the XXX CSI meeting (1997) Melbourne; Dr Jerzy Fijalkowski Award of the Polish Academy of Sciences (2010); Paschoal Senise Award, on the occasion of the Brazilian 16th National Meeting on Analytical Chemistry (2011)
Professional activities: Associate member of the IUPAC Commission on Spectrochemical Analysis (1983–1987); Member of Anal. Chem. Division of IUPAC (1987–1991); Member of Editorial Advisory Board: Zh Anal Khim (1975–1985); J Anal At Spectrosc (1987–1996); Can J Spectrosc (1978–1999); Spectrochim Acta Part B (1979–); Spectroscopy and Spectral Analysis (1981–); Full Member of Russian Academy of Natural Sciences (from 1991)
Publication record: papers (>300), books (4), patents (2), citations (>6500), h-index (38)
5 most important publications: [1] L'vov BV. Atomic Absorption Spectrochemical Analysis, Adam Hilger, London, 1970; [2] L'vov BV. Thermal Decomposition of Solids and Melts. New Thermochemical Approach to the Mechanism, Kinetics and Methodology, Springer, Berlin, 2007; [3] L'vov BV, Galwey AK. Catalytic oxidation of hydrogen on platinum: thermochemical approach. J Therm Anal Calorim 2013;112:815–822. doi:10.1007/s10973-012-2567-0; [4] L'vov BV, Galwey AK. Interpretation of the kinetic compensation effect in heterogeneous reactions: thermochemical approach. Int Rev Phys Chem 2013;32: 515–557. doi:10.1080/0144235X.2013.802109; [5] L'vov BV. Thermochemical model in kinetics of heterogeneous reactions: 100-year jubilee. J Therm Anal Calorim 2014; 116: 1041–1045. doi:10.1007/s10973-013-3580-7.

Name: Jelena Macan
Country: Croatia
Date and place of birth: 1977, Zagreb, Croatia
Present position and address: Associate professor, University of Zagreb, Faculty of Chemical Engineering and Technology, Marulićev trg 19, HR-10000 Zagreb, Croatia
Email: jmacan@fkit.hr
Education and scientific degrees: Chemical Engineer, University of Zagreb (1995–2000); Ph.D. (2006); Assistant professor (2007); Associate professor (2012)
Workplaces: University of Zagreb, Faculty of Chemical Engineering and Technology (2000–)
Main fields of interest: preparation and properties of nanocomposite and hybrid materials with epoxy resin matrix, preparation and properties of ceramics and thin coatings by sol-gel
Relevant categories in thermal analyses: fields (inorganic, materials, nano, complex, organic, polymer, ceramics); methods [TG, DTA, DSC, thermomechanical analysis, kinetics, extremely high temperature (above 1,000 °C)]
Awards and acknowledgments: State Award for Science for Young Scientists for 2006
Professional activities: member of editorial board for journal Polimeri (2007–); editor-in-chief of alumni bulletin Glasnik (2012–); member of organizing and scientific committee and editor of the book of abstracts of POLYCHAR 20 international conference in Dubrovnik (2012); secretary of Croatian Microscopy Society (2013–)
Publication record: papers (21), h-index (7)
Equipments: differential thermal calorimeter (DSC), combined differential thermal calorimeter and thermoballance (DSC-TGA)
5 most important publications: [1] J. Macan, K. Tadanaga, M. Tatsumisago, "Influence of copolymerization with alkyltrialkoxysilanes on condensation and thermal behaviour of poly (phenylsilsesquioxane) particles", Journal of Sol-Gel Science and Technology 53 (2010) 31–37; [2] J. Macan, A. Gajović, H. Ivanković, "Porous zirconium titanate ceramics synthesized by sol-gel process", Journal of the European Ceramic Society 29 (2009) 691–696; [3] L. Valentini, J. Macan, I. Armentano, F. Mengoni, J. M. Kenny, "Modification of fluorinated single-walled carbon nanotubes with aminosilane molecules", Carbon 44 (2006) 2196–2201; [4] J. Macan, I. Brnardić, S. Orlić, H. Ivanković, M. Ivanković, "Thermal degradation of epoxy-silica organic-inorganic hybrid materials", Polymer Degradation and Stability 91 (2006) 122–127; [5] J. Macan, H. Ivanković, M. Ivanković, H. J. Mencer, "Study of Cure Kinetics of Epoxy-Silica Organic-Inorganic Hybrid Materials", Thermochimica Acta 414 (2004) 219–225.

Name: Marek Maciejewski
Country: Poland
Date and place of birth: 1940, Dobre, Poland
Present position and address: Retired from Swiss Federal Institute of Technolgy (ETH), Zurich, Switzerland
Email: marek.maciejewski.pl@gmail.com
Education and scientific degrees: Chemical Engineer and M.Sc (1963), Ph.D. in chemistry (1971), D.Sc. habilitation (1988); all at Warsaw University of Technology.
Workplaces: Institute of Physical Chemistry, Polish Academy of Science, 1963–1965; Department of Fundamental Physico-Chemical Problems in Technology, 1965–1968; Department of Technological Designing, 1968–1971; Department of General Chemistry, 1971–1990; all at Chemistry Faculty of Warsaw University of Technology; Laboratory of Technical Chemistry and Institute for Chemical and Bioengineering, Swiss Federal Institute of Technology (ETH Zurich), Switzerland, 1990–2010
Main fields of interest: hyphenated techniques in thermal analysis (TA-MS and TA-FTIR), investigation of the inorganic gas-solid reactions, quantification of the spectrometric signals in TA-MS-FTIR systems, solid-state chemistry, kinetics of the solid-state reactions, application of the thermal analysis in the catalytic research, pulse thermal analysis
Relevant categories in thermal analyses: fields (inorganic, materials, catalysis, gas-solid reactions); methods (pulse thermal analysis, TA-MS-FTIR, TG, DSC, kinetics)
Awards and acknowledgments: Individual Prizes of the Polish Minister of Science and High Schools for the scientific achievements (1985, 1989); Netzch-GEFTA Award (2001), Swiss Thermal Society of Thermal Analysis and Calorimetry Award (2005), Honorary Membership of Polish Society of Thermal Analysis and Calorimetry (2007), Doctor Honoris Causa of West University of Timisoara, Romania (2006)
Professional activities: Member of the Board of the Polish Society of Calorimetry and Thermal Analysis (1984–1990), Member of the Kinetics Committee of ICTAC (1986–1995), Member of the Board of EUROSTAR (1995–2001), Member of the Board of Swiss Society of Thermal Analysis and Calorimetry STK (1995–2005), Member of Editorial Board (2002–2012) and Honorary Board (2012–) of Journal of Thermal Analysis and Calorimetry, Member of Editorial Board of Themochimica Acta (2005–)
Publication record: papers (151), book chapters (1), patents (9), citations (5020), h-index (41)
5 most important publications: [1] M. Maciejewski, Computional aspects of kinetic analysis: Part B, The ICTAC Kinetic project, Thermochimica Acta, 335 (2000) 145–154; [2] J.D. Grunwaldt, M. Maciejewski, O.S. Becker, P. Fabrizioli, A. Baiker, Comparative study of Au/TiO2 and Au/ZrO2 catalysts for low-temperature CO oxidation, Journal of Catalysis, 186 (1999) 458–469; [3] M. Maciejewski, C.A. Müller, R. Tschan, W.D. Emmerich. A. Baiker, Novel pulse thermal analysis method and its potential for investigating gas-solid reactions, Thermochimica Acta 295 (1997) 167–182; [4] M. Casapu, J.D. Grunwaldt, M. Maciejewski, M. Wittrock, U. Göbel, A. Baiker, Formation and stability of barium aluminate and cerate in NOx storage-reduction catalysts, Applied Catalysis, B; Environmental 62 (2006) 232–242; [5] M. Piacentini, M. Maciejewski, A. Baiker, Pt-Ba/alumina NOx storage-reduction catalysts Applied Catalysis, B; Environmental 62 (2005) 232–242.

Name: János Madarász
Country: Hungary
Date and place of birth: 1963, Nádudvar, Hungary
Present position and address: habil. assoc. prof. chemistry, Dept. Inorg. and Analytical Chemistry, Budapest University, Technology and Economics H–1521 Budapest, Szt. Gellért tér 4, Hungary
Email: madarasz@mail.bme.hu
Education and scientific degrees: 1994—Ph.D. in chemistry (from both Hungarian Academy of Science and Technical University of Budapest). 1985 and 1987—B.Sc. and M.Sc. in chemical engineering (from Technical University of Budapest)
Workplaces: Dept. Inorg. and Anal. Chem., Budapest University of Technology and Econ.
Main fields of interest: solid state chemistry
Relevant categories in thermal analyses: fields: inorganic, materials, minerals, complex, organic, pharmaceutical, polymer, ceramics); methods (TG, DTA, DSC, EGA-FTIR, EGA-MS, HT-XRD)
Awards and acknowledgments: Meisel-award (1987)
Professional activities: member of Committee on TA&Cal of Hungarian Academy of Sciences, member of Group for TA&Cal of Hungarian Chemical Society
Publication record: papers (96), books (2), patents (2), citations (810), h-index (15), sum of impact factors (146)
Equipments: DSC, TG, TG/DTA, TG-FTIR, TG/DTA-MS (by TA Instruments), XRD (PANanalytical), FTIR (BioRad)
5 most important publications: [1] Structure and Thermal Behaviour of Dichlorobis(thiourea)cadmium(II), a Single-Source Precursor for CdS Thin Films, M. Krunks, J. Madarász, L. Hiltunen, R. Mannonen, E. Mellikov and L. Niinistö, Acta Chem. Scand., 51 (1997) 294–301; [2] Thermal decomposition of thiourea complexes of Cu(I), Zn(II), and Sn(II) chlorides as precursors for the spray pyrolysis deposition of sulfide thin films, J. Madarász, P. Bombicz, M. Okuya, and S. Kaneko, Solid State Ionics, 141–142 (2001) 439–446; [3] Thermal, FTIR and XRD study on some 1:1 molecular compounds of theophylline, J. Madarász, P. Bombicz, K. Jármi, M. Bán, Gy. Pokol and S. Gál, J. Therm. Anal. Calorim., 69 (2002) 281–290; [4] Comparative evolved gas analyses on thermal degradation of thiourea by coupled TG-FTIR and TG/DTA-MS instruments, J. Madarász and G. Pokol, J. Therm. Anal. Calorim., 88 (2007) 329–336; [5] Co-crystal of (R,R)-1,2-cyclohexanediol with (R,R)-tartaric acid, a key structure in resolution of the (±)-trans-diol by supercritical extraction, and the related ternary phase system, P. Thorey, P. Bombicz, I. M. Szilágyi, P. Molnár, G. Bánsághi, E. Székely, B. Simándi, L. Párkányi, G. Pokol, J. Madarász, Thermochim. Acta 497 (2010) 129–136.

Name: Ján Majling
Country: Slovakia
Date and place of birth: 1942, Muránska Dlhá Lúka, Czechoslovakia
Present position and address: Retired from Slovak University of Technology, Assoc. Prof. J. Majling, 900 32 Borinka, Dlhá ul.319, Slovakia.
Email: jan.majling@stuba.sk
Education and scientific degrees: Chemical Engineer, Faculty of Chemical Technology, Slovak University of technology (SUT) (1959–1964). Ph.D., Institute of Inorganic Chemistry, Slovak Academy of Sciences (SASci) (1971). Assoc. Prof., Faculty of Chemical Technology, SUTechnology (1981)
Workplaces: Institute of Inorganic Chemistry, SASci., Bratislava (1966–1975), Faculty of Chemical and Food Technology, SUT, Bratislava (1975–2004), Retired from SUT (2004)
Main fields of interest: silicate chemistry and technology; phase diagrams; sulphoaluminate belite cements; advanced ceramics; single crystal growth
Relevant categories in thermal analyses: fields (inorganic and polymer materials, ceramics, glasses and cement, nano particulate systems, thin films); methods (TG, DTA, instrument development to carry out thermo-optical investigations)
Professional activities: Slovak Silicate Society, vice president (1995–2003)
Publication record: papers (73), books (1), patents (15), citations (288), h-index (11)
Equipments: Thermo-optical device (DS-LAB, Slovakia) to measure transmission of solid samples, Derivatograph, MOM Hungary, Previously employed, Hot stage microscope, Griffin-Telin, Great Britain
5 most important publications: [1] J. Majling, V. Kremnican, L. Pach, J. Chocholoušek, High Temperatures-High Pressures, 33 (2001) 43; [2] L. Pach, L. Bača, J. Majling, Chemical Papers, 54 (2000) 265; [3] V. Kovar, K. Bodišová, V. Kremničan, J. Majling: Journal of Thermal Analysis and Calorimetry, 79 (2005) 727; [4] J. Majling, L.Bača, J. Kozanková, M. Kocifaj, E. Fueglein, Journal of Thermal Analysis and Calorimetry, 114 (2013) 417; [5] M. Kocifaj, J. Majling, V. Kovár: Journal of Quantitative Spectroscopy and Radiative Transfer, 131 (2013) 115.

Name: Andrzej Malecki
Country: Poland
Date and place of birth: 1947, Krakow, Poland
Present position and address: full professor retired from: AGH-UST University of Science and Technology, Faculty of Materials Science and Ceramics, Department of Inorganic Chemistry Al. Mickiewicza 30, 30-059 Krakow, Poland
Email: malecki@agh.edu.pl
Website: http://home.agh.edu.pl/ ~ malecki/
Education and scientific degrees: Jagiellonian University (Krakow) (physical and theoretical chemistry). Ph.D. (1975 AGH-UST Krakow, Poland; 1985 Université de Bordeaux, France), habilitation (1987), professor of chemical sciences (1999)
Workplaces: AGH-UST University of Science and Technology, Faculty of Materials Science and Ceramics, Department of Inorganic Chemistry (1971–)
Main fields of interest: physical chemistry, solid state chemistry
Relevant categories in thermal analyses: fields (chemical kinetics, thermal decomposition of solids, mechanism of gas-solid reactions); methods (TG, DTA, EGA, DSC)
Awards and acknowledgments: Świętosławski Medal (2009)
Professional activities: President of ICTAC, Chairman of ESTAC 9 (Krakow, Poland), Past-president of Polish Society of Calorimetry and Thermal Analysis (PTKAT)
Publication record: papers (150), books (2), patents (4), citations (400), h-index (8), sum of impact factors (120)
Equipments: DTA/DSC/TG (TA Instruments) coupled with mass spectrometer (Balzers)
5 most important publications: [1] A. Koleżyński, A. Małecki, "First principles studies of thermal decomposition of anhydrous zinc oxalate", Journal of Thermal Analysis and Calorimetry; 96 (2009). 645; [2] A. Koleżyński, A. Małecki, " Theoretical approach to thermal decomposition process of chosen anhydrous oxalates", Journal of Thermal Analysis and Calorimetry, 97 (2009) 77; [3] A. Małecki, B. Małecka, "Formation of N2O during thermal decomposition of d-metal hydrates nitrates", Thermochim. Acta, 446 (2006) 113–116; [4] A.Małecki, B. Małecka, "Determination of kinetic parameters of phase transition on the basis of DTA measurements", J. Therm. Anal. Cal. 64 (2001) 837–842; [5] A. Małecki, S. Komornicki, B. Prochowska-Klisch (Małecka), "A kinetic model of gas-solid reaction under the conditions of changing energy of activation", Bull. Pol. Acad. Sci. 29 (1981) 455.

Name: Jiri Málek
Country: Czech Republic
Date and place of birth: 1959, Policka, Czechoslovakia
Present position and address: Professor, Department of Physical Chemistry, University of Pardubice, Studentská 573, 532 10 Pardubice, Czech Republic
Email: jiri.malek@upce.cz
Website: www.jiri-malek.cz
Education and scientific degrees: Chemical Engineer, Inst. Chem. Technology, Pardubice (1982); Ph.D. (1986); D.Sc. Inst. Chem. Technology, Prague (2000); Professor, University of Pardubice (2002)
Workplaces: Joint Laboratory of Solid State Chemistry, Czechoslovak Acad. Sci. (1988–2001); Department of Physical Chemistry, University of Pardubice (2002–)
Main fields of interest: kinetics of solid state processes; structural relaxation of non-crystalline materials; viscosity behavior of supercooled liquids
Relevant categories in thermal analyses: fields (inorganic, materials, polymer, glass, ceramics); methods (DTA, DSC, microscopy, thermomechanical analysis, kinetics, specific heat, calorimetry)
Awards and acknowledgments: ICTA Young Scientist Award (1992); CEEC-TAC Distinguished TA&C Researcher in Central and Eastern Europe (2013)
Professional activities: Member of Czech Group For Thermal Analysis (1985–); Editorial board of Thermochimica Acta (1997–2003); R&D Council of the Czech Government (2006–); Czech Engineering Academy (2008–); Scietific Board of Czech Science Foundation (2010); National councillor at International Confederation for Thermal Analysis and Calorimetry (2010–); Associate Editor of Journal of Thermal Analysis and Calorimetry (2011–); National commission for evaluation of research, chairman (2012)
Publication record: papers (127), book chapters (4), citations (3010), h-index (28)
Equipments: DSC (PerkinElmer, Mettler, TA-instruments), TMA (R.M.I., TA-Instruments), Optical and IR microscopy (Linkam, Olympus)
5 most important publications: [1] J. Málek, Thermochim. Acta, 138 (1989) 337–346.; [2] J. M. Criado, J. Málek, A. Ortega, Thermochim. Acta, 147 (1989) 377–385; [3] J.Málek, Thermochim. Acta, 200 (1992), 257–269; [4] J. Málek,, Thermochim. Acta, 355 (2000) 239–253.; [5] F.J. Gotor, J.M. Criado, J. Málek, N. Koga, J. Phys. Chem. A, 104 (2000) 10777–10782.

Name: Barbara Malič
Country: Slovenia
Date and place of birth: 1960, Ljubljana, Slovenia
Present position and address: Head of Electronic Ceramics Department, Jožef Stefan Institute, Jamova cesta 39, 1000 Ljubljana, Slovenia
Email: barbara.malic@ijs.si
Website: http://www-k5.ijs.si/
ORCID: 0000-0002-3438-8846
Education and scientific degrees: Chemistry, University of Ljubljana, Slovenia, Ph.D. (1995), Associate professor of chemistry of materials (2006)
Main fields of interest: lead-based and lead-free ferroelectric and piezoelectric ceramics and thin films, processing-microstructure-properties relationship
Relevant categories in thermal analyses: fields (inorganic, materials, nano, glass, ceramics, organic precursors of ceramics); methods (TG, DTA, EGA, DSC, cryo, extremely high temperature (above 1,000 °C); specific heat, calorimetry, dilatometry)
Awards and acknowledgments: Zois recognition award (2010)
Professional activities: Slovenian delegate to COST Materials, Physical and Nanosciences Domain Committee, (2010–), Symposium chair at the EMRS 2011 Fall meeting
Publication record: papers (150), patents (4), h-index (17)
Equipments: Simultaneous thermal anlysis (TG/DTA): STA 429, STA 409 PC Luxx and STA 409 (Netzsch) coupled with quadropole mass spectrometer (Balzers Thermostar), Differential scanning calorimeter DSC 204 F1 (Netzsch), Dilatometer DIL402 CD (Netzsch)
5 most important publications: [1] B. Malič, J. Bernard, J. Holc, D. Jenko, M. Kosec. J. Eur. Ceram. Soc., 2005, 25, 2707; [2] B. Malič, D. Jenko, J. Holc, M. Hrovat, M. Kosec. J. Am. Ceram. Soc., 2008, 91, 1916; [3] J. Tellier, B. Malič, B. Dkhil, D. Jenko, J. Cilenšek, M. Kosec. Solid State Sci. 2009, 11, 320; [4] A. Kupec, B. Malič, J. Tellier, E. Tchernychova, S. Glinšek, M. Kosec. J. Am. Ceram. Soc., 2012, 95, 515; [5] B. Malič, A. Kupec, M. Kosec, Thermal analysis, T. Schneller, R. Waser, M. Kosec, D. Payne, (Eds.), Springer, Wien, 2013, 163–179.

Name: João F. Mano
Country: Portugal
Date and place of birth: 1968, Sintra, Portugal
Present position: Professor, University of Minho
Email: jmano@dep.uminho.pt
Website: http://www.3bs.uminho.pt/users/jfmano
Researcher ID/ORCID: A-4418-2009/0000-0002-2342-3765
Education and scientific degrees: M.Sc. in Chemical Engineering, I.S.T. Portugal (1992); Ph.D. in Chemistry, Technical University of Lisbon (1996); D.Sc. in Tissue Engineering, Regenerative medicine and Stem Cells, University of Minho (2012)
Workplaces: University of Minho (1996–)
Main fields of interest: biomaterials and nanobiomaterials, biodegradable and biomimetic polymers, tissue engineering, thermo-physical properties of macromolecules, glass transition dynamics, smart systems and surfaces, drug delivery, surface chemistry/physics
Relevant categories in thermal analyses: fields (biomaterials, semi-crystalline polymers, biodegradable macromolecules, soft matter, nanomaterials, phase transitions, relaxation processes, crystallization); methods (DMA, DSC, dielectric relaxation, surface analysis, kinetics, non-conventional)
Awards and acknowledgments: fellow of the IUPAC 2004–; the Stimulus to Excellence Award by the Portuguese Minister for Science and Technology 2005; Federation of European Materials Societies Materials—Science and Technology Prize 2007; BES innovation award in 2010
Professional activities: Editorial board member in e.g. Tissue Eng. A, B, C; Mater. Letter.; J. Bioact. Compat. Polym.; Adv.Healthcare Mater.; organizing international meetings or symposia in major conferences; 14 Ph.D. and 30 M.Sc. thesis supervision; Former director of the Master course in Biomedical Engineering at the University of Minho, Erasmus/Sócrates responsible for the same graduation (profile in Biomaterials, Rehabilitation and Biomechanics); member of e.g. Portuguese Society for Stem Cells and Cellular Therapies, Portuguese Society of Materials (elected president of the fiscal councils) and the Federation of European Materials Societies (member of the Executive Committee since 2011); Responsible or active participant in different national and European funded projects, in the areas of polymer and biomaterials science/technology, tissue engineering, biodegradable and natural materials and nanoscience/nanotechnology for biomedicine; evaluating research projects from EC (including ERC Advanced grants), and from other agencies from Europe, Canada, Chile and New Zeeland, and reviewed manuscripts from more than 140 different journals and a series of books and special number in journals; invited talks in EU, US and Asian research groups (15) and invited lectures in international meetings (36)
Publication record: papers (400), books (3), patents (3), citations (7730), h-index (43)
Equipments: DMA, DSC, dielectric relaxation spectroscopy, QCM, microscopy
5 most important publications: [1] A.R. Bras, P. Malik, M. Dionisio, J.F. Mano, Macromolecules, 41, 6419 (2008); [2] J.F. Mano, Macromolecular Rapid Communications, 29, 1341 (2008); [3] A. Pinheiro, J.F. Mano, Polymer Testing, 28, 89 (2009); [4] G.V. Martins, J.F. Mano, N.M. Alves, Langmuir 27, 8415–8423 (2011); [5] R.R. Costa, C.A. Custódio, F.J. Arias, J.C. Rodríguez-Cabello, J.F. Mano, Nanomedicine—NBM, 9, 895 (2013).

Name: Shehab A. Mansour
Country: Egypt
Date and place of birth: 1976, Cairo, Egypt
Present position and address: Associate Professor, Basic, Engineering Science Department, Menofia University, Gamal Abd El-Nasar Street: 32511 | Menofia, Egypt.
Email: shehab_mansour@yahoo.com
Education and scientific degrees: BSc in Physics, Ain Shams University (1997); M.Sc. in Physics Ain Shams University (2003); Ph.D. in Physics, Cairo University (2008); Associate Professor (2013)
Workplaces: Faculty of Engineering, Menofia University (1998–)
Main fields of interest: glass and crystallisation kientics of chalcogenide semiconductors and thermal stabilitity of polymer nanocomposites
Relevant categories in thermal analyses: fields (chalcogenide semiconductors, quantum dots, polymer nanocomposites); methods (DTA, DSC and TG analysis)
Awards and acknowledgments: TUBITAK Postdoctoral Fellowship in Firat University (2011); ICTP grant for TRIL program in ENEA UTTP-NANO, Portici Research Centre, Italy (2012); Short Term Scientific Mission grant, COST Action MP0902 in Jozef Stefan Institute, Slovenia (2013)
Professional activities: Reviewer for some international journals such as Materials science and engineering B, Polymer engineering and science and thermal analysis and calorimetry
Publication record: papers (18), citations (71), h-index (5)
Equipments: Mettler FP80, Shimadzu TGA-50H, SETARAM LabsysTM TG-DSC16.
5 most important publications: [1] " Model-free transformation kinetics for ZnS quantum dots fabricated via colloidal reaction" Sh.A. Mansour, M.S. Al-Kotb, M.F. Kotkata Physica B 433 (2014) 127–132; [2] " Study of thermal stabilization for polystyrene/carbon nanocomposites via TG/DSC techniques" Sh.A. Mansour, Journal of Thermal Analysis and Calorimetry1 (2013) 579–583; [3] "Study of the Thermally Stimulated Depolarization Current (TSDC) For Composites of Acrylonitrile Butadiene Rubber (NBR) Loaded with HAF" M. Amen, G. M. Nasr, K.M. El-Shokrofy and Sh.A. Mansour, Iranian Journal of Science and Technology. Transaction A, Science 37 (A3), 313–318; [4] "Study of glass transition kinetics of selenium matrix alloyed with up to 10 % indium" M.F. Kotkata, Sh.A. Mansour, Journal of Thermal Analysis and Calorimetry, 103 (2011) 555; [5] "Crystallization process analysis for Se0.95In0.05 and Se0.90In0.10 chalcogenide glasses using the contemporary isoconversional models" M.F. Kotkata, Sh.A. Mansour, Journal of Thermal Analysis and Calorimetry 103 (2011) 957.

Name: Anastasia N. Mansurova
Country: Russia
Date and place of birth: 1982, Kamensk-Uralski, Russia
Present position and address: senior researcher, Institute of Metallurgy of the Ural Branch of the Russian Academy of Sciences, 620016, Yekaterinburg, 101, Amundsen St., Russia
Email: mansurova_a@list.ru, pcmlab@mail.ru
Researcher ID/ORCID: 0000-0002-5001-2662
Education and scientific degrees: Ural State Federal University (1999–2004), Ph.D. 2011.
Workplaces: Institute of Metallurgy of the Ural Branch of the Russian Academy of Sciences, Yekaterinburg (2003–)
Main fields of interest: investigation of redox processes of inorganic compounds by the thermal analysis and the mass-spectrometry
Relevant categories in thermal analyses: fields (inorganic materials, minerals, ceramics, metals, oxide, oxisulfide); methods (DTA, TGA, DSC, kinetics, extremely high temperature (above 1,000 °C), calorimetry)
Awards and acknowledgments: Diploma of the Russian Academy of Sciences (2011)
Professional activities: participations in the international conference on Diffusion in Solids and Liquids—DSL2008 in Barcelona, Spain and the XV annual conference—YUCOMAT 2013 in Herceg Novi.
Publication record: papers (12), patents (1), h-index (1), sum of impact factors (1.752)
Equipments: thermoanalazer NETZSCH STA 449C Jupiter, mass-spectrometer NETZSCH QMS 403C Aëolos
5 most important publications: [1] R. I. Gulyaeva, E. N. Selivanov, A. N. Mansurova., Kinetics of Calcium Oxysulfides Reduction by Carbon Monoxide, Diffusion in Solid and Liquids, 283–286 (2009) 539–544; [2] A. N. Mansurova, R. I. Gulyaeva, V. M. Chumarev et al. Thermochemical properties of MnNb2O6. J. Therm. Anal. Calorim., 101 (2010) 45–47; [3] A.N. Mansurova, V.M. Chumarev, R. I. Gulyaeva, V.P.Mar'evich. Thermogravimetric and mass spectrometric studies of the reduction of iron niobate by carbon. Russian Metallurgy (Metally), 1 (2012) 28–32; [4] A. N. Mansurova, V. M.Chumarev, L. I. Leont'ev, R. I. Gulyaeva, N. I. Sel'menskikh. Phase transformations during the interaction of Nb2O5 and FeNb2O6 with aluminum. Russian Metallurgy (Metally), 11 (2012) 929–934.; [5] R. I. Gulyaeva, E. N. Selivanov, A. N. Mansurova. Chemism and kinetics of the oxidation of zinc-calcium oxysulfide. Russian Metallurgy (Metally), 5 (2013), 327–331.

Name: Teresa M. R. Maria
Country: Portugal
Date and place of birth: 1969, Abrantes, Portugal
Present position and address: Assistant Professor, Department of Chemistry, University of Coimbra, Rua Larga, 3004-535 Coimbra, Portugal
Email: troseiro@ci.uc.pt
Researcher ID/ORCID: N-3024-2013/0000-0002-7143-2228
Education and scientific degrees: Degree in Chemisty, University of Coimbra (1988–1992); Ph.D., University of Coimbra (2004)
Workplaces: Chemistry Department, University of Coimbra
Main fields of interest: polymorphism of organic compounds: research is focused in the study of small organic molecules (dihydroxyl cyclohexane derivatives), and compounds with pharmaceutical interest; screening and characterization of co-crystals of organic compounds including active pharmaceutical ingredients; plastic crystals; phase transition processes
Relevant categories in thermal analyses: fields (materials, organic, pharmaceutical, glass); methods (TG, DTA, DSC, kinetics, cryo, specific heat, calorimetry)
Publication record: papers (24), h-index (9)
Equipments: DSC7 PerkinElmer; DSC Pyris1 PerkinElmer; DSC600 hot stage Linkam system.
5 most important publications: [1] Polymorphism and melt crystallisation of racemic betaxolol, a beta-adrenergic antagonist drug", Teresa M. R. Maria, Ricardo A. E. Castro, M. Ramos Silva, M. Luísa Ramos, Licínia L. G. Justino, Hugh D. Burrows, João Canotilho, M. Ermelinda S. Eusébio, J. Therm. Anal. Calorim. 111 (2013) 2171–2178; [2] 2-Quinolinecarboxaldehyde: polymorphic behavior of a small rigid molecule, Teresa M. R. Maria, M. Ermelinda S. Eusébio, J. Almeida e Silva, Abílio J. F. N. Sobral, C. Cardoso, J. A. Paixão, M. Ramos Silva, J. Mol. Struct.1030 (2012) 67–74; [3] Pyrazinamide-diflunisal: A New Dual-Drug Co-Crystal, António O.L. Évora, Ricardo A. E. Castro, Teresa M. R. Maria, Mário T. S. Rosado, Manuela R. Silva, A. Matos Beja, João Canotilho, M. Ermelinda S. Eusébio, Cryst. Growth Des., 11 (2011) 4780–4788; [4] Glass Forming Ability of Butanediol Isomers, Teresa M. R. Maria, A. J. Lopes Jesus, M. Ermelinda S. Eusébio, J. Therm. Anal. Calorim. 100 (2010) 385-390; [5] Polymorphism of trans-1, 4-cyclohexanediol: Conformational Isomorphism, Teresa M. R. Maria, Ricardo A. E. Castro, Suse S. Bebiano, Manuela R. Silva, Ana M. M. Beja, João Canotilho, M. Ermelinda S. Eusébio, Cryst. Growth Des., 10 (2010)1194–1200.

Name: Bruno Marongiu
Country: Italy
Date and place of birth: 1946, Arzana, Italy
Present position and address: Associate Professor, Physical Chemistry at Sciences Faculty of University of Cagliari, I-09042 Mouserato, Italy
Email: maronb@unica.it
Education and scientific degrees: Chemistry (1970); Researcher (1972); Associate professor (1982)
Workplaces: Department of Chemical Sciences; Cittadella Universitalia di Monserrato of Cagliari, Italy (1972–)
Main fields of interest: thermodyinamics of organic mixtures; kinetics and catalysis; technology of supercritical fluids
Professional activities: Member of SCI—Italian Society of Chemistry- Division of Chemical Physic; Interdivisional Group of Calorimetry and Thermal Analysis; Tresaurier of AICAT—ItalianAssociation of diCalorimetry iand ThermalAnalysis; Member of ISASF—International Society for the Advancement of Supercritical Fluids; Chairman del IWorkshop on Thermodynamics of Organic Mixtures (Chia Laguna, 1991); Chairman delle Giornate Mediterranee di Calorimetria ed Analisi Termica, GMEDCAT-95, (Chia Laguna, 1995);Chairman della I Scuola Nazionale sui Fluidi Supercritici (Chia Laguna, 1998); Organizing Committee Chair of 13 International Congress on Thermal Analysis and Calorimetry, ICTAC-13, (Chia Laguna, 2004). Referee for J. Chem. Eng.Data; Fluid Phase Equilib; Thermochim.Acta; J. Therm. Anal. Calorim.
Publication record: papers (120)
5 most important publications: [1] L. Forni, I. Ferino, B. Marongiu, S. Torrazza: Activity and structure of Rh/Cr2O3 catalyst for toluene steam- dealkylation, Journal of Catalysis, 9 (1984) 169; [2] H. V. Kehiaian, B. Marongiu: A comparative study of thermodynamic properties and molecular interactions in mono- and poly-chloroalkane + n-alkane or cyclohexane mixtures, Fluid Phase Equilib., 40 (1988) 23; [3] M. R. Tinè, B. Marongiu: Some recent applications of the DISQUAC group contribution model to binary liquid organic mixtures, Thermochim. Acta, 199 (1992) 63; [4] A. Caredda, B. Marongiu, S. Porcedda, C. Soro: Supercritical carbon dioxide extraction and characterization of laurus nobilis essential oil, J. Agric.and Food Chem., 50 (2002) 1492; [5] S. Lai, E. Locci, A. Piras, S. Porcedda, A. Lai, B.Marongiu: Imazalil-cyclomaltoheptaose (β-cyclodextrin) inclusion complex: preparation by supercritical carbon dioxide and 13C CP-MAS and 1H NMR characterization, Carbohydrate Research, 338 (2003) 2227.

Name: Vincent B. F. Mathot
Country: Netherlands
Date and place of birth: 1945, Haarlem, Netherlands
Present position and address: President of SciTe, Ridder Vosstraat 6, 6162 AX Geleen, the Netherlands; Guest Professor at the Katholieke Universiteit Leuven, Belgium.
Email: vincent.mathot@scite.nl
Website: www.scite.eu
Education and scientific degrees: Physics and Mathematics, University of Amsterdam
Workplaces: SciTe, the Netherlands; Chemelot, the Netherlands; Katholieke Universiteit Leuven (K.U. Leuven), Belgium
Main fields of interest: polymers; molecular structure; fractionation; crystallization and melting; morphology; thermal properties; relationships between these topics
Relevant categories in thermal analyses: methods (quantitative DSC; fast scanning calorimetry; theory)
Awards and acknowledgments: Mettler-Toledo Award of the NATAS (2002); NATAS (2009) Honorary Session 'Lifetime Achievement'
Professional activities: Heading various departments at DSM Research for more than 30 years. In 2003 start of SciTe B.V. (Science and Technology). After leaving DSM Research in 2005 full time for SciTe. Guest professorship at the K.U. Leuven. At its start, SciTe was meant to realize a commercial Fast Scanning Calorimeter based on MEMS technology, in close cooperation with the Swiss-American company Mettler-Toledo, Anatech in Sittard, the Netherlands and Xensor Integration in Delft, the Netherlands. In 2010 the Flash DSC 1 resulted as commercialized by Mettler-Toledo. With its scan rate range of 100 up to 2,400,000 °C/min in heating and 240,000 °C/min in cooling it is a unique and powerful instrument. Member of the Editorial Board of Thermochimica Acta; Former member of the Council of the International Confederation for Thermal Analysis and Calorimetry (ICTAC); Former member of the ICTAC Education Committee; Former board member and Chair holder of the Dutch Society for Thermal Analysis (TAWN); Chairman of DIN Standardization Workshop for Temperature calibration of Fast Scanning Calorimeters, DIN SPEC 91127, June 2011
Equipments: 2x Flash DSC 1 from Mettler-Toledo. DSC 8500; Diamond DSC; DSC-7, all from PerkinElmer
5 most important publications: [1] V.B.F. Mathot, M. Pyda, M. Pijpers, G. Vanden Poel, E. van de Kerkhof, S. van Herwaarden, F. van Herwaarden, A. Leenaers, Thermochimica Acta, 522 (2011) 36; [2] J. Ibarretxe, G. Groeninckx, L. Bremer, V.B.F. Mathot, Polymer, 50 (19) (2009) 4584; [3] M.G.M. Wevers, V.B.F. Mathot, M. Pijpers, B. Goderis, G. Groeninckx, Full Dissolution and Crystallization of Polyamide 6 and Polyamide 4.6 in Water and Ethanol., Lecture Notes in Physics: Progress in Understanding of Polymer Crystallization, COST, G. Reiter, G. Strobl, Eds., Springer-Verlag, Berlin Heidelberg (2007) Chapter 9, pp 151–168; [4] W. Hu, F.G. Karssenberg, V.B.F. Mathot, Polymer, 47(15) (2006) 5582; [5] M.F.J. Pijpers, V.B.F. Mathot, B. Goderis, R.L. Scherrenberg, E.W. van der Vegte, Macromolecules, 35(9) (2002) 3601.

Name: Mikhail A. Matsko
Country: Russia
Date and place of birth: 1974, Barnaul, Russia
Present position and address: A head of Laboratory of Catalytic Polymerization, Boreskov Institute of Catalysis SB RAS, 630090, Novosibirsk, Russia
Email: Matsko@catalysis.ru
Education and scientific degrees: Ph.D. (2002)
Workplaces: Boreskov Institute of Catalysis SB RAS
Main fields of interest: catalytic polymerization of olefins and polymer (polyolefin) characterization (molecular weight distribution, thermal analysis, fractionation technique)
Relevant categories in thermal analyses: fields (polymer (polyolefin) characterization); methods (DSC, thermal fractionation using DSC, DMA)
Publication record: papers (48), patents (4), citations (150), h-index (9)
Equipments: DSC Netzsch 204 F1
5 most important publications: [1] M.A. Matsko, et al., Macromol. Symp. 2009, 282 (p 157–166); [2] M.A. Matsko, et al., Journal of Applied Polymer Science Volume 126, Issue 6, 15 December 2012, Pages: 2017–2023; [3] M.A. Matsko, et al., Catalysis in Industry, 2011, Vol. 3, No. 2, pp. 109–115; [4] N.V. Semikolenova, M.A. Matsko, et al., Homogeneous Catalysts: Types, Reactions and Applications/Ed. A.C. Poehler, New York: Nova Science Publishers, 2011, pp. 97–126; [5] M.A. Matsko, et al., J Therm Anal Calorim 2013, V. 113, Is 2, pp 923–932.

Name: Said Abd-Elaziz Mazen
Country: Egypt
Date of birth: 1947
Present position and address: Prof. of Solid State Physics, Physics Department, Faculty of Science, Zagazig Univ., Zagazig, Egypt.
Email: dr.samazen@Yahoo.com, dr.saidmazen@gmail.com
Researcher ID: 7003750984
Education and scientific degrees: Ph.D.
Workplaces: Department of Physics, Faculty of Science, Zagazig Univ., Zagazig, Egypt.
Main fields of interest: solid state physics (magnetic semiconductors)
Relevant categories in thermal analyses: fields (materials, ceramics); methods [extremely high emtperature (above 1,000 °C)]
Professional activities: Editorial board member in research India Pub. Int. J. of Mat. Sci (IJOMS)
Publication record: papers (65), citations (339), h-index (11)
5 most important publications: [1] The effect of titanium on some physical properties of CuFe2O4, Phys. Stat. Sol. (a) 134 (1992) 263–271; [2] X-Ray analysis and IR absorption spectra of Li–Ge Ferrite, Mat. Chem. and Phys. 34 (1993) 35–40; [3] Structure and Magnetic Properties of Li–Cu Ferrite, Phys. Stat. Sol. (a) 172 (1999) 275; [4] IR absorption and dielectric properties of Li–Ti ferrite, J. of physics D: Applied physics 30 (1997) 1799; [5] Some physical and magnetic properties of Mg–Zn ferrite Crystal Research and Technology 38 (2003) 471.

Name: Henry G. McAdie
Country: Canada
Date and place of birth: 1930, Montreal, Canada
Present position and address: Retired, Suite 1005, 5444 Yonge Street, Toronto, Ontario, M2N 6J4, Canada
Email: HGMcAdie@aol.com
Education and scientific degrees: B.Sc. McGill University (1951); M.Sc. Queen's University (1953); Ph.D. at Queen's University (1956)
Workplaces: Ontario Research Foundation (1956–1984); Chemical Institute of Canada (1962–2002); H. G. McAdie Associates (1984–2009)
Main fields of interest: thermal properties of materials; standards creation; environmental research and monitoring, particularly with respect to air pollution
Awards and acknowledgments: Elected Fellow of Royal Institute of Chemistry (now Royal Society of Chemistry), (1966); Elected Fellow of Chemical Institute of Canada (1967); J. Charles Honey Award for service to Toronto Section, Chemical Institute of Canada (1978); Elected Fellow of North American Thermal Analysis Society (1983); N. S. Kurnakov Medal in thermal analysis (1985); Outstanding Service Award, North American Thermal Analysis Society (1986); Elected Honorary Life Member, Air and Waste Management Assoc., Ontario Section (1993); Elected Honorary Member, ICTAC (2000)
Professional activities: Chairman of the Air and Waste Management Association, Ontario Section APCA/PCAO Joint Conference (1984), Board member (1977–1980; 1984–1993); Member of the Committees D-22 and E-37 of the American Society for Testing and Materials; Canadian Standards Association Committee on Air Pollution Control; chairman of the Chemical Institute of Canada, Board of Directors (1977–1979); Various Board, Council, Subject Division and Local Section offices (1962–); chairman, Past Officers and Directors Advisory, Committee (1981–2001); ICTAC: president (1977–1980), vice-president (1974–1977), chairman, Standardization Committee (1965–1974); vice-chairman, 7th ICTAC Conference, (1982), membership secretary (1985–); chairman of InternationalOrganization for Standardization CAC/ISO/TC-146, Air Quality (1983–1989; 1974–1977), chairman, SC-4 (1977–1983)
Publication record: papers (80), books (3), patents (1)

Name: Andrew R. McGhie
Country: USA
Date and place of birth: 1940, Clydebank, Scotland
Present position and address: Associate Director, Laboratory for Research on the Structure of Matter, University of Pennsylvania, Philadelphia PA 19104-6202
Email: mcghie@lrsm.upenn.edu
Website: www.lrsm.upenn.edu
Education and scientific degrees: A.R.C.S.T., Royal College of Science and Technology, Glasgow 1962, B.Sc (with 1st Class Honors, Applied Chemistry) 1962 and Ph.D. Physical Chemistry, University of Glasgow 1965 (Advisor-Prof. John N. Sherwood)
Workplaces: Franklin Institute Research Laboratories and Drexel University, Philadelphia PA, Arthur D. Little Research Institute, Musselburgh, Scotland, Dupont Central Research Laboratory, Wilmington, DE, LRSM, University of Pennsylvania (1970–)
Main fields of interest: purification and growth of organic crystals, thermal analyis, educational outreach
Relevant categories in thermal analyses: fields (materials: nano, organic, pharmaceutical, polymer); methods (TG, DTA, DSC, EGA, instrument development)
Awards and acknowledgments: NATAS Fellow, 1989; NATAS Outstanding Service Award, 1992; NATAS Outstanding Achievement (Mettler-Toledo) Award, 2011; ICTAC Distinguished Service Award, 2004; Professor (honorary), Materials Science and Engineering, University of Pennsylvania, 1997
Professional activities: ICTAC Conf. Chair 1996, NATAS President, 2008, NATAS Conf. Chair, 2010, TAFDV, President and Symposia Chair
Publication record: papers (>84), books (1), patents (1), h-index (29)
Equipments: DSC, DTA, TGA/DTA/MS
5 most important publications: [1] Techniques of Melt Crystallization, Gilbert J. Sloan and Andrew R. McGhie, Wiley Interscience 1988 532 pp (ISBN 0-471-078751); [2] Orientational Ordering Transition in Solid C60; [3] Heiney, PA; Fischer, JE; McGhie, AR; et al., Physical Review Letters 66, 22, 2911–2914 1991; [4] Band-Hopping Transition for Electrons in Naphthalene Schein, LB, Duke, CB, McGhie, AR, Phys. Rev. Let. 40, 3, 197–200, 1978; [5] Compressibility of Solid C60, Fischer, JE, Heiney, PA, McGhie, AR; Romanow, WJ; Denenstein, AM, McCauley, JP, Smith, AB, Science, 252, 5010, 1288–1290, 1991.

Name: Gregory B. McKenna
Country: USA
Date and place of birth: 1949, Pittsburgh, USA
Present position and address: Horn Professor, Department of Chemical Engineering, Texas Tech University, Lubbock, TX 79409-3121
Email: greg.mckenna@ttu.edu
Website: www.depts.ttu.edu/CHE/
Researcher ID/ORCID: O-1134-2013/0000-0002-5676-9930
Education and scientific degrees: U.S. Air Force Academy Engineering Mechanics B.S./1970; Massachusetts Institute of Technology Composite Materials S.M./1971; University of Utah Materials Science and Engineering Ph.D./1976
Workplaces: National Institute of Standards and Technology, Texas Tech University
Main fields of interest: polymer physics, rheology, glass physics, thermodynamics and mechanics of rubber, nanomechanics, nanothermodynamics
Relevant categories in thermal analyses: fields (polymer and glass physics, nanothermodynamics); methods (calorimetry, dilatometry, rheology)
Awards and acknowledgments: 2013 ARCS, Lubbock Chapter, Scientist of the Year (co-awardee S.L. Simon); 2009 Bingham Medal of the Society of Rheology; 2007 Founders Award of the Polym. Anal. Div. Soc. Plast. Engineers; 2006 Whitby Award for Distinguished Teaching and Research, Rub. Div., ACS; 2005 Fellow, Society of Engineering Science; 2005 Paul Whitfield Horn (University) Professor, Texas Tech University; 2004 International Award, Society of Plastics Engineers; 2001 NATAS Outstanding Achievement (Mettler-Toledo) Award; 2000 SPE International Research Award (in memory of John C. Moricoli); 1999 John R. Bradford Endowed Chair in Engineering, Texas Tech University; 1998 Fellow of the Society of Plastics Engineers; 1992 U.S. Department of Commerce Silver Medal Award; 1989 Edward U. Condon Award for Distinguished Achievement in Writing; 1989 Fellow of the American Physical Society; 1985 U.S. Department of Commerce Bronze Medal; 1976 National Research Council/National Academy of Sciences Postdoctoral Fellowship; 1970 Distinguished Graduate (Summa cum laude), U.S. Air Force Academy (Third in class)
Professional activities: 2013-present President, Society of Rheology; 1990-present Member of Editorial Review Board, Journal of Rheology; 2013-present Associate Editor, Mechanics of Time Dependent Materials; 1995–1999 Member of Governing Board, American Institute of Physics; 1992–2000–2007 Editor, Journal of Polymer Science, Polymer Physics Edition; 2007–2010 Editor-in-Chief, Journal of Polymer Science, Polymer Physics Edition; 2010-present Consulting Editor, Journal of Polymer Science, Polymer Physics Edition
Publication record: papers (199), books (1), patents (1), citations (9574), h-index (48)
5 most important publications: [1] C. L. Jackson and G.B. McKenna, J. Chem. Phys., 93, 9002–9011 (1990); [2] G.B. McKenna, "Glass Formation and Glassy Behavior," in Comprehensive Polymer Science, Vol. 2: Polymer Properties, ed. by C. Booth and C. Price, Pergamon Press, Oxford, 311–362 (1989); [3] C.L. Jackson and G.B. McKenna, J. Non-Crystalline Solids, 131, 221–224 (1991); [4] P.A. O'Connell and G.B. McKenna, Science, 307, 1760–1763 (2005); [5] G.B. McKenna, J. Phys. Cond. Matter, 15, S737–S763 (2003).

Name: Maura Berger Maltez Melchert
Country: Brazil
Date and place of birth: 1971, São Paulo, Brazil
Present position and address: Researcher from Chemical Engineering Department, Polytechnic School of São Paulo University, Avenida Professor Luciano Gualberto 380, São Paulo, 05424-970, SP, Brazil.
Email: maura.melchert@ig.com.br
Education and scientific degrees: Chemical Engineer, Mackenzie University, Brazil (1990–1994); M.Sc. (2005) and Ph.D. (2012) in Chem. Eng. at University of São Paulo (USP).
Main fields of interest: study of the thermal analysis technique and their applications. thermal analysis applied to industrial products and waste thermal characterization and processing.
Relevant categories in thermal analysis: fields (stabilized and solidified wastes, inorganic, organic, industrial products, cement, catalyst wastes, hazardous wastes); methods (TG, DTG, DTA, DSC, specific heat, calorimetry, non-conventional DTA instrument)
Equipments: Simultaneous TA Instruments TG/DTA equipment; Differential Scanning Calorimeter and Non-conventional DTA instrument (NCDTA)
Publication record: papers (13), citations (16)
3 most important publications: [1] Melchert, M. B. M.; Viana, M. M.; Lemos, M. S.; Dweck, J.; Büchler, P. M. Simultaneous solidification of two catalyst wastes and their effect on the early stages of cement hydration. Journal of Thermal Analysis and Calorimetry, v. 105, pp. 625–633, 2011; [2] Viana, M. M.; Melchert, M. B. M.; Morais, L. C.; Buchler, P. M.; Dweck, J. Sewage sludge coke estimation using thermal analysis. Journal of Thermal Analysis and Calorimetry, v. 106, pp. 437–443, 2011; [3] Dweck, J.; Melchert, M. B. M.; Viana, M. M.; Cartledge, F. K.; Büchler, P. M., Importance of quantitative thermogravimetry on initial cement mass basis to evaluate the hydration of cement pastes and mortars. Journal of Thermal Analysis and Calorimetry, v. 111, pp. 1–10, 2013.

Name: Joseph D. Menczel
Country: USA
Date and place of birth: 1948, Hungary
Present position and address: Retired from Alcon, Testing lab owner; 412 Stampede Court, Fort Worth, TX 76131, USA
Email: jmenczel@yahoo.com
Education and scientific degrees: MS in Physical Chemistry, Kiev State University; Ph.D. in Physical Chemistry, Budapest University of Technology 1972; Post-doctoral Research Associate, Rensselaer Polytechnic Institure 1979–1980, 1984
Workplaces: Central Research Institute for Physics, Budapest, Hungary; Budapest Technical University, Hungary; Rensselaer Polytechnic Institute, Try, New York, USA; GAF Corporation, Wayne, NJ, USA; Hoechst Celanese, Summit, NJ, USA; Alcon Corporation, Fort Worth, TX, USA; Thermal Measurements, LLC, Fort Worth, TX, USA
Main fields of interest: glass transition in semicrystalline polymers; melting and crystallization of polymers; instrument calibration
Relevant categories in thermal analyses: fields (polymer, pharmaceutical, glass); methods (DSC, TGA, TMA, DMA, μ-TA)
Awards and acknowledgments: NATAS Fellow 2002; NATAS Outstanding Achievement (Mettler Toledo-) Award 2010; various accomplishments awards of Hoechst
Professional activities: NATAS Secretary 2009–2011; NATAS Vice-President 2012; NATAS President 2013; NATAS Executive Councilor 2014; JTAC North American Editor
Publication record: papers (85), books (3), patents (6)
Equipments: Q2000; P-E DSC7; P-E Pyris 1 DSC; P-E TGA7, P-E TMA7, P-E DMA7e
5 most important publications: [1] J.D. Menczel, B. Wunderlich, J. Polym. Sci., Polym. Lett. Ed., 1981, 19, 261; [2] J.D. Menczel. T. Leslie, Thermochimica Acta, 1990, 166, 309; [3] J.D. Menczel, R.B. Prime, Thermal Analysis of Polymers. Fundamentals and Applications, Wiley, 2009; [4] Chapter 7 (Fibers) and Chapter 8 (Films) in: Thermal Characterization of Polymeric Materials, ed. By E.A. Turi, Academic Press, 1997; [5] J.D. Menczel, M. Jaffe, C.K. Saw, T.P. Bruno, J. Thermal Anal., 1996, 46, 753.

Name: Slavko Mentus
Country: Serbia
Date and place of birth: 1946, Sarajevo, Bosnia and Herzegovina
Present position and address: Retired (2013) from the University of Belgrade, Faculty of Physical Chemistry, Prof. S. Mentus, Faculty of Physical Chemistry, Belgrade University, Studentski trg 12, 11000 Belgrade, Serbia
Email: slavko@ffh.bg.ac.rs
Website: www.ffh.bg.ac.rs
Education and scientific degrees: BSc in Physical Chemistry (1969), Ph.D. in physical Chemistry (1975), Assistant Professor in Physicochemical Methods of Analysis (1982), Professor in Electrochemistry (1995)
Workplaces: Faculty of Physical Chemistry, Belgrade University, Serbian Academy of Sciences and Arts
Main fields of interest: electrode materials, electrochemical kinetics, energy conversion, materials characterization
Relevant categories in thermal analyses: fields (inorganic, materials, nano, ceramics); methods (TG, DTA, kinetics)
Awards and acknowledgments: Award of the Association of Professors and Researchers of the University of Belgrade, Award of the Chamber of Commerce of Belgrade, Corresponding member of the Serbian Academy of Sciences and Arts (2009)
Professional activities: member of the International Society of Electrochemistry (ISE), Electrochemical Society, Inc. Penington USA, Serbian Chemical Society, Materials Research Society—Serbia, Dean of the Faculty of Physical Chemistry (2004–2009), member of the honorary scientific committee of the conferences CEEC-TAC1 2011—Craiova, Romania, and CEEC-TAC2 2013—Vilnius, Lithuania, and scientific commitees of many other conferences
Publication record: papers (130), books (9), book chapters (4), patents (11), citations (870), h-index (15), sum of impact factors (216)
Equipments: TA SDT Model 2060, for simultaneous TG-DTA/DSC Analysis
5 most important publications: [1] S. Mentus, Electrochim.Acta, 50, (2004) 27; [2] S. Mentus, Electrochim. Acta, 50, (2005) 3609; [3] B. Janković, B. Adnađević, S. Mentus, Thermochim. Acta, 456 (2007) 48; [4] S. Mentus, D. Jelić and V. Grudić, J.Therm. Anal. Calorim. 90 (2007) 393; [5] S. Mentus, G. Ćirić-Marjanović, M. Trchová, J. Stejskal, Nanotechnology, 20 (2009) 245601.

Name: Alfréd Kállay-Menyhárd
Country: Hungary
Date and place of birth: 1978, Budapest, Hungary
Present position and address: Assistant professor, H-1111 Budapest Műegyetem rkp. 3 H/1
Email: amenyhard@mail.bme.hu
Website: www.mua.bme.hu
Researcher ID: G-9701-2012
Education and scientific degrees: Ph.D.
Workplaces: Budapest University of Technology and Economics, Department of Physical Chemistry and Materials Science
Main fields of interest: thermal analysis of polymeric materials, polymorphism, crystallization kinetics, structure-property correlations in semicrystalline polymers
Relevant categories in thermal analyses: fields (crystallization kinetics, melting and crystallization of polymers); methods (calorimetry (DSC), simultaneous thermal analysis (TG-DTA), thermo-optical measurements (POM))
Awards and acknowledgments: 2010 BME Faculty of Chemical Technology and Biotechnology Paulik brothers—award (Budapest); 2011–2014 Hungarian Academy of Sciences, János Bolyai Research Scholarship (Budapest); 2013 Outstanding Young TA&C Researcher in Central and Eastern Europe (Vilnius); 2013Hungarian Academy of Sciences, József Varga University Award (Budapest)
Professional activities: Working Committee for Thermal Analysis of the Hungarian Academy of Sciences, secretary; Hungarian Academy of Sciences, public membership (ID: 17978); Journal of Thermal Analysis and Calorimetry, Editor-in-Chief; International Confederation for Thermal Analysis and Calorimetry (ICTAC), member; Central and Easter European Committee for Thermal Analysis and Calorimetry (CEEC-TAC), member
Publication record: papers (34), patents (1), citations (422), h-index (8), sum of impact factors (40.282)
Equipments: PerkinElmer DSC7, PerkinElmer DSC2, PerkinElmer Diamond DSC, PerkinElmer TGA6, PerkinElmer STA 6000, PerkinElmer Diamond DMTA
5 most important publications: [1] Menyhárd, A., Varga, J., Liber, Á., Belina, G., Polymer Blends Based on the β-modification of Polypropylene, Eur. Polym. J., 41. 669–677 (2005); [2] Menyhárd, A., Varga, J., Molnár, G., Comparison of Different β-Nucleators for Isotactic Polypropylene, Characterization by DSC and Temperature Modulated DSC (TMDSC) Measurements, J. Therm. Anal. Calorim., 83. 625–630 (2006); [3] Varga, J., Menyhárd, A., Effect of Solubility and Nucleating Duality of N,N′-Dicyclohexyl-2,6-naphthalenedicarboxamide on the Supermolecular Structure of Isotactic Polypropylene, Macromolecules, 40. 2422–2431 (2007); [4] Menyhárd, A., Dora, G., Horváth, Zs., Faludi, G., Varga, J. Kinetics of Competitive Crystallization of β- and α-modifications in β-nucleated iPP Studied by Isothermal Stepwise Crystallization Technique J. Therm. Anal. Calorim. 108. 613–620 (2012); [5] Horváth, Z., Menyhárd, A., Doshev, P., Gahleitner, M., Tranninger, C., Kheirandish, S., Varga, J., Pukánszky, B., Effect of Molecular Architecture on the Crystalline Structure and Stiffness of iPP Homopolymers: Modeling Based on Annealing Experiments, J. Appl. Polym. Sci. 130. 3365–3373 (2013).

Name: Lucildes Pita Mercuri
Country: Brazil
Date and place of birth: 1964, Salvador, Brazil
Present position and address: III adjunct professor in the Federal University of São Paulo UNIFESP.; Federal University of São Paulo, Campus Diadema, José Alencar Unit. Rua São Nicolau (Jd. Pitangueiras). Center. Zip code: 09913030, Diadema, SP, Brazil
Email: lpmercuri@gmail.com
Website: https://sites.google.com/site/latemig/
Researcher ID: C-9552-2012
Education and scientific degrees: Applied Chemical, Bahia State University (1986); M.Sc. (UFBA-1991); D.Sc. (USP-2000); Ph.D. (KSU-USA-2001); Professor (2004)
Workplaces: Institute of Environmental Sciences. Chemical and Pharmaceutical, Department of Exact Sciences and Earth-DCET. Laboratory of Engineering and Environmental Control (LENCA). Federal University of São Paulo—Campus Diadema-São Paulo-Brazil
Main fields of interest: Thermoanalytical study in characterizing of different materials and Encapsulation study and controlled release in vitro of different organic species into new materials
Relevant categories in thermal analyses: Fields (nanomaterials, biological material, pharmaceutical, food, medicinal plants, cosmetology, immunology); Methods (TG, DTA, DSC, TG/DTA-GC/MS, kinetics, specific heat and calorimetry)
Awards and acknowledgments: Best Poster Section of Biological Chemistry 22th Annual Meeting of the Brazilian Chemical Society-SBQ (1999); Best poster on "Materials Science and Soft Matter", the Brazilian Association of Crystallography (2007); Honorary Membership, Thermoanalytical Group of the Hungarian Chemical Society (2008); 2nd Place: XXIII Award Pereira Barreto (XVII Congress of Scientific Initiation—UNIFESP) Federal University of São Paulo (UNIFESP-DCC) (2009); Poster winner of 4th place, XXIII Brazilian Congress of Cosmetology (2009); Godmother the First Class of Chemistry, Federal University of São Paulo (2011); 2nd Place Prize Ivo Giolito, ABRATEC (2012)
Professional activities: Member of the Association of Thermal Analysis and Calorimetry-ABRATEC (2010–); Member of Advisory Committee of the Research Foundation of the State of São Paulo-FAPESP (2010–); Member of Advisory Committee of the CNPq (2007–); Vice-director of the Center for Technological Innovation at the Federal University of São Paulo-NIT-UNIFESP (2013–2014)
Publication record: papers (43), book chapter (1), citations (825), h-index (12)
Equipments: DTG-60 (Shimadzu) and DSC-60 (Shimadzu)
5 most important publications: [1] Mercuri, L. P. et al. Small, 2, 2, 254–256, 2006; [2] Mercuri, L. P. et al. Thermochim. Acta, 383, 79–85, 2002; [3] Mercuri, L. P. et al. J. Alloy. Comp. 344, 190–194, 2002; [4] Matos, Jivaldo Do Rosário; Kruk, M.; Mercuri, L. P.; Jaroniec, Mietek; Zhao, L.; Kamiyama, T.; Terasaki, O.; Pinnavaia, T. J.; Liu, Y., J. Am. Chem. Soc. 125, 821–829, 2003; [5] Matos, Jivaldo Do Rosário; Mercuri, L. P.; Kruk, M.; Jaroniec, Mietek., Chem. Mater. 13, 1726–1731, 2001.

Name: Katalin Mészáros Szécsényi
Country: Serbia
Date and place of birth: 1952, Mol, Serbia
Present position and address: Full professor, Department of Chemistry, Biochemistry and Environmental Protection, Faculty of Sciences, 21000 Novi Sad, Trg D. Obradovića 3, Serbia
Email: mszk@uns.ac.rs
Researcher ID/ORCID: 6602812519/0000-0002-7494-7323
Education and scientific degrees: Dipl. Chem. (1974) University of Novi Sad; M.Sc. (1982); Ph.D. (2011); Professor (2004) Department of Chemistry, Biochemistry and Environmental Protection, Faculty of Sciences, University of Novi Sad
Workplaces: Faculty of Sciences, University of Novi Sad, Serbia (1975–)
Main fields of interest: synthesis and characterization of coordination and organometallic compounds; thermal analysis of polymers/polymer composites and other materials
Relevant categories in thermal analyses: fields (inorganic, materials, complex, polymer, pharmaceutical, food); methods (TG, DTA, EGA, DSC, kinetics)
Professional activities: Serbian Chemical Society, Chemical Society of Vojvodina, Public Board of the Hungarian Academy of Sciences, Working Committee on Thermoanalytical Section of the Hungarian Academy of Sciences, Associate Editor of the Journal of Thermal Analysis and Calorimetry, Member of Editorial Board of Létünk
Publication record: papers (65), books (3), citations (278), h-index (9), sum of impact factors (>60)
Equipments: TAI SDT Q600 Simultaneous TG/DSC
5 most important publications: [1] Budinski-Simendić, J., Špirkova, M., Pavličević, J., Šomvarsky, J., Mészáros Szécsényi, K., Dušek, K., Structure and Properties of Dangling Chain Poly(urethane-isocyanurate) Model Networks, ACS—Symposium Series, 2010. Contemporary Science of Polymeric Materials, Chapter 11, pp 149–166; [2] Holló, B., Jašo, V., Leovac, V.M., Divjakovic, V., Kovács, A., Mészáros Szécsényi, K., Synthesis, structure and thermokinetic studies on perchlorate salts of metal complexes containing a formamidine-type ligand (2013) Journal of Coordination Chemistry, 66 (3), pp. 453–463; [3] Bera, O., Pilić, B., Pavličević, J., Jovičić, M., Holló, B., Mészáros Szécsényi, K., Špirkova, M., Preparation and Thermal Properties of Polystyrene/Silica Nanocomposites, Thermochim. Acta, 515 (2011) 1-5; doi:10.1016/j.tca.2010.12.006; [4] Holló, B., Rodić, M.V., Bera, O., Jovičić, M., Leovac, V.M., Tomić, Z.D., Mészáros Szécsényi, K., Cation- and/or anion-directed reaction routes. Could the desolvation pattern of isostructural coordination compounds be related to their molecular structure? (2013) Structural Chemistry, 24 (6), pp. 2193–2201; [5] Poręba, R., Špirková, M., Pavličević, J., Budinski-Simendić, J., Mészáros Szécsényi, K., Holló, B., Aliphatic polycarbonate-based polyurethane nanostructured materials. the influence of the composition on thermal stability and degradation (2014) Composites Part B: Engineering, 58, pp. 496–501.

Name: Anna Michnik
Country: Poland
Date and place of birth: 1953, Częstochowa, Poland
Present position and address: Associate Prof., University of Silesia, Department of Medical Physics, A. Chełkowski Institute of Physics, ul. Uniwersytecka 4, 40-007 Katowice, ul. 75 Pułku Piechoty 1A, 41-500 Chorzów, Poland
Email: anna.michnik@us.edu.pl
Education and scientific degrees: M.Sc. in Theoretical Physics, University of Silesia in Katowice (1977); Ph.D. in Physics, Jagiellonian University in Cracow (1988); Habilitation in Physics, University of Silesia in Katowice (2010)
Workplaces: Medical University of Silesia in Katowice, Faculty of Pharmacy, (1977–1996); University of Silesia in Katowice, A. Chełkowski Institute of Physics (1996–)
Main fields of interest: the influence of physical and chemical factors on the stability of globular proteins in aqueous solutions; microcalorimetric studies of thermal denaturation of serum and serum proteins—support for medical diagnostics and sports medicine
Relevant categories in thermal analyses: fields (organic, pharmaceutical, biology, life, proteins, medicine); methods (DSC, kinetics, specific heat, calorimetry, microcalorimetry, biocalorimetry)
Professional activities: Associate Editor in Journal of Thermal Analysis and Calorimetry (Bio-, Life- sciences and Pharmaceuticals); Member of the Organizing Committee of V, VI Symposium on Medical Physics; III, IV International Symposium on Medical Physics, 2006, 2009 and II. Symposium on Medical Physics, 1998, Szczyrk (Poland); Secretary of the Organizing Committee of IV. Symposium on Medical Physics; II. International Symposium on Medical Physics, 2003, Ustron (Poland), Treasurer of the Organizing Committee of III. Symposium on Medical Physics, 2001, Wisla (Poland), Member of the Scientific Committee at the 1st Central and Eastern European Conference on Thermal Analysis and Calorimetry, 2011, Craiova, Romania
Publication record: papers (38), books (3), citations (288), h-index (8)
Equipments: microcalorimeter VP DSC MicroCal Inc., Northampton, MA
5 most important publications: [1] A. Michnik, J Therm Anal Calorim. 2003;71:509–519; [2] A. Michnik, K. Michalik, A. Kluczewska, Z. Drzazga, J Therm Anal Calorim. 2006;84:113–117; [3] A. Michnik, K. Michalik, Z. Drzazga, J Photochem Photobiol. 2008;90:170–178; [4] A. Michnik, Z. Drzazga, A. Kluczewska, K. Michalik, Biophys Chem. 2005;118:93–101; [5] A. Michnik, chapter: "Blood plasma, serum and serum proteins microcalorimetric studies aimed at diagnosis support" in book: Thermal Analysis in Medical Application, Editor: D. Lőrinczy, Akadémiai Kiadó, Budapest; ISBN: 978 963 05 8992 5; 2011.

Name: Chiara Milanese
Country: Italy
Date and place of birth: 1974, Mortara, Italy
Present position and address: Researcher—Scientific Responsible of the Pavia Hydrogen Lab, Chemistry Department, University of Pavia, Viale Taramelli, 16, 27100 Pavia, Italy
Email: chiara.milanese@unipv.it
Website: http://h2lab.unipv.it
Education and scientific degrees: Chemistry Degree at the University of Pavia summa cum laude (1998); post-lauream degree by SAFI (Advanced School for Integrated Formation), a high quality multidisciplinary formation for Ph.D. students activated by the University of Pavia (2001); Ph.D. in Chemistry at the University of Pavia (2002); Researcher at the University of Pavia (2008)
Workplaces: Chemistry Department, Physical Chemistry Section, University of Pavia
Main fields of interest: hydrogen storage, energetics
Relevant categories in thermal analyses: fields (inorganic, materials, nano); methods (TG, DSC, High-Pressure DSC, kinetics, calorimetry, instrument development, manometric techniques, Combined High-Pressure DSC-manometric techniques)
Awards and acknowledgments: "Franzosini Award" assigned by Commission V.8 (Solubility Data Commission) of IUPAC at its 25th Annual Meeting, held in conjunction to the 40th IUPAC General Assembly (Berlin, Germany, 1999); grant by the US Department of Energy and General Motors to attend the Gordon Conference "Hydrogen—Metal Systems" (Stonehill College, Massachusetts, 2011)
Professional activities: Expert of the task 32 "Hydrogen-Based Energy Storage" of the Hydrogen Implementing Agreement (HIA) active in the frame of International Energy Agency (IEA); member of the COST Action MP1103—"Nanostructured Materials for Solid State Hydrogen Storage"
Publication record: papers (93), h-index (16)
Equipments: Gravimetric analyser; Manometric Sievert type apparatus; High Pressure Differential Scanning Calorimeter, working also in combination with the manometric instrument; Thermogravimetric instrument; Differential Scanning Calorimeter; simultaneous TG-DSC instrument
5 most important publications: [1] M. Aramini, C. Milanese, et al., "Addition of transition metals to lithium intercalated fullerides enhances hydrogen storage properties", International Journal of Hydrogen Energy, 39 5 (2014) 2124–2131; [2] C. Milanese, et al., "Thermodynamic and kinetic investigations on pure and doped $NaBH_4$–MgH_2 system", Journal of Physical Chemistry C, 115 7 (2011) 3151–3162; [3] C. Milanese, et al., "Synergetic effect of C (graphite) and Nb_2O_5 on the H_2 sorption properties of the Mg–MgH_2 system", International Journal of Hydrogen Energy, 35 (2010) 9027–9037; [4] C. Milanese, et al.,"Effect Of C (Graphite) Doping On The H_2 Sorption Performance of The Mg–Ni Storage System", International Journal of Hydrogen Energy, 35 3 (2010) 1285–1295; [5] C. Milanese, et al.,"Mg–Ni–Cu mixtures for hydrogen storage: a kinetic study", Intermetallics, 18 (2010) 203–211.

Name: Stefana Milioto
Country: Italy
Date and place of birth: 1960, Haine-Saint-Paul, Belgium
Present position and address: Full Professor of Physical Chemistry at Department of Physics and Chemistry, University of Palermo, viale delle Scienze pad 17. Palermo, Italy
Email: stefana.milioto@unipa.it
Researcher ID: A-3309-2014
Education and scientific degrees: 1989 Ph.D. degree in Chemical Science at the University of Palermo
Workplaces: 1986 Strasbourg (France) at Centre de Recherche sur les Macromolecules, Istituto C. Sadron; 1988 Saskatoon (Canada) at Department of Chemistry, University of Saskatchewan; 1992 Tennessee (USA) at Oak Ridge National Laboratory.
Main fields of interest: physico-chemical characterization of colloidal systems; recently, interest has been addressed to polymer-nanoparticle interactions
Relevant categories in thermal analyses: fields (materials, nano, polymer, clay, cyclodextrin, cultural heritage); methods (TG, DSC, specific heat, microcalorimetry, DMA)
Awards and acknowledgments: 1988. A. Lucci Award, Italian Association of Calorimetry and Thermal Analysis; 1989. W. F. Giauque Award, International Calorimetry Conference; 1989. Italian Chemical Society, Sicily Section
Professional activities: 1992–1996 member of the scientific committee of the Italian Association of Calorimetry and Thermal Analysis; 2009–2011 member of the scientific committee of the Italian Council of Physical Chemistry; organizer and chair of the International Congress MEDICTA 2007 held in Palermo (Italy), 25–29 September 2007; 2005–2010 Head of Department of Physical Chemistry at the University of Palermo; 2012- Coordinator of Master Course in Conservation and Restoration of Cultural Heritage at the University of Palermo
Publication record: papers (106), books (5), citations (1800), h-index (26)
Equipments: TGA Q5000 IR (TA Instruments), nano-ITC200 (MicroCal), micro-DSC III (SETARAM), DSC 2920 CE (TA Instrument), DMA Q800 (TA Instruments)
5 most important publications: [1] De Lisi, R.; Fisicaro, E.; Milioto, S. J. Solution Chem 1988, 17, 1015–1041; [2] De Lisi, R.; Inglese, A.; Milioto, S.; Pellerito, A. Langmuir 1997, 13, 192–202; [3] De Lisi, R.; Lazzara, G.; Milioto, S.; Muratore, N. J. Phys. Chem. B 2003, 107, 13150–13157; [4] Cavallaro, G.; Lazzara, G.; Milioto, S. Langmuir 2011, 27, 1158–1167; [5] Cavallaro, G.; Donato, D. I.; Lazzara, G.; Milioto, S. J. Therm. Anal. Calorim. 2011, 104, 451–457.

Name: Dragica M. Minić
Country: Serbia
Date and place of birth: 1947, Brus, Serbia
Present position: Professor at University of Belgrade, Faculty for Physical Chemistry, Belgrade, Serbia
Email: dminic@ffh.bg.ac.rs
Website: www.dragicaminic.info
ORCID: 0000-0001-5055-2039
Education and scientific degrees: B.Sc. in Physical Chemistry, Faculty of Science and Mathematics, University of Belgrade, Yugoslavia (1970), M.Sc. in Physical Chemistry, Faculty of Science and Mathematics, University of Belgrade, Yugoslavia (1974), Ph.D. in Physical Chemistry, Faculty of Science and Mathematics, University of Belgrade, Yugoslavia (1983)
Workplaces: Department of Physical Chemistry, University of Belgrade, Belgrade, Serbia (1977–2014), College of Agriculture, University of Belgrade, Zemun, Yugoslavia (1973–1977), College of Mining and Metallurgy, University of Belgrade, Bor, Yugoslavia (1970–1973)
Main fields of interest: thermodynamics and kinetics of reactions in solid phase; correlation of structural and functional properties; thermal stability and kinetics of thermal degradation of organometallic compounds; hydrogen energy.
Relevant categories in thermal analyses: fields (thermal stability, mechanism, thermodynamics and kinetics of solid state reactions); methods (DTA, DSC, TG)
Publication record: papers (142), books (15), patents (5), citations (592), h-index (13), sum of impact factors (142.86)
5 most important publications: [1] Vladimir A. Blagojević, Milica Vasić, Bohumil David, Dušan M. Minić, Naděžda Pizúrová, Tomáš Žák, Dragica M. Minić, Thermally Induced Crystallization of Fe73.5Cu1Nb3Si15.5B7 Amorphous Alloy, Intermetalics, 45 (2014) 53–59; [2] Vladimir A. Blagojević, Dušan M. Minić, Milica Vasić, Dragica M. Minić, Effect of Thermal Treatment on Structural and Physical Properties of Fe89.8Ni1.5Si5.2B3C0.5 Amorphous Alloy, Mater. Phys.Chem, 142 (1) (2013) 207–212; [3] Vladimir A. Blagojević, Milica Vasić, Dušan M. Minić, Dragica M. Minić, Kinetics and thermodynamics of thermally induced structural transformations of amorphous Fe75Ni2Si8B13C2 alloy, Thermochimica Acta, 549 (2012) 35–41; [4] Maja Šumar Ristović, Maja Gruden Pavlović, Matija Zlatar, Vladimir Blagojević, Katarina Anđelković, Dejan Poleti, Dragica M. Minić, Kinetics, mechanism and DFT calculations of thermal degradation of Zn(II) complex with N-benzyloxycarbonylglycinato ligand, Monatshefte für Chemie, 143 (8) (2012) 1133–1139; [5] Dragica M. Minić, Vladimir. A. Blagojević, Dušan.M. Minić, Aleksandra. Gavrilović, Lidija Rafailovic, Tomas. Žak, Influence of microstructure on microhardness of Fe81Si4B13C2 amorphous alloy after thermal treatment, Metallurgical Materials Transactions A, 42A (2011) 4106–4112.

Name: Ratikanta Mishra
Country: India
Date of birth: 1968
Present position: Scientific Officer (G), Chemistry Division, Trombay, Mumbai-400085, India
Email: mishrar@barc.gov.in
Education and scientific degrees: M.Sc., Ph.D.
Workplaces: Chemistry Division, Trombay, Mumbai-400085, India
Main fields of interest: chemical thermodynamics of alloys, oxides, phase diagram, calorimetry and vapor pressure measurement
Relevant categories in thermal analyses: fields (phase diagram delineation, vapor pressure measurement and calorimetry); methods (TG-DTA, DSC, room temperature solution calorimetry, HT-calorimetry, Knudsen effusion mass spetrometry, transiration technique)
Awards and acknowledgments: TA-ITAS-Young Scientist Award for Excellence in Thermal Analysis, Alexander-von-Humboldt Fellowship, Marie Curie International Incoming Fellowship
Professional activities: Executive member of Indian Thermal Analysis Society, Member of several Professional bodies like Indian Science Academy, Indian Nuclear Society, Indian thermal Society, Odisha Academy of Science
Publication record: papers (99), books (1), citations (>700), h-index (10), sum of impact factors (3)
Equipments: TG-DTA (Setram Setsys and 92-16.18), DSC (Mettler & Toledo), Knudsen effusion Mass loss (Setaram Micro-balance), Calvet Calorimter, Isoperibol solution calorimeter, Kundsen effusion mass spectrometer
5 most important publications: [1] Ales Kroupa, Ratikanta Mishra, Divakar Rajamohan, Hans Flandorfer, Andrew Watson, Herbert Ipser, Phase Equilibria in the Ternary Ni-Sb-Sn System: Experiments and Calculations, CALPHAD: Computer Coupling of Phase Diagrams and Thermochemistry 45 (2014) 151–166; [2] R. Mishra, A. Zemanova, A. Kroupa, H. Flandorfer, H. Ipser, Synthesis and characterization of Sn-rich Ni–Sb–Sn nanosolders, Journal of Alloys and Compounds 513 (2012) 224–229; [3] M. Basu (Ali), R. Mishra, S.R. Bharadwaj and D. Das, Review on Thermodynamic and transport properties of thoria-urania fuel of Advanced Heavy Water Reactor Journal of Nuclear Materials 403 (2010) 204–215; [4] R. Mishra, W Hermes, Ute Ch. Rodewald, R-D Hoffmann, R. Pöttgen, Trivalent-Intermediate Valent Cerium Ordering in Ce2RuZn4., Z. Anorg. Allg. Chem. 634(2008) 470–474; [5] R. Mishra, M. Basu, S.R. Bharadwaj, A.S. Kerkar, S.R. Dharwadkar, D. Das, Thermodynamic stability of barium thorate, BaThO3 from Knudsen effusion study, J. Alloys and Compounds. 290 (1999) 97.

Name: Subhash C. Mojumdar
Country: Canada
Date and place of birth: 1971, Feni, Bangladesh
Present position and address: Special Graduate Faculty, University of Guelph, 177 Springfield Place, Waterloo, ON, N2T2M6 Canada
Email: scmojumdar@yahoo.com
Education and scientific degrees: Chemical Engineer, Slovak University of Technologies, Bratislava, SR (1990–1995), Ph.D. (1999)
Workplaces: University of Waterloo, Canada (2007–2008), University of Toronto, Canada (2006–2009), National Research Council of Canada, Ottawa (2004–2006), Mount Alison University, Canada (2002–2004), Slovak Academy of Sciences, Bratislava (1998–2002)
Main fields of interest: thermal analysis and calorimetry, spectroscopy, X-ray, chromatography, nanotechnology, construction materials, organometallics, biotechnology
Relevant categories in thermal analyses: fields (inorganic, materials, nano, minerals, complex, organic, pharmaceutical, polymer, food, biology, life, glass, ceramics, cement, antibiotics, antioxidants); methods (TG, DTA, EGA, DSC, thermomechanical analysis, kinetics, specific heat, calorimetry, microcalorimetry, instrument development, DMA, thermal conductivity, light beam thermal analysis)
Professional activities: Past President, Canadian Thermal Analysis Society (CTAS), Local Sections Chair: North American Thermal Analysis Society (NATAS), Executive Council Member: Association of the Chemical Profession of Ontario (ACPO), Member of the Editorial Board of Journal of Thermal Analysis and Calorimetry, Research Journal of Chemistry and Environment, Advances in Bioscience and Biotechnology, Journal of Environmental Protection, Global Journal of Analytical Chemistry, International Journal of Chemistry and Chartered Chemist News (Editor-in-Chief), Program Chair: Annual Workshops and Exhibitions of CTAS, Session Chair: Annual Conferences of NATAS and Canadian Society of Chemistry (CSC)
Publication record: papers (401), citations (2005)
Equipments: Simultaneous TG-DTG-DTA (T.A.I. SDT 2960), simultaneous TG-DTG-DTA-DSC (T.A.I. SDT Q600), TG-DTG-DTA (T.A.I. Q500), DSC (T.A.I. Q1000), DSC (T.A.I. Q2000), DSC (Setaram C 80), DMA (T.A.I. DMA Q800), DMA V (Rheometric Scientific), DSC-Dupont 910
5 most important publications: [1] S. C. Mojumdar, M. Melník and E. Jóna, J. Anal. Appl. Pyrol., 46 (1998) 147; [2] S. C. Mojumdar, M. Melník and E. Jóna, Thermochim. Acta, 352 (2000) 129; [3] S. C. Mojumdar, J. Therm. Anal. Calorim., 64 (2001) 1133; [4] S. C. Mojumdar, J. Therm. Anal. Calorim., 64 (2001) 629; [5] S. C. Mojumdar, D. A. Becker, G. A. DiLabio, J. J. Ley, L. R. C. Barclay and K. U. Ingold, J. Org. Chem., 69 (2004) 2929.

Name: Gurli Mogensen
Country: Denmark
Date and place of birth: 1949, Copenhagen, Denmark
Present position and address: Research Scientist, Haldor Topsoe, Nymollevej 91, DK-2800 Lyngby, Denmark
Email: gum@topsoe.dk
Website:www.topsoe.com
Education and scientific degrees: M.Sc. (Chem. Eng.) (1972), Ph.D. (Ceramics) (1975)
Workplaces: (1972–1975) DTU (Technical University of Denmark), (1975–1989) Rockwool International A/S, Denmark, (1989–) Haldor Topsoe, Denmark
Main fields of interest: materials science, catalysis
Relevant categories in thermal analyses: fields (materials research, catalyst preparation, reduction, decomposition, sintering); methods (TG-DSC-MS, dilatometry)
Professional activities: Vice Chairman of the 12th ICTAC, 2000, in Copenhagen, President of NoSTAC (2000–)
Publication record: patents (6), internal reports in R&D projects (>150)
Equipments: Netzsch, Mettler
Most important publication: Mogensen, Mogens Bjerg; Lindegaard, Thomas; Hansen, Uffe Rud; Mogensen, Gurli., Physical Properties of Mixed Conductor Solid Oxide Fuel Cell Anodes of Doped CeO2., Electrochemical Society. Journal, Vol. 141, No. 8, 1994, pp. 2122–2128.

Name: Elisabeth Ermel Da Costa Monteiro
Country: Brazil
Date and place of birth: 1940, São Paulo, Brazil
Present position and address: Guest Professor, Macromolecules Institute, Professor Eloisa Mano, Rio de Janeiro Federal University Av. Horácio Macedo 2030, Bl. J, Technology Center—UFRJ, Rio de Janeiro, RJ, Brazil, CEP 21941-598
Email: ecmonteiro@ima.ufrj.br, elisermel@yahoo.com.br
Website: http://www.ima.ufrj.br
Education and scientific degrees: Chemical Engineer, Rio de Janeiro Federal University (UFRJ), Brazil (1958–1962), D.Sc. at Chemistry Institute/UFRJ (1981), Adjunct Professor UFRJ (1981–1999), Guest Professor at IMA/UFRJ (1999–), Brazilian National Council (CNPq) Researcher (1982–2010)
Workplace: Rio de Janeiro Federal University at Macromolecules Institute Professor Eloisa Mano
Main fields of interest: spectroscopy, thermal analysis, polymer characterization
Relevant categories in thermal analyses: fields (natural and synthetic polymers, organic materials, composite materials, nanocomposites, clays and organoclays, pozzolans, industrial solid wastes and sludges solidified and stabilized wastes); methods (TG, DTG, DTA, DSC, thermomechanical analysis, kinetics, specific heat, calorimetry)
Professional activities: Research Group Leader 1988-IMA/UFRJ, (1997–). Advisor teacher accredited by the Brazilian National Council (CNPq) (1998); Professor of Graduate Course in Science and Technology of Polymers at IMA/UFRJ (1981–) teaching classes of spectrometry applied to polymers, physical chemistry of polymers in solution, physical methods applied to polymers; Committee member of the J. Therm. Anal. Calorim. (JTAC) (1997–2001); Committee member of Polymer: science and technology (1991–); M.Sc and Ph.D. and Post Doct. Advisor (1981–)
Publication record: papers (52), books (1), patents (1), citations (150), h-index (7)
5 most importqant publications: [1] Thaumaturgo, C.; Monteiro, E.E.C., Thermal Stability and Miscibility in PVC/EVA Blends. J. Therm. Anal., 49 (1997), 247–254; [2] Thaumaturgo, C.; Monteiro, E.E.C., Thermal Transitions and Polymer/Polymer Miscibility.II. PVC/EVA Miscible Systems. J. Therm Anal., 49 (1997) 235–245; [3] Thaumaturgo, C.; Monteiro, E.E. C., Thermal Transitions and Polymer/Polymer Miscibility.I. PVC/EVA Immiscible Systems. J. Therm Anal., 49 (1997) 227–234; [4] Monteiro, E. E. C.; Mansur C.R.E. Characterization of methacrylic polymers by calorimetry and infrared analyses. J. Appl. Polym. Sci., 68 (1998) 345–354; [5] Mansur C.R.E., Tavares, M.I.B.; Monteiro, E.E.C., Thermal analysis and NMR studies of methyl methacrylate (MMA)-methacrylic acid by an unusual polymerization of MMA. J. Appl. Polym. Sci., 75 (2000) 495–507.

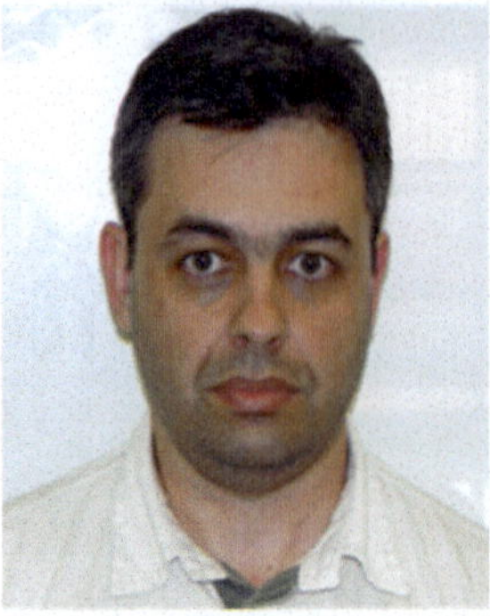

Name: Leandro Cardoso de Morais
Country: Brazil
Date and place of birth: 1974, São Paulo, Brazil
Present position and address: Doctor Professor, Environmental Engineering Department, São Paulo State University —UNESP, Avenue 3 de Março, 511, Sorocaba, SP, Brazil, Zip Code 18087-180
Email: leandro@sorocaba.unesp.br, leandrocmorais@yahoo.com.br
Education and scientific degrees: Chemical Engineer, Faculty of Industrial Engineering (FEI), Brazil (1992–1997), M.Sc. Eng. in Chem. Eng. at EPUSP/USP (2001–2003), and Dr. Eng. in Chem. Eng. at EPUSP/USP (2003–2006) all in São Paulo University, Brazil (USP).
Workplace: São Paulo State University: Environmental Engineering Department.
Main fields of interest: thermal analysis applied to industrial, agricultural and forests produtcts and to their effluent and waste thermal characterization and processing
Relevant categories in thermal analyses: fields (inorganic and organic materials, industrial products, coal, biochar, biomass, bioenergy, biofuel, syngas, industrial solid wastes and sludges, lactose, agricultural and forest waste, wood, bio-oil, pyrolysis and combustion); methods (TG, DTG, DTA, kinetics, specific heat, calorimetry, conventional TG and DTA instrument)
Professional activities: Referee of the Journal of Hazardous Materials (2009–), Journal of Oceanography and Marine Science (2010–), Environmental Engineering Science (2010–), Acta Amazonica (2011–) and Journal of Polymer Engineering (2013–)
Publication record: papers (87), books (3), book chapters (4), patents (1), citations (60), h-index (4)
5 most important publications: [1] Morais, L.C.; Dweck, J.; Diaz, F.R.V.; Buchler, P. M. Characterization of lactose and derived products in dairy industry effluents processing. J. Therm. Anal. Calorim., 82 (2005) 315–18; [2] Morais, L.C.; Dweck, J.; Gonçalves, E. M.; Buchler, P. M. An experimental study of sewage sludge incineration. Environ. Technol., 26 (2006) 1047–51; [3] Dweck, J.; Morais, L.C.; Fonseca, M.V.A.; Campos, V.; Buchler, P.M. Calcined sludge sintering evaluation by heating microscopy thermal analysis. J. Therm. Anal. Calorim., 95 (2009) 985–89; [4] Morais, L.C.; Dweck, J.; Campos, V.; Rosa, A.H.; Fraceto, L.F.; Buchler, P. M. Combustion and Pyrolysis of a Sludge from Wastewater Treatment Plant. Mater. Sci. For., 660 (2010) 1009–14; [5] Viana, M.M.; Melchert, M.B.M.; Morais, L.C.; Buchler, P.M.; Dweck, J. Sewage sludge coke estimation using thermal analysis. J. Therm. Anal. Calorim., 106 (2011) 437–43.

Name: Cheila Gonçalves Mothé
Country: Brazil
Date and place of birth: 1952, Rio de Janeiro, Brazil
Present position and address: Titular Professor of the School of Chemistry, Federal University of Rio de Janeiro. Av. Horacio Macedo, 2030, Escola de Química/Sala E-204, Ilha do Fundão, Rio de Janeiro, Brazil
Email: cheila@eq.ufrj.br
Website: www.eq.ufrj.br/docentes
Workplace: Federal University of Rio de Janeiro, Rio de Janeiro, Brazil
Education and scientific degrees: Chemical Engineer (1975); Professor (1976); M.Sc. (1979); D.Sc. (1992) University of São Paulo/University of the Air—Japan; Postdoctoral (1998) Cornell University/USA; Postdoctoral (2003) Cleveland State University/USA; Titular Professor (2005–)
Main fields of interest: technological characterization by thermal analysis; polymer composites with natural fibers; recovery of natural products; kinetic study by thermal analysis; rheological behavior of polysaccharides; recycling materials; energetic potential associated with residues from biomass; energy recovery and treatment of industrial waste; thermal characterization and application of asphalt, biodiesel and heavy oil
Relevant categories in thermal analyses: fields (polymer and composite, organic, biomass, biofuel, nano, pharmaceutical and cosmetics, food, ceramics, crude oil and asphalt binder); methods (TG/DTG, DTA, DSC, DMA, EGA, kinetics, extremely high temperature (above 1,000 °C); specific heat, calorimetry, microcalorimetry)
Awards and acknowledgments: Prêmio Metanol/Copenor de Química (1986); Honorable Mention (1989, 1994, 1996, 1997, 2000, 2002); Chemical Award of the Year and Medal (2006) by the CRQ-Brazil; Honorary Member of the Group of the Hungarian Chemical Thermoanalytical Society, Budapest, Hungary (2010); 1st place XI Award Abrafati-Petrobras Science in Paint (2008); Leopoldo Hartmnn Award (2010); Prêmio Oscar Niemeyer de Trabalhos Científicos e Tecnológicos- CREA-RJ (2013)
Professional activities: Regional Editor of J. Therm. Anal. Calorim.; NATAS Member and Volunteer (2001–); CNPq ad hoc consulter for 10 years; Fellow of the African Scientific Institute (2011–); President of the Brazilian Society of Science and Technology RJ Food Regional (2007–); President of the Brazilian Association of Rheology (2010–); Vice-President of Brazilian Association of Thermal Analysis and Calorimetry (2008–). Coordinator of three laboratories: Natural and Synthetic Polymers Technology, Rheology Laboratory, Thermal Analysis Laboratory at UFRJ/Brazil
Publication record: papers (108), books (10), patents (7), citations (589)
Equipments: SDT Q600 (TA Instruments); DSC 1 (Mettler Toledo); STA 409P6/4 16 Luxx (NETZSCH); SDT 2960 (TA Instruments); DSC 2010 (TA Instruments)
5 most important publications: [1] Mothé, C.G., Azevedo, A.D., "Thermal Analysis of Materials—Análise Térmica de Materiais", 2^a ed., editora Art liber Ltda, 324 pp, ISBN 978-85-88098-49-7, (2009); [2] Mothé, C. G., Correia, D. Z., França, F. P.: J. Therm. Anal. Calorim., v. 85, pp. 31–36, (2006); [3] Mothé, C. G., Araújo, C. R., Thermochim. Acta, v. 357–358, pp. 321–325, (2000); [4] Mothé, C. G., Rao, M. A., Food Hydrocolloids, v. 13, pp. 501–506, (1999); [5] Mothé, M.G., Mothé, C.G., Carvalho, C.H.M., Oliveira, M.C.K., J. Therm. Anal. Calorim., doi:10.1007/s10973-014-3891-3, (2014).

Name: Dionysios E. Mouzakis
Country: Greece
Date and place of birth: 1971, Patras, Greece
Present position and address: Vice-Chair, Assistant Professor, Department of Mechanical Engineering, Technological Educational Institute of Larisa, GR 41110, TEI of Thessaly-Larisa, Greece.
Email: mouzakis@upatras.gr;mouzakis@teilar.gr
Website: http://scholar.google.gr/citations?sortby=pubdate&hl=el&user=ejfsdrsAAAAJ&view_op=list_works
ORCID: 0000-0001-5842-5202
Education and scientific degrees: Diploma in Mechanical Engineering, Department of Aeronautics and Mechanical Engineering, University of Patras, Greece, Ph.D. in Engineering, TU Kaiserslautern, Germany (Doktor des Ingenieurwisseschaftens)
Workplaces: University of Kaiserslautern, University of Patras, Technological Educational Institute of Larisa, Hellenic Military Academy.
Main fields of interest: polymers, composites, nanocomposites, biomaterials
Relevant categories in thermal analyses: fields (polymers, biomaterials, composites, nanocomposites, sensors); methods (DMA, DSC, MDSC)
Publication record: papers (42), books (1), book chapters (4), citations (>500), h-index (14)
Equipments: Q800 DMA, Q200 m-DSC (TA Instruments)
5 most important publications: [1] DE Mouzakis, DG Dimogianopoulos, S Zaoutsos, Damage assessment of carbon fiber reinforced composites under accelerated aging and validation via stochastic model-based analysis, International Journal of Damage Mechanics, 1056789513508798; [2] DE Mouzakis, GC Papanicolaou, K Giannadakis, I Zuburtikudis, On the Toughness Response of iPP and sPP/MWNT Nanocomposites, Strain 49 (4), 348–353; [3] DE Mouzakis, TD Papadopoulos, GL Polyzois, PG Griniari, Dynamic mechanical properties of a maxillofacial silicone elastomer incorporating a ZnO additive: the effect of artificial aging, Journal of Craniofacial Surgery 21 (6), 1867–1871; [4] N Bouropoulos, A Stampolakis, DE Mouzakis, Dynamic Mechanical Properties of Calcium Alginate-Hydroxyapatite Nanocomposite Hydrogels, Science of Advanced Materials 2 (2), 239–242; [5] DE Mouzakis, N Bouropoulos, P Kallidonis, D Karnabatidis, K Katsanos, Viscoelastic property mapping along encrusted polymeric urinary catheters, Journal of Endourology 22 (8), 1761–1770.

Name: Jules Mullens
Country: Belgium
Date and place of birth: 1944, Hoeselt, Belgium
Present position and address: Universiteit Hasselt, Agoralaan, Gebouw D, BE 3590 Diepenbeek
Email: jules.mullens@uhasselt.be
Website: www.uhasselt.be
Education and scientific degrees: M.Sc (1967 Cath. Univ. Leuven); Aggr. Sec. Education (1968 Cath. Univ. Leuven); Ph.D. (1971 Cath. Univ. Leuven); Aggr. Univ. Education (1985 Limb. Univ. Cent)
Workplaces: Catholic University Leuven (1971–1972); Hasselt University (1972–)
Main fields of interest: characterization and identification of the properties of a variety of materials (inorganic, polymers, waste products)
Relevant categories in thermal analyses: Coupling of thermal analysis with gas analysis: TGA-MS; TGA-FTIR.
Professional activities: Full Professor Limburg University Center/Hasselt University; Head of Chemistry Department 1986–1990; Dean of Faculty of Science 1992–1997
Publication record: papers (>200), books (3)
5 most important publications: [1] J. Mullens and L.C. Van Poucke: Investigation of the stability of the minimum in a minimization program for the simultaneous determination of K and ΔH by calorimetric measurements, Journal of Physical Chemistry 89 (1985) 294; [2] J. Mullens, A. Vos, R. Carleer and L.C. Van Poucke: The decomposition of copper oxalate to metallic copper is well suited for checking the inert working conditions of TA equipment, Thermochimica Acta, 207 (1992) 337; [3] K. Van Werde, G. Vanhoyland, D. Nelis, D. Mondelaers, M.K. Van Bael, J. Mullens and L. C. Van Poucke: Phase formation of ferroelectric perovskite 0.75 $Pb(Zn_{1/3}, Nb_{2/3})O_3$–0.25 $BaTiO_3$ prepared by aqueous solution-gel chemistry, Journal of Materials Chemistry 11 (2001) 1192; [4] I. Truijen, A. Hardy, M.K. Van Bael, H. Van den Rul, J. Mullens: Study of the decomposition of aqueous citratoperoxo-Ti(IV)-gel precursors for titania by means of TGA-MS and FTIR, Thermochimica Acta 456, 1 (2007) 38; [5] De Dobbelaere, Christopher; Hardy, An; D'Haen, Jan; Van den Rul, Heidi; Van Bael, Marlies and Mullens, Jules: Morphology of water-based chemical solution deposition (CSD) lead titanate films on different substrates: Towards island formation, Journal of the European Ceramic Society, 29(9).(2009) 1703.
Books:[1] J. Mullens: Evolved Gas Analysis Chapter 12 in Handbook of Thermal Analysis and Calorimetry, Principles and Practice, Elsevier Science, Ed. M. Brown, P. Gallagher, (1998). [2] H. Van den Rul, M.K. Van Bael, A. Hardy, K. Van Werde, J. Mullens: Aqueous solution based synthesis of nanostructured metal oxides review chapter in Handbook of Nanoceramics and Their Based Nanodevice, Ed. H.S. Nalwa, American Scientific Publishers (2008). [3] Marlies K. Van Bael, An Hardy, and Jules Mullens: Aqueous Precursor Systems, review chapter 5 in Chemical Solution Deposition of Functional Oxide Thin Films, T. Schneller et al. (eds.), Springer-Verlag Wien (2014).

Name: Viorica Musat
Country: Romania
Date and place of birth: 1952, Bîrlad, Romania
Present position and address: Professor at Dunărea de Jos University of Galați, vice-Rector in charge of the Ph.D. activities in UDJG, 47 Domneasca street, 800008 Galati, Romania
Email: viorica.musat@ugal.ro
Website: www.cnmf.ugal.ro
Researcher ID: C-7198-2014
Education and scientific degrees: Chemist/Physical chemistry- Polytechnical Institute of Bucharest (1977), Ph. D. (1997), Full Professor (2002–)
Workplaces: Dunărea de Jos University of Galați
Main fields of interest: nanotechnologies and nanomaterials; solution-based preparation (precipitation, sol-gel, hydrothermal/solvothermal, chemical bath deposition) of magnetic oxides (spinel ferrites), semiconductive oxides (ZnO, CuO, TiO2) and hubride oxide-organics materials (nanoparticles, nanowires, thin films; multifunctional nanomaterials for optoelectronics, transparent/flexible electronics, protective coatings, photocatalytic and antimicrobial applications
Relevant categories in thermal analyses: fields (thermal behaviour based on TGA, DTA, DSC, FTIR and XRD measurements of oxides based materials and on the non-isothermal kinetic investigation of these materials, using isoconversional methods (Friedman (FR), Flynn-Wall-Ozawa (FWO) and Kissinger-Akahira-Sunose (KAS)) and the invariant kinetic parameter (IKP) method); methods (modulated thermogravimetry, modulated DSC)
Professional activities: Member of the National Council for the Recognition of Degrees, Diplomas and Certificates (CNATDCU).; Member of the National Council of Scientific Research (CNCS)—Materials Science.; Director of Doctoral Studies University "Dunarea de Jos" University of Galati; Director Research Centre Nanostructures and Functional Materials —NCCS, University Dunarea de Jos Galati, CNCSIS accredited in 2006, www.cnmf.ugal.ro; Director of Master Program Nanotechnology and Multifunctional Materials; Co-organizer of the International Symposium European Sociery of Materials E- MRS K entitled: Solution-Derived Electronic- Oxide Films, Nanostructures and patterning: from Materials to Devices, Warsaw, September 2011; Organizer of Symposium UgalNano-1/2010, UgalNano2/2011, UgalNano2/2013.; Co-Organizer of Romphyschem 1988, Galati, 23–25 Sept. 1998
Publication record: papers (85), books (4), book chapters (3), patents (1), citations (380), h-index (11), sum of impact factors (50.30)
Equipments: modulated Q5000IR Thermogravimetry balance, modulated Differential Scanning Calorimeter Q20DSC Tzero, TA Instruments
5 most important publications in thermal anaysis: [1] V. Muşat, P. Budrugeac and C. Gheorghies, J Therm Anal Calorim 94 (2008) 373–377; [2] V. Musat, P. Budrugeac, R. C. C. Monteiro, E. Fortunato, E. Segal, J Therm Anal Calorim 89(2) (2007) 505–509; [3] Budrugeac, V. Musat and E. Segal, J Therm Anal Calorim 88/3 (2007) 699–702; [4] V. Musat, L. Frangu, E. Segal, J Therm Anal Calorim 68 (2002) 787–801, ISSN 1388-6150; [5] V. Musat, E. Segal, Solid State Sci 4 (2001) 407–415, ISSN 1293-2558.

Name: Adina Magdalena Musuc
Country: Romania
Date and place of birth: 1977, Alexandria, Romania
Present position and address: Ph.D. Senior Researcher 3rd degree at "Ilie Murgulescu" Institute of Physical Chemistry of Romanian Academy, 202 Spl. Independentei, 060021 Bucharest, Romania
Email: amusuc@icf.ro; musucadina@yahoo.com
Website: www.icf.ro
Researcher ID: C-2485-2011
Education and scientific degrees: BSc in Physical Chemistry (1999); M.Sc. in Environmental Quality Control (2001); Ph.D. in Chemistry (2007); Assistant Researcher (1999–2002), Scientific researcher (2002–2008); Senior researcher 3rd degree (2008–)
Workplaces: "Ilie Murgulescu" Institute of Physical Chemistry of Romanian Academy (1999–)
Main fields of interest: thermal analysis and calorimetry; chemical kinetics and theory of runaway reactions, chemical kinetics in homogeneous and heterogeneous systems; combustion and flames of gases
Relevant categories in thermal analyses: fields (inorganic, materials, nano, minerals, complex, organic, pharmaceutical, polymer, food, ceramics, composites); methods (TG, DTA, DSC, DIL, STA, kinetics, specific heat, calorimetry, microcalorimetry, instrument development—isoperibolic calorimeter)
Professional activities: Member of the CATCAR (Commission of Thermal Analysis and Calorimetry of the Romanian Academy), member of the Romanian Chemical Society, member of the Romanian Catalysis Society, vice-president of the Romanian Chemical Society- Bucharest 3, member in the conference secretariat of International Conference of Physical Chemistry—RomPhysChem 15, Alternate Member of Scientific council of "Ilie Murgulescu" Institute of Physical Chemistry. reviewer at Journal of Thermal Analysis and Calorimetry and Journal of Material Research
Publication record: papers (30), books (1), patents (1), citations (75), h-index (5), sum of impact factors (32.108)
Equipments: Mettler Toledo TGA/sDTA 851^{e} thermal analyzer, DSC 823^{e} Mettler Toledo calorimeter, Setaram MicroDSC VII Evo calorimeter, Netzch STA 449 F1 Jupiter Simultaneous Thermal Analyzer
5 most important publications: [1] A.M. Musuc, D. Razus, D. Oancea, Thermochim. Acta 448 (2) (2006)130; [2] J. Pandele Cusu, A.M. Musuc, M. Matache, D. Oancea, J. Therm. Anal. Calorim. 110(3) (2012) 1259; [3] S. Preda, V. Teodorescu, A. Musuc, C. Andronescu, M. Zaharescu, J. Mater. Res. 28(3) (2013) 294; [4] A.M. Musuc, M. Badea-Doni, L. Jecu, A. Rusu, V.T. Popa, J. Therm. Anal. Calorim. 114 (1) (2013) 169; [5] G. Patrinoiu, J. M. Calderon-Moreno, S. Somacescu, N. Spataru, A. M. Musuc, R. Ene, R. Birjega, O. Carp, Eur. J. Inorg. Chem. (6) (2014), 1103–1115.

Name: Maxim S. Nazarov
Country: Russia
Date and place of birth: 1981, Belgorod, Russia
Present position and address: Researcher, Russian Federation, 141313, Sergiev Posad, Moscow Region, Academic Silin street, 3
Email: nazarovmaksi@yandex.ru
Website: www.niiph.ru
Education and scientific degrees: Belgorod National Research University (2003, honours degree)
Workplaces: Applied Chemistry R&D Institute JSC, Russian Federation, Sergiev Posad
Main fields of interest: leading specialist by chemical compatibility of the components of pyrotechnic compositions, physical and chemical stability of the pyrotechnic compositions, thermal analysis and calorimetry pyrotechnic compositions and components, including effect of influence relative humidity, thermal ageing, heat and humidity ageing
Relevant categories in thermal analyses: fields (pyrotechnic compositions and components, high-energy materials, inorganic and construction materials, nano-sized powders and ultra-fine inorganic oxidants and metal fuels); methods (TG, DTA, DSC, STA, kinetics, extremely high temperature (1,500 °C); Calvet isothermal microcalorimetry, Calvet-DSC)
Awards and acknowledgments: NETZSCH (2007) for scientific and technical achievements in the research of high temperature processes by thermal analysis, presidential scholarship (2013)
Publication record: papers (16)
Equipments: simultaneous thermal analysis instrument STA 449 F3 Jupiter NETZSCH, simultaneous TGA/Calvet-DSC instrument TG-DSC111 SETARAM, derivatograph Q-1500D, isothermal Calvet microcalorimeter DAC-1-7 M (Russia)
3 most important publications: [1] Nazarov M.S., Mozgunov N.E., Emelianov M.V. Study of chemical compatibility and physicochemical stability of pyrotechnic compositions by Calvet isothermal/differential scanning calorimetry and simultaneous thermal analysis., Abstracts of the XIV International Conference on Thermal Analysis and Calorimetry in Russia (RTAC-2013), 23–28 September, 2013 Saint-Petersburg, Russia. pp. 366–369; [2] Nazarov M.S., Mozgunov N.E. Hardware-software complex «Microcalorimeter DAC-1-7M» research of chemical compatibility of pyrotechnic compositions components., Abstracts of the XVII International Conference on Chemical Thermodynamics in Russia (RCCT-2009), June 29–July 3, Kazan, Russia, Volume I. p. 292; [3] Nazarov M.S., Mozgunov N.E., Fedin V.I., Konuhof B.N., Sarabev V.I. Research ultradisperse high energy substances., Abstracts of the XVI International Conference on Chemical Thermodynamics in Russia (RCCT-2007), 1-6 July, Suzdal, Russia, Volume II. p. 435.

Name: Alex Neves Junior
Country: Brazil
Date and place of birth: 1980, Cuiabá, Brazil
Present position and address: Adjunct Professor, Federal University of Mato Grosso; Rua Los Angeles n 775 Bairro Jardim Califórnia, Cuiabá-MT
Email: alexnevesjr@hotmail.com, alexnevesjr@gmail.com
Education and scientific degrees: Civil Engineer, Federal University of Mato Grosso (UFMT), Brazil (2002), M.Sc. at São Paulo University, Brazil (USP) (2006), and Ph.D. at Federal University of Mato Grosso, Brazil (COPPE-UFRJ) (2014).
Workplace: Mato Grosso Federal University: College of Architecture, Engineering and Technology.
Main fields of interest: thermal analysis applied to cementitious materials and cement chemistry.
Relevant categories in thermal analyses: fields (cements, pozzolans, natural fibres, composite materials, CO2 capture in cementitious materials, hydration reactions and pozzolanic activity); methods (TG, DTG, calorimetry, conventional and non-conventional TG and DTA instrument and system development)
Professional activities: Guest reviewer of the J. Therm. Anal. Calorim. (JTAC)
Publication record: papers (8), citations (8)
4 most important publications: [1] Neves Junior, Alex; Lemos, M.S; Toledo Filho, R.D; Dweck, J.; Fairbairn, E. M. R.; Early stages hydration of high initial strength Portland cement part II: NCDTA and Vicat analysis. J. Therm. Anal. Calorim., v. 113, pp. 659–665, 2013; [2] Neves Junior, Alex; Toledo Filho, R.D; Fairbairn, E.M.R; Dweck, J; CO2 sequestration by high initial strength Portland cement pastes. J. Therm. Anal. Calorim., v. 113, pp. 1577–1584, 2013; [3] Neves Junior, Alex; Toledo Filho, R.D; Fairbairn, E.M.R; Dweck, J.; Early stages hydration of high initial strength Portland cement Part I. thermogravimetric analysis on calcined mass basis. J. Therm. Anal. Calorim., v. 108, pp. 725–731, 2012; [4] Neves Junior, Alex; Toledo Filho, R.D; Fairbairn, E.M.R; Dweck, J.; A study of the carbonation profile of cement pastes by thermogravimetry and it's effect on the compressive strenght. J. Therm. Anal. Calorim., doi:10.1007/s10973-013-3556-7 2014.

Name: Lauri Niinistö
Country: Finland
Date and place of birth: 1941, Helsinki, Finland
Present position and address: Retired from the Helsinki University of Technology (Aalto University) Professor emeritus, postal address: Viittakuja 12, FI-01640 Vantaa, Finland
Education and scientific degrees: M.Sc., (Inorganic Chemistry) Helsinki University of Technology (1968), D. Techn. (1973), Professor (1977)
Workplaces: Laboratory of Inorganic and Analytical Chemistry, Helsinki University of Technology, also worked in Sweden, France, USA, Austria and Hungary
Main fields of interest: preparation and characterization of thin films for optoelectronic and electronic devices, thermal analysis of precursors, thermoanalytical techniques in the thin film technology, rare earths
Relevant categories in thermal analyses: fields (thermoanalytical techniques in the thin film technology and nanotechnology); methods (TG, DTA, EGA, DSC)
Awards and acknowledgments: Cannizzaro gold medal of the Italian Chemical Society (1996), FECS award (2000), honorary doctorates from Tallinn Technical University (1990) and Budapest University of Technology and Economics (2002)
Professional activities: Federation of European Chemical Societies (FECS), President 1995–1999.; FECS Working Party on Analytical Chemistry, Chairman 1987–1993, 1997–1998.; European Rare Earth Research Society (ERES), Chairman 1990–2003.; Chairman of Program and/or Organization Committee of eight international conferences and symposia; Member of Organising/Scientific/Advisory committee of several international conferences; Acta Chemica Scandinavica, Editor 1982–90, Managing Board member 1991–2000.; Journal of Thermal Analysis and Calorimetry, Regional Editor for Scandinavia and the Baltic countries Fresenius' Journal of Analytical Chemistry, Member of International Editorial Board, 1996–2004, Journal of Materials Chemistry, Member of International Advisory Editorial Board, 1995–2003; Journal of the Chemical Society, Dalton Transactions, Member of International Advisory Editorial Board, 2000–2003.; Periodica Polytechnica (Budapest), Editorial Board Member
Publication record: papers (370), citations (8400), h-index (45)
5 most important publications: [1] Niinistö, L., Thermal Analysis, In Analytical Chemistry, 2nd Ed, (2004) Mermet, J.M., et al. (ed.) Wiley, VCH, Weinheim, pp. 374–391; [2] Niinistö, L., From precursors to thin films—thermoanalytical techniques in the thin film technology, Journal of Thermal Analysis and Calorimetry (1999), 56, 7-15; [3] Knez, M., Nielsch, K., Niinistö, L., Advanced Materials (2007) 19, 3425–3438; [4] Niinistö, L., Karppinen, M. Journal of Thermal Analysis (1989), 35, 319–27; [5] Leskelä, M., Leskelä, T., Niinistö, L., Journal of Thermal Analysis (1993), 40, 1077–88.

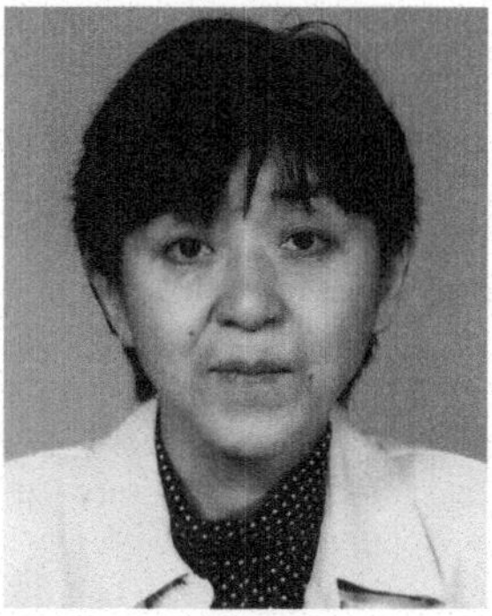

Name: Yuko Nishimoto
Country: Japan
Date and place of birth: Shimane, Japan
Present position and address: Professor, Faculty of Science, Kanagawa University, Tsuchiya, Hiratsuka, Kanagawa 259–1293, Japan
Email: y24moto@kanagawa-u.ac.jp
Website: http://www.sci.kanagawa-u.ac.jp/english/chem/index.html
Education and scientific degrees: Chiba University: M.S. in Chemistry, B.S. in Chemistry; Ph.D.: Tokyo University, June 1991
Workplaces: Kanagawa University
Main fields of interest: analytical chemistry, thermal analysis, environmental analysis, water analysis, polymer analysis
Relevant categories in thermal analyses: fields (environmental analysis, water analysis, polymer analysis); methods (TG, DTA, DSC, TMA, EGA)
Professional activities: 2012–2014: Treasurer of The Japan Society for Analytical Chemistry; 2009–2011: Secretary of The Japan Society of Calorimetry and Thermal Analysis; 2009-Present : Member, ISO/TC61 –Plastics SC5 Physical Chemistry Properties; 2005–2007: Editorial Board Member of "Netsu Sokutei" (Calorimetry and Thermal Analysis); 2005–2009: Associate Editors of "Journal of Thermal Analysis and Calorimetry"; 2003–2005: Editorial Advisory Board Member of "Journal of Thermal Analysis and Calorimetry"; 1999–2002: Editorial Board Member of "Bunseki" (The Japan Society for Analytical Chemistry); 2000–2002: Secretary of The Japan Society of Calorimetry and Thermal Analysis; 1996–1998: Planner of The Japan Society of Calorimetry and Thermal Analysis; 1994–1996: Editorial Board Member of "Analytical Sciences"; 1993–1999: Member, ISO/TC61 –Plastics SC5 Physical Chemistry Properties WG5B Thermal Analysis; 1997-Present: Member, ISO/TC45—Rubber WG1 Chemical Tests; 2001–2011: Member, Japanese Industrial Standards Committee; JISC, General Chemistry, Chemical Products
Publication record: papers (78), books (39)
Equipments: TG/DTA, DSC, TMA, EGA-MS
5 most important publications: [1] Transaction of the Material Research Society of Japan, 38, 589–592 (2013); [2] Transaction of the Material Research Society of Japan, 31, 937–940 (2006); [3] Analytical Sciences, 20, 1079–1082 (2004); [4] Thermochimica Acta, 399, (139–144) (2003); [5] Analytical Sciences, 8, 873–874 (1992).

Name: Wiesława Nocuń-Wczelik
Country: Poland
Date and place of birth: 1951, Jaworzno, Poland
Present position and address: Professor; AGH University of Science and Technology, Faculty of Material Science and Ceramics, 30-059 KRAKÓW, al. Mickiewicza 30
Email: wiesia@agh.edu.pl
Education and scientific degrees: D.Sc., eng.
Workplaces: AGH University of Science and Technology, Faculty of Material Science and Ceramics, Faculty of Mining and Geo-engineering
Main fields of interest: building materials, especially cement and concrete, supplementary cementing materials, waste disposal in building materials technology, hydration process and characterization of hydration products
Relevant categories in thermal analyses: fields (calorimetry and microcalorimetry in the studies of cements/binders hydration); methods (from microcalorimetry to standard calorimetric methods (EC standards))
Awards and acknowledgments: many awards from local university authorities (education, books, etc.)
Professional activities: lecturer, supervisor of Eng., M.Sc. and Ph.D. student's projects, head of calorimetry laboratory, member of Polish Society of Thermal Analysis and Calorimetry, Polish Ceramic Society, Committee of Civil Engineering associated with Polish Academy of Science
Publication record: papers (170), books (7), patents (2), citations (230), h-index (8), sum of impact factors (35)
Equipments: Microcalorimeters and standard calorimeters
5 most important publications: [1] W. Nocuń-Wczelik, B. Trybalska, The studies of Al-doped tricalcium silicate hydration, Journal of Thermal Analysis, 32, 1719–1722, 1987; [2] W. Nocuń-Wczelik, "Structure and properties of hydrated calcium silicates", book edited by Polish Academy of Science in Kraków 1999. in Polish; [3] W. Nocuń-Wczelik, "Silica fume, the properties and application in concrete technology" book edited by Polski Cement in Kraków 2005; [4] CEMENT. Properties. Applications. Editor of book and author of 3 chapters, edited by AGH University of Science and Technology Scientific Edition Kraków 2010; ISBN 987-83-7464-313-0; [5] W. Nocuń-Wczelik, B. Trybalska, E. Żugaj, Application of calorimetry in evaluation the effect of carbonate additives on cement hydration, Journal of Thermal Analysis and Calorimetry, 113(1), 351–356, 2013.

Name: Dumitru Oancea
Country: Romania
Date and place of birth: 1941, Vulturesti, Arges, Romania
Present position and address: Professor Emeritus University of Bucharest, Department of Physical Chemistry, University of Bucharest, Bulevardul Regina Elisabeta 4-12, 030018 Bucharest, Romania
Email: doan@gw-chimie.math.unibuc.ro, dumitruoancea2012@gmail.com
Education and scientific degrees: Chemist (Graduate of Faculty of Chemistry, University of Bucharest 1964); Ph.D. (1972); Assistant Professor (1964–1976), Lecturer (1976–1990), Associate Professor (1990–1994), Full Professor (1994–2012), Professor Emeritus (2012–) at Department of Physical Chemistry, University of Bucharest
Workplaces: Department of Physical Chemistry, University of Bucharest
Main fields of interest: chemical kinetics (in solution, in gases, in solid/gas heterogeneous systems), enzyme catalysis, ion exchange, combustion kinetics, catalysis
Relevant categories in thermal analyses: fields (stability and decomposition of energetic materials, catalytic combustion); methods (DSC, micro-calorimetry)
Awards and acknowledgments: Award "Gheorghe Spacu" of the Romanian Academy (1987), Medal "Gheorghe Spacu" of the Romanian Chemical Society (2009), Corresponding Member of the Romanian Academy (2009)
Professional activities: Dean of the Faculty of Chemistry (2004–2012), Member of the Editorial Board of the journal Revue Roumaine de Chimie, Member of the Roumanian Chemical Society, Member of the Romanian Catalysis Society, Member of the Scientific Committee of ROMPHYSCHEM Conferences
Publication record: papers (170), books (5), citations (550*), h-index (12*) (*cf. Google Scholar)
Equipments: DSC, noncommercial isothermal micro-calorimeter for catalytic combustion
5 most important publications: [1] J. Pandele Cusu, A. M. Musuc, D. Oancea, J. Therm. Anal. Calorim., 109, 255–263 (2012); [2] D. Oancea, V. Munteanu, D. Razus, M. Mitu, J. Therm. Anal. Calorim., 103, 911–916 (2011); [3] D. Oancea, V. Munteanu, D. Razus, J. Therm. Anal. Calorim., 102, 993–1000 (2010); [4] D. Oancea, A. Stuparu, M. Nita, M. Puiu, A. Raducan, Biophys Chemistry, 138, 50–54 (2008); [5] D. Oancea, O. Staicu, V. Munteanu, D. Razus, Catal. Lett. 121, 247–254 (2008).

Name: Nina Obradović
Country: Serbia
Date and place of birth: 1977, Belgrade, Serbia
Present position and address: Senior Research Associate, Institute of Technical Sciences of Serbian Academy of Sciences and Arts, Knez Mihajlova 35/IV, 11000 Belgrade
Email: nina.obradovic@itn.sanu.ac.rs
Website: http://www.itn.sanu.ac.rs/ninaobradovic_eng.html
Researcher ID: 23134986000
Education and scientific degrees: Doctor of Philosophy (Ph.D.) in Materials Science, Physical Chemistry, Faculty of Physical Chemistry, Belgrade, Serbia (Dec. 2007); Master of Science (M.Sc.) in Materials Science, Physical Chemistry, Faculty of Physical Chemistry, Belgrade, Serbia (Apr. 2005); Bachelor of Science (BS) in Physical Chemistry Faculty of Physical Chemistry, Belgrade, Serbia (Dec. 2001)
Workplaces: Institute of Technical Sciences of Serbian Academy of Sciences and Arts
Main fields of interest: nanostructured materials, powder technology, ceramics, materials characterization, thermal analysis, sintering.
Relevant categories in thermal analyses: fields (inorganic, materials, nano, minerals, complex, organic, pharmaceutical, polymer, food, glass, ceramics, cement); methods (DTA, DSC)
Awards and acknowledgments: For the best competent and scientific-research paper in 2000/2001. by Belgrade University at the Faculty of natural Sciences and mathematics; "The top of synthesis and sintering" award for success in sintering under extreme conditions for paper presented at the International conference "Mechanochemical synthesis and sintering" held on 14–18, 2004 at Novosibirsk, Russia.; Award for the best M.Sc. thesis given by the Yugoslav Materials Research Society in 2005.; February 2009—Belgrade Chamber of Commerce Annual Award for the best doctoral thesis in the 2007/8 school year
Professional activities: Member of Serbian Ceramic Society, Journal Science of Sintering Editorial Board Secreteriate
Publication record: papers (37), patents (1), h-index (6)
Equipments: SHIMADZU DTA-50, SHIMADZU DSC-50
5 most important publications: [1] S. Filipovic, N. Obradovic, J. Krstic, M. Šcepanovic, V. Pavlovic, V. Paunovic, M.M. Ristic, "Structural characterization and electrical properties of sintered magnesium–titanate ceramics", J. of Alloys and Compounds 555 (2013) 39–44; [2] N. Obradović, N. Djordjević, S. Filipović, N. Nikolić, D. Kosanović, M. Mitrić, S. Marković, V. Pavlović "Influence of mechanical activation on sintering od cordierite ceramics in the presence of Bi_2O_3 as a functional additive", Powd. Technol. 218 (2012) 157–161; [3] N. Obradović, A. Terzić, Lj. Pavlović, S. Filipović, V. Pavlović "Dehydration investigations of a refractory concrete using DTA method", J. Therm Anal Calorim, 110 (2012) 37–41; [4] N. Obradović, S. Filipović, V. Pavlović, M. Mitrić, S. Marković, V. Mitić, N. Đorđević, M. M. Ristić "Isothermal sintering of barium-zinc-titanate ceramics", Ceram. Int. 37 (2011) 21–27; [5] Nina Obradovic, Miodrag Mitric, Maria Vesna Nikolic, Dragica Minic, Nebojsa Mitrovic, Momcilo M. Ristic "Influence of MgO addition on the synthesis and electrical properties of sintered zinc-titanate ceramics", J. Alloys Compd. 471:1–2 (2009) 272–277.

Name: Marianne Odlyha
Country: UK
Date and place of birth: 1945, Kiel, Germany
Present position and address: Senior lecturer and Manager of the Thermal Methods and Conservation Science Centre, Birkbeck, University of London, Malet St. London WC1E 7HX.
Email: m.odlyha@bbk.ac.uk
Education and scientific degrees: B.Sc Hons, B.A, M.Sc., Ph.D. (University of London). Training in analytical methods applied to cultural materials Doerner Institute, Munich.
Workplaces: University of Adelaide (1968–1970), Institute of Technology, South Australia, Australia (1974–1987). Courtauld Institute of Art: student assistant (1988/9) radiography and analysis of paintings.
Main fields of interest: application of thermal methods with spectroscopic, surface analytical and surface imaging techniques (atomic force microscopy) for damage assessment and preservation of indoor movable cultural heritage; preventive conservation and development of portable devices for monitoring corrosive environments in display cases and microclimate frames containing paintings
Relevant categories in thermal analyses: fields (materials, nano, organic, polymer, cellulose, textiles, wood, leather, parchment, collagen, composite materials, accelerated ageing and life time predictions); methods (TGA, DSC (oxidative degradation studies of materials), thermomechanical analysis (glass transition temperature measurements), controlled environment RH-DMA, nanogravimetric devices for monitoring corrosivity of environments)
Professional activities: ICTAC awards committee (2004–2008), ICTAC Nomenclature Group (1991–1998), Committee Member U.K Thermal Methods Group (TMG) (1991–1994), Member ICOM-CC (International Council of Museums), Member of IIC (International Institute for Conservation). Guest editor Thermochim Acta (Vol 365 (1-2) 2000 and special chapter on conservation in J Therm Anal Calorim (2011) 104, Teaching at EC-funded Cost Action schools and courses in cultural heritage. Coordinator of 2 FP4 projects and project partner in 6 projects (FP5-FP7) on cultural heritage including damage assessment, preventive conservation, and evaluation of nanoformulations for conservation treatment of cellulosic (paper, painting canvases) and collagen-based materials
Publication record: papers (108), citations (690), book chapters (5), patents (2)
Equipments: Controlled Environment DMA, TG, DSC, TMA
5 most important publications: [1] Odlyha, M. et al., "Thermoanalytical (macro to nanoscale) techniques and Non-invasive Spectroscopic Analysis for Damage Assessment of Parchment" in "Improved Damage Assessment of Parchment IDAP" ed. R. Larsen, European Commission pp. 73–85 (2007) ISBN 978-92-79-05378-8; [2] Odlyha, M. The Application of Thermoanalytical techniques to the Preservation of Art and Archaeological Objects " Chapter 2, Vol 2, Handbook of Thermal Analysis and Calorimetry, ed. M.E.Brown and P.K.Gallagher, Elsevier 47–92 (2003); [3] Odlyha M. et al. Dosimeters for Indoor Microclimate Monitoring for Cultural Heritage. In: Padfield T, Borchersen K, editors. Museum Microclimates. Copenhagen: National Museum of Denmark. 2007. pp. 73–9; [4] Odlyha, M. et al. 1997 J Therm Anal Calorim, 49, 1571–1584; [5] Odlyha, M. et al. 1993. J Therm Anal Calorim 40, 285–302.

Name: Lucia Odochian
Country: Romania
Date and place of birth: 1938, Botosani, Romania
Present position and address: Retired from "Alexandru Ioan Cuza" University of Iasi, Consultant Professor Lucia Odochian, 14 Neculau Street, 700522 Iasi, Romania
Email: lodochian@yahoo.com
Education and scientific degrees: Chemist, Faculty of Chemistry, "Alexandru Ioan Cuza" University of Iasi (1955–1960); Ph.D. (1973); Professor (1992).
Workplaces: Faculty of Chemistry, "Alexandru Ioan Cuza" University of Iasi (1960–2005); Sci. Advisor (2005–)
Main fields of interest: non-isothermal kinetics—thermal behavior of micro- and macro-molecular compounds; isothermal kinetics—chain reactions
Relevant categories in thermal analyses: fields (inorganic, materials, organic, polymer, biology, ceramics); methods (TG, DTA, DSC, kinetics, specific heat, calorimetry, TG-FTIR, TG-FTIR-MS)
Professional activities: Member of the Chemical Society of Romania (SChR); member of Scientists Association from Romania; member of International Confederation of Thermal Analysis and Calorimetry (1997–2000)
Publication record: papers (110), books (9), patents (5), citations (167), h-index (8)
Equipments: TG/DTA Diamond (PerkinElmer) thermo-balance and a FTIR spectrometer, Spectrum 100 (PerkinElmer), provided with a TG-FTIR (PerkinElmer) gas transfer accessory with a gas cell of 100 mm length and KBr windows, heated at 150 °C
5 most important publications: [1] L. Odochian: J. Thermal Anal., 45 (1995) 1437; [2] M. Dumitras, L. Odochian: J. Thermal Anal. Calorim., 69 (2002) 599; [3] L. Odochian, A.M. Mocanu, C. Moldoveanu, G. Carja, C. Oniscu: J. Therm. Anal. Calorim., 93 (2008) 907; [4] L. Odochian, C. Moldoveanu, A.M. Mocanu, G. Carja: Thermochimica Acta, 526 (2011) 205; [5] L. Odochian, C. Moldoveanu, G. Carja: Thermochimica Acta, 558 (2013) 22.

Name: Rodica Olar
Country: Romania
Date and place of birth: 1963, Deda, Romania
Present position and address: University of Bucharest, Faculty of Chemistry, Department of Inorganic Chemistry, 90-92 Panduri Str., 050663 Bucharest, Romania
Email: rodica_m_olar@yahoo.com
Education and scientific degrees: M.Sc. Inorganic Chemistry (Polytechnical Institute of Bucharest, 1986), Ph.D. Chemistry (University of Bucharest, 1999)
Workplaces: Mecanical factory Zarnesti (1986–1990), Metalurgical Research Institute Bucharest (1990), University of Bucharest (1990–)
Main fields of interest: thermal characterisation of complexes with biological activity (antimicrobial, cytotoxic, insulin mimetic), complexes (synthesis and characterisation), bioinorganic chemistry, inorganic materials
Relevant categories in thermal analyses: fields (inorganic, materials, nano, complex, pharmaceutical); methods (TG, DTA, DSC, EGA)
Professional activities: Biointerface Research in Applied Chemistry Editorial Board member, ICTAC member, Roumanian Chemical Society member, CEEC-TAC member, CEEC-TAC Projects Department Coordinator, CEEC-TAC 1 National Organising Committee member, CEEC-TAC 2 Scientific Committee member
Publication record: papers (69), books (2), patents (1), citations (350), h-index (11), sum of impact factors (127.111)
Equipments: Labsys 1200 SETARAM thermobalance, DSC 550 Cahn instrument
5 most important publications: [1] C. Bucur, R. Cerc Korošec, M. Badea, L. Calu, M.C. Chifiriuc, M.N. Grecu, N. Stanică, D. Marinescu, R. Olar: Investigation of thermal stability, spectral, magnetic and antimicrobial behavior for new complexes of Ni(II), Cu(II) and Zn(II) with a bismacrocyclic ligand, J. Therm. Anal. Calorim., 113 (2013) 1287; [2] R. Olar, M. Badea, D. Marinescu, R. Mardale: Thermal behaviour of new Cu(II) complexes with Schiff bases functionalised with 1,3,5-triazine moieties as potential antibacterial agents, J. Therm. Anal. Calorim., 105 (2011) 553; [3] R. Olar, M. Badea, D. Marinescu, C. Chifiriuc, C. Bleotu, N. Grecu, E.E. Iorgulescu, M. Bucur, V. Lazar, A. Finaru: Prospects for new antimicrobials based on N,N-dimethylbiguanide complexes as effective agents on both planktonic and adherent microbial strains. Eur. J. Med. Chem., 45 (2010) 2868; [4] R. Olar, M. Badea, D. Marinescu, C. Chifiriuc, C. Bleotu, N. Grecu, E.E. Iorgulescu, V. Lazar, N,N-Dimethylbiguanide complexes displaying low citotoxicity as potential large spectrum antimicrobial agents, Eur. J. Med. Chem., 45 (2010) 3027; [5] R. Olar, M. Badea, D. Marinescu, V. Lazar, C. Chifiriuc: Thermal behaviour of new Ni(II) and Cu(II) complexes with macrocyclic ligands functionalised with 1,2,4-triazole, J. Therm. Anal. Calorim., 97 (1) (2009) 315.

Name: Riko Ozao
Country: Japan
Date and place of birth: 1957, Kyoto, Japan
Present position and address: Professor, Department of Informatics and Media Technology, Shohoku College, SONY Institute of Higher Education, Atsugi, Kanagawa, 243-8501, Japan
Email: ozao@shohoku.ac.jp
Website: http://ozaolab.sakura.ne.jp/index.html
Education and scientific degrees: B. Sc. (1980), M.Sc. (1982), Ph.D. (Dr Eng) (1986), Waseda University, Department of Science and Engineering (Tokyo, Japan)
Workplaces: SONY Institute of Higher Education, Waseda University, Meiji University
Main fields of interest: materials characterization, information science, databases, education (environmental science)
Relevant categories in thermal analyses: fields (materials characterization); methods (DSC, TG-DTA, TMA)
Awards and acknowledgments: NATAS (North American Thermal Analysis Society) Fellow Award (2012)
Professional activities: ICTAC: Membership Secretary (Executive Board) since 2006, Affiliate Society (Japan) Councillor 2004–2006)/NATAS (Symposium Program Chair (2013), Session Organizer (2006, 2008, 2009, 2012))/Materials Research Society of Japan: Executive Committee Member (2005–2012), Chairman of Annual Symposium Sessions every year 2004–2011, Session Chairs for IUMRS-ICEM2012 (Yokohama, Japan), IUMRS-ICA2008 (Nagoya, Japan), IUMRS-ICMAT2005 (Singapore), IUMRS-ICAM2003 (Yokohama, Japan), The Japan Society of Calorimetry and Thermal Analysis (JSCTA): Treasurer (Board of Organization Committee) (1995–1999 (2 terms)), Chairman of Standardization Working Group (2004–2006), Organization Committee, CATS-2005 (4th International and 6th Japan-China Joint Symposium on Calorimetry and Thermal Analysis), Chairman, Website Working Group (1999–2004), Organization Committee, CATS (2nd International and 4th Japan-China Joint Symposium on Calorimetry and Thermal Analysis) (1999), Organization Committee, Annual Meeting of JSCTA: No.28 (1992), No.34 (1998), Eco-materials Forum: Database Subcommittee chair (2006–13), Journal of Thermal Analysis and Calorimetry: Associate Editor, Guest Editor for Vol.116 (Issue 3) (2014)
Publication record: papers (150), books (12), patents (1)
5 most important publications: [1] "Comprehensive Handbook of Calorimetry and Thermal Analysis" (Editorial board), Wiley, UK (2004); [2] "Novel Thermal Analysis" ("Netsu-bunseki" in Japanese), Kodansha-Scientific, Tokyo, Japan (2005); [3] Ozao, W–P. Pan, N.Whitely, T. Okabe; Coal-like Thermal Behavior of Carbon-based Environmentally Benign New Material, Woodceramics (2004) Energy Fuels, 18, pp. 638–43; [4] Riko Ozao and M. Ochiai; Fractal Nature and Thermal Analysis of Powders (1993) J. Thermal Analysis, 40, pp. 1331–40; [5] R.Ozao, H.Yoshida, Y.Ichimura, T.Inada, M.Ochiai; Crystallization of Anodic Alumina Membranes Studied by Simultaneous TG-DTA/FTIR (2001) J. Therm. Anal. Calorim., 64, pp. 915–22.

Name: Barbara Pacewska
Country: Poland
Date and place of birth: 1946, Konstantynówka, Poland
Present position and address: Professor, Head of Department of Fundamental Chemistry, Warsaw University of Technology, Faculty of Civil Engineering, Mechanics and Petrochemistry, Institute of Chemistry, Łukasiewicza 17 St., PL-09-400 Płock, Poland
Email: bpacewska@pw.plock.pl
Education and scientific degrees: M.Sc. (1969), Ph.D. (1978) Warsaw University of Technology, D.Sc. (1993) Łódź University of Technology, Professor (2006)
Workplaces: Warsaw University of Technology, Institute of Chemistry, Płock, Poland (1970-present)
Main fields of interest: thermal analysis, chemistry of solids, inorganic chemistry
Relevant categories in thermal analyses: fields (inorganic, materials, minerals, cement, sorbents); methods (TG, DTA, DSC, calorimetry, complementary methods e.g. XRD, FTIR, SEM)
Awards and acknowledgments: First Award of the JTAC (2006), Three Awards of Education Ministry, Award in 4th International Competition EKO 2001, Sixteen Scientific Awards of Rector of Warsaw University of Technology
Professional activities: ICTAC Affiliate Councillor for Poland (2008–), Member of Board of Polish Society of Calorimetry and Thermal Analysis (1994–) and President of this society (2006–2012, 2013–), Chairman of Płock Division of Polish Chemical Society (1996–2007), Member of Board of Płock Scientific Society (2002–2005), Member or Chairman of Organising or Scientific Committees of conferences: Seminars to the memory Prof. St. Bretsznajder (1977–2011), Conferences on Calorimetry and Thermal Analysis (1994–2012) and others, Guest co-editor or editor of six special issues of JTAC. Regional Editor of JTAC (2000–)
Publication record: papers (97), book (1), patents (3), citations (351), h-index (10)
Equipments: SDT 2960 thermoanalyser (TA Instruments); BMR calorimeter constructed at Institute of Physical Chemistry, Polish Academy of Sciences
5 most important publications: [1] Pacewska B., Szychowski D., Kluk O., Adsorption and structural properties of mineral-carbon sorbents. J. Therm. Anal. Calorim. 2002; 67:773–787; [2] Pacewska B., Bukowska M., Wilińska I., Swat M., Modification of the properties of concrete by a new pozzolan—a waste catalyst from the catalytic process in a fluidized bed. Cement and Concrete Research, 2002; 32:145–152; [3] Pacewska B., Blonkowski G., Wilińska I., Investigations of the influence of different fly ashes on cement hydration. J. Therm. Anal. Calorim. 2006; 86:179–186; [4] Pacewska B., Kluk- Płoskońska O., Szychowski D., Influence of aluminium precursor on physico-chemical properties of aluminium hydroxides and oxides. Part I–IV. J. Therm. Anal. Calorim. 2006; 85:351–359, 2006; 86:751–760, 2007; 87: 383–393, 2007; 90:783–793; [5] Pacewska B., Nowacka M., Wilińska I., Kubissa W., Antonovich V., Studies on the influence of spent FCC catalyst on hydration of calcium aluminate cements at ambient temperature. J. Therm. Anal. Calorim. 2011; 105:129–140.

Name: Cornelia Păcurariu
Country: Romania
Date and place of birth: 1952, Timişoara, Romania
Present position and address: Professor at the Faculty of Industrial Chemistry and Environmental Engineering, POLITEHNICA University Timişoara, 6 Pîrvan Blv., RO-300223 Timisoara, Romania,
Email: cornelia.pacurariu@upt.ro
Website: http://www.chim.upt.ro/Facultatea-de-Chimie-Industriala-si-Ingineria-Mediului-Cadru-Didactic_Pacurariu-Cornelia_olK_sQ.html
Education and scientific degrees: Chemical Engineer, POLITEHNICA University Timişoara, Faculty of Chemical Engineering (1971–1976); Ph.D.(1997); Professor (2004)
Workplace: University POLITEHNICA Timişoara, Faculty of Industrial Chemistry and Environmental Engineering
Main fields of interest: teaching and research activities within the field of: chemical kinetics, applied physical chemistry, physical chemistry of interfaces, synthesis of nanomaterials, thermal analysis (DTA, DSC, TG), spectroscopic (UV-Vis, FT-IR) methods, environmental protection
Relevant categories in thermal analyses: fields (inorganic, materials, nano, glass, ceramics, polymer, cement); methods [TG, DTA, DSC, kinetics, extremely high temperature (above 1,000 °C)]
Professional activities: Member of the Scientific Committee of the 2nd Central and Eastern European Conference on Thermal Analysis and Calorimetry, 27–30 August, 2013, Vilnius, Lithuania; Member of the Scientific Committee of the 1st Central and Eastern European Conference on Thermal Analysis and Calorimetry, 7-10 September, 2011, Craiova, Romania; Editorial board member of: Romanian Journal of Materials and Chemical Bulletin of the POLITEHNICA University Timisoara; Membership in professional bodies: Romanian Ceramic Society and Romanian Chemical Society
Publication record: papers (>100), books (15), patents (1), citations (220), h-index (8)
Equipments: Netzsch STA 449 C
5 most important publications: [1] Pacurariu C., Lazău R., Lazău I., Ianoş R., Vlase T., Influence of the specific surface area on crystallization process kinetics of some silica gels, J. Therm. Anal. Calorim., 97 (2009) 409–414; [2] Pacurariu C., Lazau R. I., Lazau I., Tita D., Dumitrel A. Non-isothermal crystallization kinetics of some aventurine decorative glaze, J. Therm. Anal. Calorim., 105 (2011) 435–441; [3] Pacurariu C., Lazau I., Non-isothermal crystallization kinetics of some glass-ceramics with pyroxene structure, J. Non-Cryst Solids, 358 (2012) 3332–3337; [4] Mihoc G., Ianos R., Pacurariu C., Lazau. I., Combustion synthesis of some iron oxides used as adsorbents for phenol and p-chlorophenol removal from wastewater, J. Therm. Anal. Calorim., 112 (2013) 391–397; [5] Pacurariu C., Mihoc G. Popa A., Muntean S., Ianos R., Adsorption of phenol and p-chlorophenol from aqueous solutions on poly(styrene-co-divinylbenzene) functionalized materials, Chem Eng. J., 222(2013) 218–227.

Name: Mrinal R. Pai
Country: India
Date and place of birth: 1973, New Delhi, India
Present position and address: Scientist, Chemistry Division, 3-193 H, Modular Laboratories, Bhabha Atomic Research Centre, Trombay Mumbai-400085.
Email: mrinalpai9@gmail.com, mrinalr@barc.gov.in
Education and scientific degrees: Ph.D. (Chemistry)
Workplaces: Chemistry Division, Bhabha Atomic Research Centre
Main fields of interest: mixed oxide catalysts for reactions related to energy and environment; H_2 generation via thermochemical cycles and photocatalysis; CO oxidation, methanol decomposition, abatement of NO_x; H_2SO_4 decomposition reaction up to 850 °C.
Relevant categories in thermal analyses: fields (solid state synthesis and characterization of active species present on fresh and spent catalysts by temperature programmed reduction/ oxidation/desorption; monitoring of lattice oxygen mobility and absorbed H_2 by storage material as a function of temperature; calculation of oxygen nonstoichiometry of oxide catalysts by TG in H_2 atmosphere; evalution of temperature dependent catalytic activity for various reactions mentioned above; mechanistic aspects were studied by insitu-FTIR as a function of temperature.); methods (sample preparation by various routes ceramic as well as solution routes and characterization by relevant techniques; evaluation of oxides as catalysts for different reactions mentioned above)
Awards and acknowledgments: TA instruments ITAS- Young Scientist Award (2008), "Young associateship", by Maharashtra Academy of Sciences in year 2010. DAE group achievement award, 2011 for development of catalysts for hydrogen technology conferred by Department of atomic Energy, India
Professional activities: Life member of e.g. ITAS, SMC, CSI, IWSA. Treasurer for Indian Thermal Analysis Society (ITAS). Co-editor for Bulletin of Indian Thermal Analysis Society (ITAS). As Secretary, Technical Committee, editor of the proceedings of THERMANS-2010, INDIA. Orginized a two day National workshop on Catalytic Materials (NWMC 2013) in Bhabha Atomic Research Centre. Reviewer of several scientific papers in national and international journals. Guide for M-Tech, Ph.D. students, registered with Pune University and Homi Bhabha National Institute
Publication record: papers (30), books (1), Conference (>45)
Equipments: TPDRO, EGA, GC, N_2 sorption measurements, High temperature furnaces, Vapor phase catalytic reactors (indigenously fabricated) and fabrication of various instrumentation facilities for in situ FTIR, GC
5 most important publications: [1] A. M. Banerjee, M. R. Pai, R. Tewari, N. Raje, A. K. Tripathi, S. R. Bharadwaj, D. Das, Applied Catalysis B; Environmental, 162 (2015) 327–337; 127 (2012) 36–46; [2] M. R. Pai, A. M. Banerjee, K. Kartha, R. V. Pai, V. S. Kamble, S. R. Bharadwaj, J. Phys. Chem. B, 114 (2010), 6943–6953; [3] M. R. Pai, J. Majeed, A. M. Banerjee, A. Arya, Rekha rao, S. Bhattacharyaa, S. R. Bharadwaj, J. Phys. Chem. C, 116 (2012) 1458; [4] M. R. Pai, B. N. Wani, B. Sreedhar, S. Singh and N. M. Gupta, J. Molec. Catal. A: Chem., 246 (2006) 128; [5] M. R. Pai, A. M. Banerjee, S. R. Bharadwaj and S. K. Kulshreshtha, J of Mater. Res., 22 (2007) 1787.

Name: Martin T. Palou
Country: Slovakia
Date and place of birth: 1962, Chad
Present position and address: Slovak Academy of Sciences, SAV, Dúbravská cesta 9, 845 03 Bratislava 45, Slovakia; Brno University of Technology, BUT, Purkyňova 464/118 612 00 Brno, Czech Republic; Slovak University of Technology, Radlinského 9, 812 37 Bratislava, Slovakia
Email: martin.palou@stuba.sk, martin.palou@savba.sk, palou@fch.vutbr.cz
Researcher ID: 616025
Education and scientific degrees: Ph.D.
Workplaces: Slovak University of Technology, Slovak Academy of Sciences, Brno University of Technology
Main fields of interest: inorganic materials (Inorganic binders and Inorganic biomaterials)
Relevant categories in thermal analyses: fields (hydration inorganic binders, crystallization of glass); methods (conduction calorimetry, DSC and TGA)
Awards and acknowledgments: 2010, Honorable Mention of Award Ján Bahýľ from Industrial Property Office of the Slovak Republic in connection with the invention "patent SK 286943—Method of production of Portland clinker using alternative raw materials, especially crystalline blast/furnace slag. 1992, Award of the Rector of Slovak University of Technology
Professional activities: Lecturer at Slovak University of Technology and Brno University of Technology, senior researcher at Slovak Academy of Sciences
Publication record: papers (43), patents (1), citations (172), h-index (8)
Equipments: Conduction Calorimeter ZIAC, Simultaneous Thermal Analysis Mettler Toledo Star TGA/DSC, Differential Scanning Calorimeter DSC Mettler Toledo Star DSC
5 most important publications: [1] Palou M.T.; Živiva V.; Ifka T., Boháč M.; Zmrzly M. Effect of hydrothermal curing on early hydration of G-Oil well cement. J Therm Anal Calorim, 2013, Page 1–7; [2] Palou M.T., Bágeľ Ľ., Živica V., Kuliffayová M., Ifka T. Hydration of high alumina cement–silica fume composite with addition of Portland cement or sodium polyphosphate under hydrothermal treatment. J Therm Anal Calorim (2013) 113:385–394; [3] Palou, M.T.; Kuzielova, E.; Vitkovic, M.; et al. Mechanism and kinetics of glass-ceramics formation in the LiO_2-SiO_2-CaO-P_2O_5-CaF_2 system. Cent Eur J Chem, 2009, 7 (2), 228–233; [4] Palou M.T.; Majling J. Hydraulic activity of C_4A_3Cr in presence of C_4A_3 J Therm Anal Calorim, 2003, 71 (2), 367–373; [5] Palou M.T.; Majling J. Effects of sulfate, calcium and aluminium ions upon the hydration of sulphoaluminate belite cement. J Therm Anal 1996, 46 (2), 549–556.

Name: Wei-Ping Pan
Country: USA
Date and place of birth: 1954, Taiwan
Present position and address: Assistant to the President, Western kentucky University, 2413 Nashville Rd, C-2, Bowling Green, KY 42101, USA
Email: wei-ping.pan@wku.edu
Website: www.wku.edu/icset
Education and scientific degrees: Michigan Technological University (Ph.D., 1986)
Workplaces: Western Kentucky University, Sumpter Professor
Main fields of interest: catalysts, coal combustion, and emission control
Relevant categories in thermal analyses: fields (material characterization and kinetic); methods (simultaneous TA methods (TG-FTIR-MS system))
Awards and acknowledgments: Fellow of the Society, North American Thermal Analysis Society, 1997. NATAS Outstanding Achievement (Mettler-Toledo) Award, 2004. Distinguished College/University Scientist Superlative Award, Kentucky Academy of Science, 2006. Mettler Award, North American Thermal Analysis Society, 2008. Western Kentucky University's Ten Million Dollars Grant Club, 2010
Professional activities: American Chemical Society, Combustion Institute, International Confederation for Thermal Analysis and Calorimetry, Kentucky Academic of Science, American Society of Mechanical Engineers, American Society of Test Method, North American Thermal Analysis Society (President, 2001)
Publication record: papers (214), books (1), patents (1), citations (5283), h-index (34)
Equipments: HPTGA, TGA, HPDSC, DSC, SDT, TMA, DMA, TGA-FTIR-MS system, Dilatometer, Viscometer, Flash Diffusivity System and Guarded Heat Flow meter
5 most important publications: [1] Wei Xie, Zongming Gao, Wei-Ping Pan, Doug Hunter, Anant Singh, and Richard Vaia, "Thermal Degradation Chemistry of Alkyl Quaternary Ammonium Montmorillonite," Chemical Materials, 2001, 13, 2979–90; [2] Wei Xie, Rongcai Xie, Wei-Ping Pan, Doug Hunter, Bryan Koene, Loon-Sent Tan, and Richard Vaia, "Thermal Stability of Quaternary Phosphonium Modified Montmorillonite," Chemical Materials, 2002, 14, 4837–45; [3] Wei Xie, Z Gao, K Liu, Wei-Ping Pan, R Vaia, D Hunter, A Singh, Thermal Characterization of organically modified montmorillonite, Thermochimica Acta, 2001, 367, 339–50; [4] W. Xu, G. Liang, H. Zhai, S. Tang, G. Hand and W.-P. Pan, "Preparation and Crystallization Behavior of PP/PP-g-MAH/Org-MMT Nanocomposite,"European Polymer Journal, 2003, 39, 1467–1474; [5] Yan Cao, Bianca Casenas, and Wei-Ping Pan, "Investigation of Chemical Looping Combustion by Solid Fuels 2. Redox Reaction Kinectics and Product Characterization with Coal, Biomass and Solid Waste as Solid Fuels and CuO as Oxygen Carrier," Energy and Fuels, 2006, 20, 1845–1854.

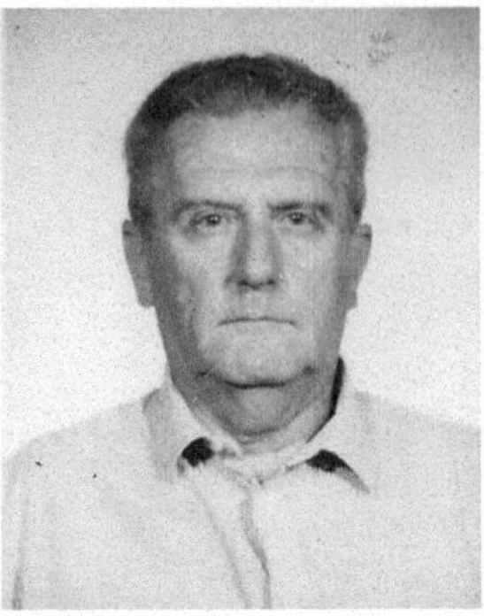

Name: Piero G. Paoletti
Country: Italy
Data and place of birth: 1931, Sesto Fiorentino, Italy
Present position and address: Emeritus Professor of university of Florence (Italy). Dipartimento Hugo Schiff, Via della lastruccia 3, 50019 Sesto Fiorentino, Firenze (Italy).
Email: piero.plt@tiscali.it
Education and scientific degrees: Doctor in Chemistry, University of Florence (1949–1954), Libero Docente (1960), Professor of inorganic chemistry (1965), University of Cagliari
Workplaces: Institutes of general chemistry, Universities of Palermo, Cagliari and Florence. Past Dean of department Hugo Schiff, University of Florence
Main fields of interest: solution calorimetry, potentiometry, spectrophotometry, building of an electrical microcalorimeter (1959)
Awards and acknowledgments: First President of the group on thermodynamics of complexes (GTC), ICTA membership, IUPAC member of a commission of stability constants and author of a critical review of ethylenediamine complexes
Professional activities: Chairman in many I.C.C.C.congresses, Organizer of two international symposia on macrocyclic chemistry. Invited to give a course at KTH of Stockholm (1978)
Publication record: papers (221), books (2)
5 most important publications: [1] P. Paoletti et al. Nature, 1960, 18, 880; [2] P. Paoletti et al. I.C.C.C. 1961, Detroit 303; [3] P. Paoletti and A. Vacca, Trans. Faraday Soc. 1964, 60, 50; [4] P. Paoletti et al. Chem. Comm., 1969, 513; [5] P. Paoletti: Thermochemistry of metal-polyamine complexes, in Thermochemistry and its application to chemical and biochemical Systems, 339–352, 1984 By D. Reidel Publishing Co. M.A.V. Ribeiro da Silva (Ed.).

Name: Paweł Pasierb
Country: Poland
Date and place of birth: 1968, Kraków, Poland
Present position and address: assistant professor, AGH University of Science and Technology, Faculty of Materials Science and Ceramics, al. Mickiewicza 30, 30-059 Kraków, Poland
Email: ppasierb@agh.edu.pl
Education and scientific degrees: M.Sc. in Materials Science at AGH University of Science and Technology, Faculty of Materials Science and Ceramics, Kraków, Poland (1992), Ph.D. in Chemistry (1997) Habilitaton in Chemistry (2011)
Workplaces: AGH University of Science and Technology, Faculty of Materials Science and Ceramics, Kraków, Poland (1997–)
Main fields of interest: solid state electrochemistry, ionic conductors, chemical stability of ceramic materials, reactivity of solids and corrosion mechanisms, practical aspects of thermal properties of ceramic materials, temperature dependence of electrochemical properties of materials
Relevant categories in thermal analyses: fields (inorganic materials, ionic conductors, electrochemical properties, solid-solid and gas–solid reactions); methods (DTA-TG-EGA, DSC)
Awards and acknowledgments: Numerous Awards of the Rector of AGH University of Science and Technology for scientific achievements (2003, 2004, 2007–2011)
Publication record: papers (33), citations (360), h-index (11)
Equipments: DTA-TG-DSC (TA Instruments) MS (Oxford Instruments), High temperature Eletrochemical Impedance Spectroscopy EIS (Solartron Analytical, Norecs sample holder)
5 most important publications: [1] P. Pasierb, M.Wierzbicka, S. Komornicki, M. Rekas, “Electrochemical impedance spectroscopy of $BaCeO_3$ modified by Ti and Y”, J Power Sources 194 (2009) 31–37; [2] P. Pasierb, M. Rękas, “Solid-state potentiometric gas sensor —current status and future trends” J Solid State Electrochemistry 13 (2009) 3–15; [3] P. Pasierb, E. Drożdż-Cieśla, R. Gajerski, S. Łabuś, S. Komornicki and M. Rękas “Chemical stability of $Ba(Ce_{(1-x)}Ti_x)_{(1-y)}Y_yO_3$ proton-conducting solid electrolytes” J Therm Anal Calorim 96 (2009) 2, 475–480; [4] P. Pasierb, M. Osiadły, S. Komornicki, M. Rękas, “Structural and electrical properties of $BaCe_{(Ti,\ Y)}O_3$ protonic conductors”, J Power Sources 196 (2011) 6205–6209; [5] Łącz, P. Pasierb, “Synthesis and properties of $BaCe_{1-x}Y_xO_{3-d}$–$BaWO_4$ composite protonic conductors”, J Therm Anal Calorim 113 (2013) 405–412.

Name: Kashinath C. Patil
Country: India
Date and place of birth: 1937, Savanur, India
Present position and address: Retd. Professor, Department of Inorganic and Physical Chemistry, Indian Institute of Science, Bangalore-560012, India
Email: kcpatil@ipc.iisc.ernet.in, kcpatil37@yahoo.co.in
Researcher ID: A-1747-2014
Education and scientific degrees: Ph.D:1968 (Indian Institute of Technology, Kanpur), D.Sc:1991 (Indian Institute of Science, Bangalore)
Workplaces: Indian Institute of Technology, Kanpur; St. Francis Xavier University, Antigonish (Nova Scotia), Canada; University of Florida, Gainesville, Florida, Jet Propulsion Laboratory, California Institute of Technology, Pasadena, California, USA and Indian Institute of Science, Bangalore, India.
Main fields of interest: high energy materials based on metal-hydrazine derivatives, combustion synthesis of advanced ceramics: alumina to zirconia, synthesis of nanocrystalline oxide materials by the controlled combustion of aqueous redox mixtures and combustible compounds (solution combustion synthesis), catalysis: metal/oxide nanocomposites
Relevant categories in thermal analyses: fields (inorganic chemistry; solid state chemistry; propellant chemistry and chemistry of hydrazine derivatives); methods (TG-DTG-DTA; DSC, dilatometry and temperature programmed reduction/oxidation)
Awards and acknowledgments: Indian Thermal Analysis Society –Netzsch Germany award (1989), Materials Research Society of India medal (1993), Diploma (Scientific center of Russian Academy of Sciences in Chernogolovka and International Association "Self-propagating High-temperature synthesis" (SHS-AS) award with jubilee medal for contribution to R&D in SHS (2007); Who is who in SHS (Self propagating High temperature Synthesis), Published by ISMAN, Chernoglovokia, Russia (2006)
Professional activities: Material Research Society of India, Indian Ceramic Society, Indian Thermal Analysis Society, Magnetic Society of India, National Institute of Advanced Studies
Publication record: papers (300), books (2), patents (3), citations (3848), h-index (33)
5 most important publications: [1] "Inorganic Hydrazine Derivatives" Eds. K.C.Patil and Tanu Mimani Rattan, Wiley; London (2014); [2] M.S. Hegde, Giridhar Madras and K.C. Patil, "Noble Metal Ionic Catalysts' Accounts of Chemical Research 42 (2009) 704–712; [3] J.J.Kingsley and K.C. Patil, "A novel combustion process for th synthesis of fine particle alpha alumina and related xide materials", Mater Letters 6 (988) 427–432; [4] P. Ravindranathan and K.C. Patil," A low temperature path to the preparation of ultrafine ferrites" Am.Ceram.Soc.Bull 66 (1987) 688–692; [5] K.C. Patil, S.T. Aruna and T. Mimani "Combustion synthesis: An update", Current Opinion in Solid State and Material Science 6 (2002) 507–512.

Name: Yoncho Pelovski
Country: Bulgaria
Date and place of birth: 1945, Bulgaria
Present position and address: University of Chemical Technology and Metallurgy; Bulgaria, PC 1756, Sofia, 8 Kliment Ohridski Blvd.
Email: pelovsky@uctm.edu; yonchop@gmail.com
Website: www.uctm.edu; www.bcci2001.com
Education and scientific degrees: Professor, Ph.D.
Workplaces: University of Chemical Technology and Metallurgy; Bulgarian Chamber of the Chemical Industry
Main fields of interest: thermochemical processes and technologies –thermal decomposition of solids—studies on thermo mechanochemistry, mechanism and kinecks; new materials and methods for production of inorganic chemicals; environmental sciences—utilization of wastes, clean up systems for industrial wastes waters and gases; sustainable development
Relevant categories in thermal analyses: fields (solid—thermal processes-decomposition mechanism and kinetics; solid type processes and mechanochemistry; thermal treatment of inorganic wastes); methods (TG, DTA, DSC and simultinious TG_FTIR and TG- solid electrolyte system measurements partial pressure of oxygen, high temperature microscopy)
Awards and acknowledgments: Prize for Science and Technology in Bulgaria (1984); Honoured inventor of Bulgaria (1988); NATO Awards (1994, 1999); Prize for best patent in Bulgaria (2013)
Professional activities: Head of Central Research Laboratory of the University of the Chemical Technology and Metallurgy (1985–1991); Chairman of the Bulgarian Society of Thermal Analysis and Calorimetry, ICTAC Council-Councillor at large-(1988–2001); ICTAC Council-Affiliate Councilor (2001–2004); Member of the Inion of Chemists in Bulgaria; Vice President of the Bulgarian Chamber of the Chemical Industry (2004–2014)
Publication record: papers (348), books (14), patents (95), total citations (237), sum of impact factors (24.6)
Equipments: Derivatograph 1500Q with inlet flow control and FTIR outlet gases, Shimadzu TG-DTA system with gas medium control, High temperature microskopy, STA PT1600 TG-DTA/DSC (STA Simultaneous Thermal Analysis), LINSEIS Messgeräte GmbH
5 most important publications: [1] Y. Pelovski, Thermal decomposition of Inorganic Sulphates, ICTAC News, Vol. 27, No 2 (1994), 95; [2] Y. Pelovski, V.Petkova, S. Nikolov, Study of the mechanism of the thermochemical decomposition of ferrous sulphate monohydrate, Thermochimica Acta, 274, (1996), 273; [3] Y. Pelovski, V. Petkova, Thermal decomposition of FeSO4-H2O-BaO2 mixtures, Thermochimica Acta, 284 (1996), 407; [4] Kotsilkova, R., V. Petkova, Y. Pelovski, 2001. Thermal analysis of polymer-silicate nanocomposites. Journal of Thermal Analysis and Calorimetry, v. 64, 591–598; [5] Low-dos irradiation by electron beam for the treatment of high SOx flue gas on the semi-pilot scale, Safety and Environmental Protection, vol, 87, No 2, (2009), 135–143.

Name: Luis A. Pérez-Maqueda
Country: Spain
Date and place of birth: 1968, Aberdeen, UK
Present position and address: Research Scientist, Instituto de Ciencia de Materiales de Sevilla (CSIC-Universidad de Sevilla), C. Americo Vespucio 49 Sevilla 41092 Spain
Email: maqueda@cica.es
Researcher ID: H-1686-2012
Education and scientific degrees: Chemist, Ph.D. in Chemistry, University of Seville
Workplaces: 1991–1995 Universidad de Sevilla (Spain), 1996–1997 Clarkson University (USA), 1998—Institute of Materials Science of Seville (Spain)
Main fields of interest: material science, solid-state chemistry, reactivity of solids, kinetics of heterogeneous processes, ceramics, polymer degradation, mechanochemsitry
Relevant categories in thermal analyses: fields (inorganic, materials, nano, minerals, polymer, glass, ceramics); methods (TG, DTA, EGA, DSC, thermomechanical analysis, kinetics, instrument development, other (sample controlled thermal analysis))
Professional activities: Member of the editorial board of Journal of Thermal Analysis and Calorimetry
Publication record: h-index (31)
Equipments: TG, DTA, DSC, TMA, EGA, SCTA
5 most important publications: [1] L. A. Perez-Maqueda; L. F. Wang; E. Matijevic. Langmuir 1998, 14(16), 4397–4401; [2] L. A. Perez-Maqueda; J. M. Criado; C. Real; J. Subrt; J. Bohacek. Journal of Materials Chemistry 1999, 9(8), 1839–1845; [3] L. A. Perez-Maqueda; J. M. Criado; F. J. Gotor; J. Malek. Journal of Physical Chemistry A 2002, 106 (12), 2862–2868; [4] S. Vyazovkin; A. K. Burnham; J. M. Criado; L. A. Perez-Maqueda; C. Popescu; N. Sbirrazzuoli. Thermochimica Acta 2011, 520(1-2), 1-19; [5] A. Perejon; N. Murafa; P. E. Sanchez-Jimenez; J. M. Criado; J. Subrt; M. J. Dianez; L. A. Perez-Maqueda. Journal of Materials Chemistry C 2013, 1(22), 3551–3562.

Name: Vilma Petkova
Country: Bulgaria
Date and place of birth: 1961, Bulgaria
Present position and address: Associate Professor in New Bulgarian University, Department "Natural Sciences", 21 Montevideo Str., 1618 Sofia, Bulgaria; Associate Professor in Institute of Mineralogy and Crystallography, Bulgarian Academy of Sciences, Acad. G. Bonchev Str., bldg.107, 1113 Sofia, Bulgaria
Email: vpetkova@nbu.bg, vilmapetkova@gmail.com
Website: http://nbu.bg/entrance.php?lang=1, http://www.imc.bas.bg/
Education and scientific degrees: University of Chemical Technology and Metallurgy, Sofia, Bulgaria
Workplaces: Associate Professor in New Bulgarian University, Department "Natural Sciences" (2013–), Institute of Mineralogy and Crystallography, Bulgarian Academy of Sciences (2000–)
Main fields of interest: study of inorganic materials, nano materials, minerals, cement, solids wastes and their applications with thermal, XRD, spectroscopic methods
Relevant categories in thermal analyses: fields (mineral resources and materials sciences; tribological-chemical activation, ecology; nanosciences and nanomaterials; cement and composites); methods (TG, DTA, DTG, DSC, kinetics, extremely high temperature (above 1,000 °C); specific heat, calorimetry)
Professional activities: Participation in Organizing Committee of 1st (2011) and 2nd (2013) Central and Eastern European Conference on Thermal Analysis and Calorimetry
Publication record: papers (104), citations (242), h-index (8)
Equipments: SETARAM SETSYS2400 (France), Stanton Redcrofd (England)
5 most important publications: [1] Pelovski, Y., V. Petkova, S. Nikolov, Study of the mechanism of the thermochemical decomposition of ferros sulphate monohydrate., Thermochimica Acta, ISSN 0040-6031, v. 274, (1996), 273–280; [2] Petkova V., Y. Pelovski, D. Paneva, I. Mitov, Influence of gas media on the thermal decomposition of second valence iron sulfates, Journal of Thermal Analysis and Calorimetry, 105, (2011), 793–803; [3] Petkova V., E. Serafimova, N. Petrova, Y. Pelovski, Thermochemistry of triboactivated Natural and NH4-exchanged Clinoptilolite mixed with Tunisian Apatite, Journal of Thermal Analysis and Calorimetry, Volume 105, Issue 2, (2011), 535–544, doi:10.1007/s10973-010-1224-8; [4] Koleva V., V. Petkova, IR spectroscopic study of high energy activated Tunisian phosphorite, Vibrational Spectroscopy, ISSN: 0924-2031, 58, (2012), 125–132; [5] Petkova V., V. Stoyanov, Y. Pelovski, TG–DTG–DTA in studying white self-compacting cement mortars, Journal of Thermal Analysis and Calorimetry, 109, (2012), 797–806.

Name: Luigi Petraccone
Country: Italy
Date and place of birth: 1972, Napoli, Italy
Present position and address: Assistant Professor of Physical Chemistry, Department of Chemical Sciences, University of Naples Federico II, Via Cintia 4, 80126 Naples, Italy
Email: luigi.petraccone@unina.it
Website: www.docenti.unina.it/luigi.petraccone
Education and scientific degrees: Degree in Chemistry, University of Naples Federico II (1998); Ph.D. in Chemical Sciences (2002); Assistant Professor in Physical Chemistry (2006).
Main fields of interest: conformational stability of biological macromolecules and their interactions with small molecule or other bio-macromolecules
Relevant categories in thermal analyses: fields (pharmaceutical, biology, life, nano); methods (DSC and ITC)
Awards and acknowledgements: "Lucci" Award of the Italian Association of Calorimetry and Thermal Analysis (A.I.C.A.T.)
Professional activities: Member of the Ph.D. Board in Chemical Sciences of the University of Naples Federico II. Member of the Italian Chemical Society (S.C.I.), Division of Physical Chemistry. Member of the Board of the Interdivisional Group of Calorimetry and Thermal Analysis (G.I.C.A.T.). Member of the Italian Association of Calorimetry and Thermal Analysis (A.I.C.A.T.). Member of the organizing committee of the Third International Meeting on G-quadruplex in Sorrento (Italy 2011). Member of the scientifc comitee of the XXXVI National Congress on Calorimetry, Thermal Analysis and Applied Thermodynamics in Cagliari (Italy 2014). Reviewer for prominent scientific journals published by the American Chemical Society (ACS), Elsevier and Wiley. Reviewer for the Czech Science Foundation.
Publication record: papers (48), books (2 book chapters), citation (710), h-index (17)
Equipments: Nano ITC Low Volume—TA Instruments and Nano DSC—TA Instruments
5 most important publications: [1] Fotticchia I, Giancola C, Petraccone L. " G-quadruplex unfolding in higher-order DNA structures". Chem Commun (2013) 49 (82) 9488–90; [2] Petraccone L, Malafronte A, Amato J, Giancola C "G-Quadruplexes from Human Telomeric DNA: How Many Conformations in PEG Containing Solutions?" J Phys Chem B. (2012) 116 (7) 2294–2305; [3] Petraccone L, Spink C, Trent JO, Garbett NC, Mekmaysy CS, Giancola C, Chaires JB "Structure and stability of higher-order human telomeric quadruplexes" J Am Chem Soc. (2011) 133 (51) 20951–61; [4] Cummaro A, Fotticchia I, Franceschin M, Giancola C, Petraccone L. "Binding properties of human telomeric quadruplex multimers: a new route for drug design" Biochimie (2011) 93 (9) 1392–1400, Top 25 Hottest Articles: Biochemistry, Genetics and Molecular Biology. Biochimie—July–September 2011; [5] Petraccone L, Trent JO, Chaires JB "The Tail of the Telomere" J. Am. Chem. Soc. (2008) 130 (49) 16530–156532.

Name: Henryk Piekarski
Country: Poland
Date and place of birth: 1945, Konstantynów, Poland
Present position and address: Full professor of Chemistry, Head of the Department of Physical Chemistry, Fac. of Chemistry, University of Łódź, ul. Pomorska 165, PL-90-236 Łódź, Poland
Email: kchfpiek@uni.lodz.pl
Education and scientific degrees: M.Sc. (1967), Ph.D. (1975), Dr. habil. (1987), Full Professor (1997) all at the University of Łódź,
Workplaces: University of Łódź, Fac. of Chemistry, Dept. of Physical Chemistry (1967–); Vrije Universiteit Amsterdam (17 months in 1979, 1986 and 1990)
Main fields of interest: physical chemistry of liquids; experimental thermodynamics; thermochemistry; solution calorimetry; biological calorimetry; micellar systems
Relevant categories in thermal analyses: fields (complex, organic, biology, liquid systems); methods (DSC, specific heat, calorimetry, microcalorimetry)
Awards and acknowledgments: Award of Ministry of Science and Higher Education (1976), Bene Merenti Medal University of Regensburg (2000), W. Świętosławski Medal of the Polish Society of Calorimetry and Thermal Analysis (2006), Medal Universitas Lodzensis Merentibus (2009)
Professional activities: Dean of the Faculty of Physics and Chemistry University of Łódź (1999–2002), Prorector in Charge of Research (2002–2008), Head of the Dept. of Physical Chemistry (1993–); Member of the Polish Society of Calorimetry and Thermal Analysis Board (1988–2003), vice-president (1991–1994 and 2000–2003), president (1994–2000); member of Editorial Advisory Board JTAC (1996–2000), Editor for Calorimetry JTAC (2000–2003); organiser and chairman of several international calorimetric conferences
Publication record: papers (108), conference contributions (130), citations (1050), h-index (19)
Equipments: TG-DSC 111, Micro DSC III, MS 80D—all "SETARAM", VP-ITC Micro Cal titration microcalorimeter, solution microcalorimeters "isoperibol" type
5 most important publications: [1] Effect of Non-electrolyte Properties on the Enthalpic Interaction Coefficients for NaCl/NaI –Non-electrolyte Pairs in Water, J. Chem. Soc. Faraday Trans., 87 (1991) 3661–3666; [2] Calorimetry-an important tool in solution chemistry, Thermochimica Acta, 420 (2004) 13–18; [3] Calorimetric study as a potential test for choosing treatment of B-cell chronic lymphocytic leukemia, Leukemia Research, 33 (2009) 308–314; [4] "Modification of the two-point scaling theory for the description of the phase transition in the solution. Analysis of the sodium octanoate aqueous solutions. J. Solution Chem., 41 (2012) 318–334; [5] A microcalorimetric titration study on the micelle formation of alkanediyl-α, ω-bis(dimethylalkylammonium bromide) surfactants at a 283.15–343.15 temperature range, J. Thermal Analysis and Calorimetry, 110 (2012) 263–271.

Name: Krzysztof Pielichowski
Country: Poland
Date and place of birth: 1969, Kędzierzyn, Poland
Present position and address: Head of Department of Chemistry and Technology of Polymers, CracowUniversity of Technology, ul. Warszawska 24, 31-155 Kraków, Poland
Email: kpielich@usk.pk.edu.pl
Website: www.pk.edu.pl
Education and scientific degrees: chemical technology, Cracow University of Technology (1992), Ph.D. (1995), D. Sc. (1999), Professor (2006)
Workplaces: Cracow University of Technology (1992–)
Main fields of interest: thermal decomposition of polymer nanocomposites, thermal energy storage materials, organic-inorganic hybrid materials
Relevant categories in thermal analyses: fields (polymer, nano); methods (TG, DSC, DMA, EGA)
Awards and acknowledgments: Award of the Rector of the Cracow University of Technology (2000, 2006, 2010, 2011)
Professional activities: Vice-chairman of the Polish Society of Calorimetry and Thermal Analysis (PTKAT, 2013); member of the Editorial Board of Polimery, Journal of Soft Matter, Technical Transactions—Chemistry (Editor); member of the Committee of Chemistry of the Polish Academy of Sciences (PAN); member of the Commission of Technical Sciences of the Polish Academy of Sciences and Arts (PAU)
Publication record: papers (95), books (3), patents (2)
Equipments: TG, DSC, DSC-TOPEM, DMA, TG/DSC
5 most important publications: [1] K. Pielichowski, J. Njuguna, Thermal Degradation of Polymeric Materials, Rapra Technology, Shawbury 2005 (ISBN 1-85957-498-X); [2] K. Pielichowski, Kinetic analysis of the thermal decomposition of polyaniline, Solid State Ionics, 104, 1997, 123; [3] K. Pielichowska, S. Głowinkowski, J. Lekki, D. Biniaś, K. Pielichowski, J. Jenczyk, PEO/carboxylic acid blends for thermal energy storage materials. Structural/morphological features and hydrogen interactions, European Polymer Journal, 44 (2008) 3344; [4] K. Raftopoulos, Ch. Pandis, L. Apekis, P. Pissis, B. Janowski, K. Pielichowski, J. Jaczewska, Polyurethane—POSS Hybrids: Molecular Dynamics Studies, Polymer, 51 (2010) 709; [5] K.N. Raftopoulos, M. Jancia, D. Aravopoulou, E. Hebda, K. Pielichowski, P. Pissis, POSS along the Hard Segments of Polyurethane. Phase Separation and Molecular Dynamics, Macromolecules, 46 (18) (2013) 7378–7386.

Name: Denis P. Pishchur
Country: Russia
Date and place of birth: 1984, Kokchetav, Kazakhstan
Present position and address: researcher, Nikolaev Institute of Inorganic Chemistry Siberian Branch of Russian Academy of Sciences. 3, Acad. Lavrentiev Ave., Novosibirsk, 630090, Russia
Email: denispischur@ngs.ru
ORCID: 0000-0002-0532-6478
Education and scientific degrees: Bachelor of physics, Novosibirsk State University (2001–2005); Master of physics, Novosibirsk State University (2005–2007); Ph.D. (2012).
Workplaces: Nikolaev Institute of Inorganic Chemistry Siberian Branch of Russian Academy of Sciences. (2007–)
Main fields of interest: spin crossover phenomena in Fe(II) complexes, chemical thermodynamic of coordination and cluster compounds
Relevant categories in thermal analyses: fields (inorganic, materials, nano, complex); methods (DSC, specific heat, calorimetry, instrument development)
Awards and acknowledgments: Netzsch award on XVII international conference on chemical thermodynamics in Russia (2009), award on XLV International Scientific Student Conference (Novosibirsk, 2007)
Equipments: Netzsch DSC 204 F1 Phoenix, precision low-temperature adiabatic calorimeter
5 most important publications: [1] G. A. Berezovskii, M. B. Bushuev, D. P. Pishchur and L. G. Lavrenova: Heat capacity of polynuclear $Fe(HTrz)_3(B_{10}H_{10})H_2O$ and trinuclear $[Fe_3(PrTrz)_6(ReO_4)_4(H_2O)_2](ReO_4)_2$ complexes (HTrz = 1,2,4-triazole, PrTrz = 4-propyl-1,2,4-triazole) manifesting $^1A_1 \leftrightarrow {}^5T_2$ spin transition. Journal of Thermal Analysis and Calorimetry, Vol. 93 (2008), 999–1002; [2] S. G. Kozlova, S. P. Gabuda, G. A. Berezovskii, D. P. Pischur, Y. V. Mironov, A. Simon, V. E. Fedorov: Quantum chemical study and low-temperature calorimetry of phase transition in $V_4S_9Br_4$. Journal of Solid State Chemistry 181 (2008) 2877–2881; [3] G. A. Berezovskii, V. A. Daletskii, D. P. Pishchur, A. D. Strekalova, L. G. Lavrenova: Thermodynamic properties of complex compounds of iron(II) nitrate with tris(3,5-dimethylpyrazol-1-yl)methane. Russian Journal of Physical Chemistry A, Vol. 87 (2013) , No. 8, 1272–1275; [4] Sergey A. Adonin, Maxim N. Sokolov, Pavel A. Abramov, Svetlana G. Kozlova, Denis P. Pishchur, Lilya A. Sheludyakova, Vladimir P. Fedin: Thermochromic behavior and phase transition of new octanuclear polyiodobismuth(III)ate. Inorganica Chimica Acta 419 (2014), 19–25; [5] M. B. Bushuev, V. A. Daletsky, D. P. Pishchur, Y. V. Gatilov, Ilya V. Korolkov, E. B. Nikolaenkova and V. P. Krivopalov: Unprecedented bistability domain and interplay between spin crossover and polymorphism in a mononuclear iron(II) complex. Dalton Transactions, Vol. 43 (2014), 3906–3910.

Name: Polycarpos Pissis
Country: Greece
Date and place of birth: 1947, Cyprus
Present position and address: Professor, National Technical University of Athens, Department of Physics, Zografou Campus, 15780 Athens, Greece
Email: ppissis@central.ntua.gr
Website: http://dielectricsgroup.physics.ntua.gr/pissis
Education and scientific degrees: Diploma (1973) and Ph. D. (1977) in Physics, University of Goettingen, Germany
Workplaces: University of Goettingen, Germany (1973–1977), National Technical University of Athens (1978–)
Main fields of interest: structure—property relationships in polymers and composites, complex polymeric systems, interfacial effects in polymer nanocomposites, conductive polymers and composites, biomaterials, hydration properties of polymers and biopolymers
Relevant categories in thermal analyses: fields (nano, organic, polymer, biology); methods (DSC, thermomechanical analysis, specific heat)
Professional activities: editorial membership of journals, organization of conferences and workshops, coordination of national and international projects
Publication record: papers (251), books (11), citations (3600), h-index (32)
Equipments: TA Q200 Modulated Differential Scanning Calorimeter, PerkinElmer Pyris 6
5 most important publications: [1] G. Barut, P. Pissis, R. Pelster and G. Nimtz, Glass transition in liquids: two versus three-dimensional confinement, Phys. Rev. Lett. 80, 3543–3546 (1998); [2] Ye. P. Mamunya, V. V. Davydenko, P. Pissis, E. V. Lebedev, Electrical and thermal conductivity of polymers filled with metal powders, Eur. Polym. J. 38, 1887–1897 (2002); [3] A. Espadero Berzosa, J. L. Gomez Ribelles, S. Kripotou, P. Pissis, Relaxation spectrum of polymer networks formed from butyl acrylate and methyl methacrylate monomeric units, Macromolecules 37, 6472–6479 (2004); [4] D. Fragiadakis, P. Pissis, L. Bokobza, Glass transition and molecular dynamics in poly(dimethylsiloxane)/silica nanocomposites, Polymer 46, 6001–6008 (2005); [5] A. Panagopoulou, A. Kyritsis, N. Shinyashiki, P. Pissis, Protein and water dynamics in bovine serum albumin—water mixtures over wide ranges of composition, J. Phys. Chem. B 116, 4593–4602 (2012).

Name: Alfonz Plško
Country: Slovakia
Date and place of birth: 1953, Skalica, Slovak Republik
Present position and address: Ass. Prof. University of Alexander Dubcek in Trencin, Studentska 2, 91150 Trencin, Slovak Republic
Email: alfonz.plsko@tnuni.sk
Education and scientific degrees: Inorganic chemistry, Doc. CSc. Ing.
Workplaces: University of Alexander Dubcek in Trencin, Slovak Republik
Main fields of interest: glass, sol-gel method, thermal properties of materials
Relevant categories in thermal analyses: fields (inorganic, materials, nano, glass, ceramics); methods (TG, DTA, DSC, kinetics)
Publication record: papers (110), patents (4), citations (72), h-index (4)
Equipments: NETZSCH STA 449 F1 Jupiter
5 most important publications: [1] Alfonz Plško, Marek Liška, Jana Pagáčová, Crystallization Kinetics of Al2O3-Yb2O3 Glasses, Journal of Thermal Analysis and Calorimetry. ISSN 1388-6150., 2012 Vol.108, No.2, pp. 505–509; [2] Katarína Moricová, Eugen Jóna, Alfonz Plško, S.C. Mojumdar, Thermal Stability of Li2O-SiO2-TiO2 gels evaluated by the Induction Period of Crystallization, Journal of Thermal Analysis and Calorimetry, 2010 Vol.100, No.3, pp. 817–820; [3] Alfonz Plško, Katarína Faturíková, Marek Liška, Jana Pagáčová, Iveta Papučová, Jana Šulcová, Dielectric Properties of SiO2-TiO2-Al2O3 thin Films prepared by Sol-Gel Method for humidity Sensors, Physics and Chemistry of Glasses—European Journal of Glass Science and Technology part B. ISSN 1753-3562.—2012 Vol.53, No.1, pp. 1–6; [4] Marek Liška, Jana Holubová, Eva Černošková, Zdeněk Černošek, Mária Chromčíková, Alfonz Plško, Nucleation and Crystallization of an As(2)Se(3) undercooled melt, Physics and Chemistry of Glasses—European Journal of Glass Science and Technology part B. ISSN 1753–3562. 2012 Vol.53, No.6, pp. 289–293; [5] Peter Simon, Katarína Nemčeková, Eugen Jóna, Alfonz Plško, Darina Ondrušová, Thermal stability of glass evaluated by the induction period of crystallization, Thermochimica Acta, 2005 Vol.428, No.1–2 (2005), pp. 11–14.

Name: Pavel Plyusnin
Country: Russia
Date and place of birth: 1974, Kurgan, USSR
Present position and address: Senior Researcher of the Nikolaev Institute of Inorganic Chemistry Siberian Branch of Russian Academy of Sciences.; 3, Acad. Lavrentiev Ave., Novosibirsk, 630090, Russia
Email: plus@niic.nsc.ru
Researcher ID: B-2062-2014
Education and scientific degrees: chemist, Kurgan State University (1998–2003); Ph.D. (2009)
Workplaces: Nikolaev Institute of Inorganic Chemistry Siberian Branch of Russian Academy of Sciences (2003–); Novosibirsk State University (2007–)
Relevant categories in thermal analyses: fields (inorganic compounds, complex compounds); methods [TG, DTA, DSC, EGA, extremely high temperature (above 1,000 °C)]
Publication record: papers (43), h-index (5)
Equipments: Q-1000 derivatograph modified, TG 209 F1 Iris® thermobalance (NETZSCH), STA 449A1 Jupiter® combined with QMS 403D Aëolos® (NETZSCH)
List of the 5 most important publications: [1] P. Plyusnin, I. Baidina, Y. Shubin, S. Korenev $[M(NH_3)_5Cl][AuCl_4]Cl \cdot nH_2O$ (M = Rh, Ru, or Cr): Synthesis, crystal structure, and thermal properties Russian Journal of Inorganic Chemistry, 2008, Vol. 53, № 11, pp. 1724–1732; [2] T. Nedoseykina, P. Plyusnin, Y. Shubin, S. Korenev. XAFS investigation of $[Pd(NH_3)_4][AuCl_4]_2$ and its thermolysis products, JTAC, 2010. V. 102, № 2. pp. 703–708; [3] A. Zadesenets, E. Filatov, P. Plyusnin, I. Baidina, V. Dalezky, Y. Shubin, S. Korenev Bimetallic single-source precursors $[M(NH_3)_4][Co(C_2O_4)_2(H_2O)_2] \cdot 2H_2O$ (M = Pd, Pt) for the one run synthesis of CoPd and CoPt magnetic nanoalloys, Polyhedron 30 (2011) pp. 1305–1312; [4] E. Semitut, P. Plyusnin, Yu. Shubin, S. Veniaminov, S. Korenev Investigation of thermal properties of double complex salts $[M(NH_3)_5Br][AuBr_4]_2 \cdot nH_2O$, M = Rh, Ir, JTAC 2012. V. 109, № 2 pp. 901–905; [5] Y. Shubin, P. Plyusnin, M. Sharafutdinov In Situ Study of Au-Pd Nanoporous Alloy Formation by Single-Source Precursor Thermolysis Nanotechnology 23 (2012) 405302.

Name: György Pokol
Country: Hungary
Date and place of birth: 1950, Budapest, Hungary
Present position and address: Budapest University of Technology and Economics (BME), Department of Inorganic and Analytical Chemistry; 1521 Budapest, P.O.Box 91, Hungary
Email: pokol@mail.bme.hu
Website: http://iaachem.bme.hu
ORCID: 0000-0003-1597-9808
Education and scientific degrees: Chemical engineer, Technical University of Budapest (BME), 1973; C.Sc. (equivalent to Ph.D.), Hungarian Academy of Sciences (HAS), 1987; Dr. Habil., BME, 1994; D.Sc., HAS, 1996
Workplaces: Budapest University of Technology and Economics (BME), Department of Inorganic and Analytical Chemistry 1973–
Main fields of interest: analytical chemistry, chemistry of solids
Relevant categories in thermal analyses: fields (inorganic, materials, nano, pharmaceutical, ceramics); methods (TG, DTA, TG-MS, TG-FTIR, DSC, temperature programmed powder XRD)
Awards and acknowledgments: Corresponding Member of the Finnish Chemical Society, 1999; Náray-Szabó Scientific Award (Hungarian Chemical Society), 2010
Professional activities: European Symposium on Thermal Analysis and Calorimetry (ESTAC), Scientific Committee, Chairman: 1998–2002; Secretary: 2002–2006; HAS Scientific Committee of Analytical and Environmental Chemistry, co-chair, 2011–; Hungarian Chemical Society, Analytical Section, Chairman, 2009–
Publication record: papers (155), book chapters (3), patents (5), citations (915), sum of impact factors (171.8)
Equipments: TG-DTA-MS; TG-FTIR; DSC (all TA Instruments); powder XRD (Panalytical)
5 most important publications: [1] Pokol, G., Várhegyi, G. Kinetic aspects of thermal analysis. Critical Reviews in Analytical Chemistry 19:(1) pp. 65–93. (1988); [2] Kozma, D., Pokol, G., Ács, M.: Calculation of the efficiency of optical resolutions on the basis of the binary phase diagram for the diastereoisomeric salts. Journal of the Chemical Society-Perkin Transactions 2:(3), pp. 435–439. (1992); [3] Kök, M.V., Pokol, G., Keskin, C., Madarasz, J., Bagci, S. Combustion characteristics of lignite and oil shale samples by thermal analysis techniques. Journal of Thermal Analysis and Calorimetry 76:(1), pp. 247–254. (2004); [4] Okuya, M., Shiozaki, K., Horikawa, N., Kosugi, T., Kumara, G.R.A., Madarasz, J., Kaneko, S., Pokol, G.: Porous TiO_2 thin films prepared by spray pyrolysis deposition (spd) technique and their application to UV sensors. Solid State Ionics 172:(1–4), pp. 527–531. (2004); [5] Szilágyi, I.M., Madarász, J., Pokol, G., Király, P., Tárkányi, G., Saukko, S., Mizsei, J., Tóth, A.L., Szabó, K., Varga-Josepovits, K.: Stability and controlled composition of hexagonal WO_3. Chemistry of Materials 20:(12), pp. 4116–4125. (2008).

Name: Vlad Tudor Popa
Country: Romania
Date and place of birth: 1952, Bucharest, Romania
Present position and address: Director, Romanian Academy, "Ilie Murgulescu" Institute of Physical Chemistry Splaiul Independentei 202, Bucharest, 060021, Romania
Email: vtpopa@icf.ro
Website: www.icf.ro
Education and scientific degrees: B.Sc.—Chemistry (1975), M.Sc.—Physical Chemistry (1976), Ph.D.—Chemistry (1998)
Workplaces: IAMC Otopeni (1976–1980); Syracuse University (1994–1996); Institute of Physical Chemistry (1980–)
Main fields of interest: statistical mechanics; chemical kinetics; complex reaction theory; homogeneous, heterogeneous and enzyme catalysis; materials science; microbiology
Relevant categories in thermal analyses: fields (materials, nano, complex, organic, pharmaceutical, polymer, biology, life); methods (TG, DSC, kinetics, microcalorimetry, complex thermal signal analysis)
Awards and acknowledgments: "Ilie Murgulescu" award of the Romanian Academy, 1995
Professional activities: Co-chair and organizer of the International Conference of Physical Chemistry: ROMPHYSCHEM-13 (2008), -14 (2010), -15 (2013); Member of the Romanian Academy Commission of Thermal Analysis and Calorimetry; Romanian Catalysis Society; Romanian Chemical Society; American Association for the Advancement of Science (AAAS)
Publication record: papers (49), citations (420), h-index (10), sum of impact factors (69)
Equipments: MicroDSC III and VII (Setaram)
5 most important publications: [1] V. T. Popa and E. Segal, "Shape Analysis of DSC Ice Melting Endotherms: Towards an Estimation of the Instrumental Profile", J. Therm. Anal. Calorim., 69 (2002) 149; [2] M. Constantinescu, L. Dumitrache, D. Constantinescu, E. M. Anghel, V. T. Popa, A. Stoica., M. Olteanu, "Latent heat nano composite building materials", Eur. Polym. J., 46(12) (2010) 2247; [3] D. C. Zaharia, C. Iancu, A. T. Steriade, A. A. Muntean, O. Balint, V. T. Popa, M. I. Popa, M. A. Bogdan, "MicroDSC study of Staphylococcus epidermidis growth", BMC Microbiology, 10:322 (2010); [4] A. M. Musuc, M. Badea-Doni, L. Jecu, A. Rusu, V. T. Popa, "FTIR, XRD, and DSC analysis of the rosemary effect on polyethylene structure and biodegradability", J. Therm. Anal. Calorim., 114 (1) (2013) 169; [5] D. C. Zaharia, A. A. Muntean, M. G. Popa, A. T. Steriade, O. Balint, R. Micut, C. Iftene, I. Tofolean, V. T. Popa, C. Baicus, M. A. Bogdan, and M. I. Popa, "Comparative analysis of Staphylococcus aureus and Escherichia coli microcalorimetric growth", BMC Microbiology, 13:171 (2013).

Name: Crisan Popescu
Country: Romania
Date and place of birth: 1953, Lugoj, Romania
Present position and address: Professor at University "Aurel Vlaicu" Arad, Romania; research scientist at DWI at the RWTH Aachen University, Aachen, Germany; DWI, Forckenbeckstrasse 50, D-52064 Aachen, Germany
Email: popescu@dwi.rwth-aachen.de
Researcher ID: C-1207-2013
Education and scientific degrees: Chemical Engineer, Polytechnic University of Iasi, Romania and Shanghai Textile Engineering Institute, China (1972–1977); Ph.D., University of Bucharest, Romania (1990–1993); Professor (1996–).
Workplaces: University "Aurel Vlaicu" Arad, Romania (1996–), DWI at the RWTH Aachen University, Aachen, Germany (2001–); LACECA research Centre Bucharest, Romania (1983–1989; 1990–2000); Wool Centrale (1977–1983; 1989)
Main fields of interest: physics and chemistry of materials; bio-polymers (keratins); heterogeneous processes.
Relevant categories in thermal analyses: fields (materials, biology); methods (kinetics)
Awards and acknowledgments: Honorific member of CEEC-TAC for contribution to the Thermal Analysis and Calorimetry
Professional activities: Founding member of the Romanian Committee of TAC; Councillor at large in ICTAC (1998–2006); Chairman of the Committee for "Life-time prediction of materials" of the ICTAC (2004–2013); Chairman of the Scientific Commission of the ICTAC (since 2012); Co-chairman of the 1st Central and East European Conference on TAC (2011); Member of the Editorial Board of "Thermochimica Acta"
Publication record: papers (105), books (15), patents (11), citations (1227), h-index (14)
Equipments: TGA, TGA-FTIR, DSC, ITC
5 most important publications: [1] S. Vyazovkin, A. K. Burnham, J. M. Criado, L. A. Perez-Maqueda, C. Popescu, N. Sbirrazzuoli: ICTAC Kinetics Committee recommendations for performing kinetic computations on thermal analysis data, Thermochimica Acta, 520 (2011) 1-19; [2] D. Istrate, C. Popescu, M. Moeller: Non-Isothermal Kinetics of Hard alpha-Keratin Thermal Denaturation, Macromolecular Bioscience, 9 (2009) 805–812; [3] I. Berndt, C. Popescu, F. J. Wortmann, W. Richtering: Mechanics versus thermodynamics: Swelling in multiple-temperature-sensitive core-shell microgels, Angewandte Chemie-International Edition, 45 (2006) 1081–1085; [4] C. Popescu, E. Segal: Critical considerations on the methods for evaluating kinetic parameters from nonisothermal experiments, International Journal of Chemical Kinetics, 30 (1998) 313–327; [5] C. Popescu: Integral method to analyze the kinetics of heterogeneous reactions under non-isothermal conditions—A variant on the Ozawa-Flynn-Wall method, Thermochimica Acta, 285 (1996) 309–323.

Name: Arun Pratap
Country: India
Date and place of birth: 1959, Chapra, India
Present position and address: Professor of Condensed Matter Physics, Applied Physics Department, Faculty of Technology and Engineering, The Maharaja Sayajirao University of Baroda Vadodara-390 001 (Gujarat State), India
Email: apratapmsu@yahoo.com, apratapmsu@gmail.com
Education and scientific degrees: M.Sc. (Physics) 1984, Bihar University, Muzaffarpur; Ph.D. (Condensed Matter Physics) 1989, Rajasthan University, Jaipur
Workplaces: Assistant Professor of Physics, Rajasthan University, Jaipur (1988–1994); Reader in Applied Physics, The M.S. University of Baroda (1994–2002); Professor of Applied Physics, The M.S. University of Baroda (2002–2008); Professor of Condensed Matter Physics, The M.S. University of Baroda (2008–)
Main fields of interest: atomic dynamics of liquid and amorphous alloys; kinetics of crystallization of multi-component amorhous alloys through thermal analysis; thermodynamics of amorphous phase formation; nano-strucured materials and nano-quasi crystals, polymers, superconductors
Relevant categories in thermal analyses: fields (experimental studies of crystallization kinetics, glass transition and theoretical investigation of glass forming ability and estimation of Kauzmann temperature); methods (DSC, Modulated DSC, TGA)
Awards and acknowledgments: Netzsch-ITAS award in 2008 conferred by Indian thermal analysis society (ITAS); IIM-KK award 2012 by Indian Institute of Metals (IIM)
Professional activities: Vice President of Indian Thermal Analysis Society (ITAS) 2008–2011; Charman of Indian Physics Association (IPA), Baroda Chapter 2013–; Secretary, IPA, Baroda Chapter 2002–2004; Vice Chairman, IPA, Baroda Chapter 2004–2006; Fellow, Gujarat Science Academy (GSA); Secretary, GSA, Baroda Chapter; Life member of National Academy of Sciences, India (NASI), Materials Society of India (MRSI), Indian Institute of Metals (IIM), Indian Society of Technical Education (ISTE), Thermophysical Society of India (TPSI)
Publication record: papers (110), books (4), patens (1), citations (459), h-index (13)
Equipments: DSC Shimadzu- DSC50, Simultaneous TGA-DTA Linseis—Germany
5 most important publications: [1] A Pratap, T Lilly Shanker Rao, KN Lad, HD Dhurandhar, Isoconversional vs. Model fitting methods Journal of thermal analysis and calorimetry 2007, 89 (2), 399–405; [2] SR Joshi, A Pratap, NS Saxena, MP Saksena, A Kumar, Heating rate and composition dependence of the glass transition temperature of a ternary chalcogenide glass Journal of materials science letters 1994, 13 (2), 77–79; [3] KG Raval, KN Lad, A Pratap, AM Awasthi, S Bhardwaj, Crystallization kinetics of a multicomponent Fe-based amorphous alloy using modulated differential scanning calorimetry, Thermochimica Acta 2005, 425 (1), 47–57; [4] Rehani, PB Joshi, KN Lad, A Pratap, Crystallite size estimation of elemental and composite silver nano-powders using XRD principles, BR Indian Journal of pure and applied physics 2006, 44 (2), 157; [5] A Pratap, K Sharma, Applications of some thermo-analytical techniques to glasses and polymers, Journal of thermal analysis and calorimetry 2012, 107 (1), 171–182.

Name: Duncan M. Price
Country: UK
Date and place of birth: 1964, Bath, England
Present position and address: Senior Development Chemist, Edwards Ltd., Kenn Road, CLEVEDON, North Somerset, UK BS21 6TH
Email: duncan.price@edwardsvacuum.com
Education and scientific degrees: BSc (Hons) Chemistry (1985), Ph.D. Materials Engineering (2002)
Workplaces: Courtaulds plc. (1988–1998), Loughborough Unversity (1998–2006), Edwards Ltd. (2007–)
Main fields of interest: all aspects of materials science and engineering, semiconductor processing
Relevant categories in thermal analyses: fields (materials, nano, pharmaceutical, polymer, metallurgy, semiconductors); methods (TG, EGA, DSC, thermomechanical analysis, instrument development, dielectric techniques, thermo-microscopy and scanning probe techniques)
Professional activities: former regional editor of J. Therm. Anal. Calorim., committee member of RSC Thermal Methods Group (UK)
Publication record: papers (83), books (6), patents (4), citations (1113), h-index (20)
5 most important publications: [1] Price, Duncan M.; Reading, Michael; Hammiche, Azzedine; Pollock, Hubert M.; Micro-thermal analysis: scanning thermal microscopy and localised thermal analysis. International Journal of Pharmaceutics 192 (1) (1999) 85–96; [2] Hammiche, A.; Bozec, L.; Conroy, M.; Pollock, H. H.; Mills, G.; Weaver, J. M. R.; Price, D. M.; Reading, M.; Hourston, D. J.; Song, M.; Highly localized thermal, mechanical and spectroscopic characterization using miniaturized thermal probes. Journal of Vacuum Science and Technology B 18 (3) May/Jun (2000) 1322–1332; [3] Price, Duncan M.; Hawkins, Michael; Calorimetry of two disperse dyes using thermogravimetry. Thermochimica Acta 315 (1) (1998) 19–24; [4] Price, Duncan M.; Jarratt, Mark; Thermal conductivity of PTFE and PTFE composites. Thermochimica Acta 392–393 (2002) 229–234; [5] Price, D.M.; Reading, M.; Hammiche, A.; Pollock, H. M.; Branch, M. G.; Localised thermal analysis of a packaging film. Thermochimica Acta 332 (2) (1999) 143–149.

Name: R. Bruce Prime
Country: USA
Date and place of birth: 1943, Los Angeles, California
Present position and address: Retired from IBM. 4018 Shona Ct, San Jose, CA 95124
Email: rbprime@sbcglobal.net
Website: www.primethermosets.com
Education and scientific degrees: BS Chemistry Loyola University, Los Angeles (now Loyola Marymount University). Ph.D. Chemistry, Rensselaer Polytecnic Institute, Troy, NY (with Bernhard Wunderlich)
Workplaces: IBM Endicott, NY and San Jose, CA
Main fields of interest: thermal analysis of polymers and composites; cure and characterization of thermosets
Relevant categories in thermal analyses: fields (materials, nano, polymer, thermosets, composites); methods (DSC, TGA, EGA, thermomechanical analysis, dynamic mechanical analysis, kinetics)
Awards and acknowledgments: NATAS Fellow (1985). NATAS Outstanding Achievement (Mettler-Toledo) Award (1989). SPE Fellow (1997)
Professional activities: NATAS: Emeritus, Publications Chair (1977), Annual Conference Technical Program Chair (1981), Awards Chair (1983). SPE: Emeritus, Polymer Analysis Division Chair (1987–1988). ACS, Emeritus. Golden Gate Polymer Forum (GGPF): Founder (1980), Board of Directors
Publication record: papers (68), books (1), patents (8), citations (1690), h-index (17)
5 most important publications: [1] Chapter on “Thermosets” in Thermal Characterization of Polymeric Materials (E. A. Turi, ed.) Academic Press, (1981), pp. 453–569. 2nd ed. Academic Press, (1997), pp. 1379–1766; [2] Thermal Analysis of Polymers: Fundamentals and Applications (J. D. Menczel, R. B. Prime, eds.) Wiley (2009); [3] R. B. Prime, Differential scanning calorimetry of the epoxy cure reaction, Polymer Engineering and Science 13 (5), 365–371 (1973); [4] R. B. Prime, B. Wunderlich, L. Melillo, Extended-chain crystals. V. Thermal analysis and electron microscopy of the melting process in polyethylene, Journal of Polymer Science Part A□2: Polymer Physics 7 (12), 2091–2097 (1969); [5] R. B. Prime, B. Shushan, Thermogravimetric analyzer/atmospheric pressure chemical ionization tandem triple quadrupole mass spectrometer system for evolved gas analysis, Analytical Chemistry 61 (11), 1195–1201 (1989).

Name: Marek Pyda
Country: Poland, USA
Date and place of birth: 1950, Wagrowiec, Poland
Present position and address: Rzeszow University of Technology, Al. Powstancow Warszawy 6, 35-959 Rzeszow, Poland
Email: mpyda@utk.edu
Education and scientific degrees: A. Mickiewicz University Poznan, Poland (Ph.D.-1982); Professor (2006)
Workplaces: University of Tennessee, Department of Chemistry. Knoxville, TN, USA (1992–2005); University of Technology, Department of Chemistry. Rzeszow, Poland (2005–)
Main fields of interest: thermal analysis and physical chemistry of synthetic and biological polymers and small molecules advanced thermal analysis of polymers
Relevant categories in thermal analyses: fields (polymer, materials, pharmaceutical, food, other: biomaterials); methods (DSC, TG, specific heat, calorimetry, other: temperature-modulated DSC (TMDSC), fast scanning calorimetry (FSC))
Awards and acknowledgments: from TA Instuments (2001)
Professional activities: Chairman of Polymer Session at ACS Meeting 2005; Chairman of Thermoplastic Polymers Session at NATAS 2003 and 2004; Member of the North American Thermal Analysis Society (NATAS), the American Physics Society (APS), and the American Chemical Society (ACS).; Journal Reviewer for: Macromolecules; J.Polymer Science; Polymers Physics; Polymer; Thermochimica Acta; Langmiur; J.Colloid Science; Journal of Agricultural and Food Chemistry; European Polymer Journal; Journal of Thermal Analysis and Calorimetry
Publication record: papers (190), book chapters (5), citations (1600), h-index (23), sum of impact factors (120)
Equipments: DSC, TMDSC, optical microscopy with hot stage
5 most important publications: [1] M. Pyda, M. Bartkowiak, and B. Wunderlich, "Computation of Heat Capacities of Solids Using a General Tarasov Equation" J. Thermal Analysis, 52, 631–656 (1998); [2] M. Pyda, A. Boller, J. Grebowicz, H. Chuah, B. V. Lebedev, and B. Wunderlich, "Heat Capacity of Poly(trimethylene Terephthalate)" J. Polymer Sci., Part B: Polymer Phys., 36, 2499–2511 (1998); [3] M.Pyda and B. Wunderlich "Heat Capacities of High Polymers"; in J. Brandrup, E. H. Immergut, and E. A. Grulke, eds., "Polymer Handbook," Wiley, 4th, revised ed. Pages VI/1–VI/38, 1999; [4] M. Pyda, Melting, Chapter 9 in Handbook of Polymer Crystallization, Eds. Ewa Piorkowska-Galeska, and Gregory Rutledge by Wiley, July 2013 pages: 265–285. ISBN: 978-0-470-38023-9; [5] M. Pyda Conformational Heat Capacity of Interacting Systems of Polymer and Water, Macromolecules, 35, 4009–4016 (2002).

Name: Janusz J. Pysiak
Country: Poland
Date and place of birth: 1933, Wilno, Poland
Present position and address: Professor, Warsaw University of Technology, School of Technical and Social Sciences, Institute of Chemistry, PL-09-400 Płock, 17 Łukasiewicza St., Poland
Email: ich@zto.pw.plock.pl
Education and scientific degrees: Warsaw Technical University of Technology (1952); Ph.D. (1964), D.Sc. (1978)
Workplaces: Warsaw Technical University of Technology (1958–1960); Institute of Physical Chemistry Polish Academy of Sciences (1960–1962); Warsaw Technical University of Technology (1962–1970); Warsaw University of Technology, School of Technical and Social Sciences in Plock (1970–)
Main fields of interest: inorganic chemistry; thermal analysis; chemistry of solids
Awards and acknowledgments: Sci. Awards of Polish Society of Chemistry (1967); Honour Member of Polish Society of Thermal Analysis and Calorimetry; Two Scientifical Awards of Education Ministry (1981, 1982)
Professional activities: Member of International Committee 8th ICTA (1985); member of International confederation for Thermal Analysis and Calorimetry (ICTAC); chairman of Sci. Comm. of I-VIII Polish Seminars to the memory of Prof. St. Bretsznajder (1977–2001); member of Polish Chemical Society; member of Polish Society of Thermal Analysis and Calorimetry; guest editor of special issues of J. Therm. Anal. Calorim., 43 (1995); 60 (2000)
Publication record: papers (248), patens (10)
5 most important publications: [1] J. J. Pysiak, B. Sabalski: Compensation effect and isokinetics temperature an thermal dissociation reactions of type A(solid) = B(solid) + C(gas) interpretation of the Arrhenius equation as a projection correlation, J. Thermal Anal., 17 (1979) 243; [2] J. J. Pysiak, B. Pacewska: Studies on thermal dissociation of basic aluminium–ammonium sulphate, J. Thermal Anal., 19 (1980) 79; 19 (1980) 89; 19 (1980) 243; [3] J. J. Pysiak, A. Glinka: Thermal dissociation of basic aluminium-potassium sulphate, Thermochim. Acta, 44 (1981) 21; 44 (1981) 22; 44 (1981) 101; [4] J. J. Pysiak: On some errors committed in the studies of kinetics of thermal dissociation of solids, J. Thermal Anal., 43 (1995) 9; [5] J. J. Pysiak, A. Y. Al.-Badwi: Kinetics equations for thermal dissociation processes, J. Therm. Anal. Calorim., 63 (2001); 68 (2002) 169, 70 (2002)429. Books: J. J. Pysiak: General chemistry, Warsaw University of Technology, Institute of Chemistry, Płock (2001).

Name: Stanislaw L. Randzio
Country: Poland
Date and place of birth: 1941, Płaska, Poland
Present position and address: Professor Emeritus, Institute of Physical Chemistry, Polish Academy of Sciences, Kasprzaka 44/52, 01-224 Warszawa, Poland
Email: randzio@ichf.edu.pl
Education and scientific degrees: Master of Chemistry, Warsaw University (1965), Ph.D. (1975), D.Sc. (1985)
Workplaces: Polish Academy of Sciences, Institute of Physical Chemistry (1967–); Several long term stays at BYU (USA), Blaise, Pascal University (France), University of Lund (Sweden) and University of Paris (France)
Main fields of interest: pVT-controlled calorimetry and transitiometry; pressure effects on phase transitions in various substances and materials (dense sub-, near- and supercritical fluids, polymers, biopolymers and food materials, molecules of pharmaceutical importance)
Awards and acknowledgments: Four Polish Academy of Sciences Awards (1972, 1976, 1985, 1988), CalvetMedal (AFCAT 1993), Applied Chemical Thermodynamics Award (STK1995), Christensen Award (American Calorimetry Conference 1996)
Professional activities: Associated Member of the IUPAC Commission on Chemical Thermodynamics (1987–1995); Member of the Editorial Board of Thermochimica Acta (1996–2003)
Publication record: papers (85), books (3), patens (14), citations (274)
5 most important publications: [1] S. L. Randzio: A Pressure-scanning calorimeter, J. Phys., E 16 (1983) 691.; [2] S. L. Randzio: Scanning calorimeters controlled by an independent thermodynamic variable: definitions and some metrological problems, Thermochim. Acta, 89 (1985) 215; [3] S. L. Randzio: From calorimetry to equations of state, Chem. Soc. Rev., 24 (1995) 359; [4] S. L. Randzio: Scanning transitiometry, Chem. Soc. Rev., 25 (1996) 383; [5] S. L. Randzio: Recent developments in calorimetry, Ann. Rep. Prog. Chem., (The Royal Society of Chemistry), Sect. C, 98 (2002) 157.

Name: Jonjaua Ranogajec
Country: Serbia
Date and place of birth: 1953, Kustilj, Serbia
Present position and address: Full professor, Faculty of Technology, University of Novi Sad, Bulevar cara Lazara 1, 21000 Novi Sad, Serbia
Email: janjar@uns.ac.rs
Education and scientific degrees: 1986 Ph.D. in Technical Sciences–Materials Engineering, Faculty of Technology, University of Novi Sad; 1981 Magister of Technical Sciences, Faculty of Technology, University of Novi Sad; 1976 Engineer of Technology, Faculty of Technology, University of Novi Sad. Specializations: Ecole Nationale Suprieure de Ceramique Industrielle, Limoges, France (2004); Ecole Polytechnique Fédérale de Lausanne, Lausanne, Switzerland (1987); Université "Pierre et Marie Currie", Laboratoire d' Etude et de Synthese des Microstructures, Paris, France (1979, experimental part of master thesis).
Workplaces: Faculty of Technology, University of Novi Sad: Head of the Department of Materials Engineering (2003–2006, 2012–); Vice Dean for Science and Development (2006–2008); Fulltime Professor (1996–)
Main fields of interest: brick, roofing tile, ceramic tile and cement industry, nanomaterials in cultural heritage protection
Relevant categories in thermal analyses: fields (raw clay materials, ceramics, cement and historic materials); methods (TGA/DTA, DIL, DSAC analysis/scope: desing of firing process and final properties of ceramics, deterioration phenomena of historical materials)
Awards and acknowledgments: Serb. Chem. Soc. (promotion of science in ceramic indystry), 2002
Professional activities: Coordination of more than 120 successfully implemented projects oriented to application of scientific results in brick and cement industry. Reviewer of scientific journals (J. Am. Ceram.Soc., J. Eur. Ceram. Soc., National Congresses of Bricks and Clay Roofing Tiles (2011, 2008, 2005, 2002), Appl. Surf. Sci., J. Civil Eng. Manag., Appl. Clay Sci.). Coordinator of projects: FP7 HEROMAT (www.heromat.com), EUREKA 3969!, EUREKA 5861!, WUS CDP + 003/2004 project leader. President of the Scientific board for Interdisciplinary Integral Science projects of the Ministry of Science, Serbia. Project leader for over 20 projects in the field of cultural heritage related to material testing embedded in objects of cultural heritage and their degradation mechanisms
Publication record: papers (40), books (2), chapters in books (2), patents (1), citation index (146), h-index (7)
Equipments: TG/DTA: STA-503 BAEHR, DSC Q20 TA Instruments, Dilatometric analysis: Dilatronic Theta Industries
5 most important publications: [1] J. Ranogajec, M. Radeka, Self-Cleaning Surface of Clay Roofing Tiles- In Self-Cleaning Materials and Surfaces: A Nanotechnology Approach, Wiley, (2013); [2] T. Vulic, M. Hadnadjev-Kostic, O. Rudic, M. Radeka, R. Marinkovic-Neducin, J. Ranogajec, Cem. Concr. Compos. 36, (2013); [3] J. Kurski, J. Ranogajec, S. Vučetić, D. Zorić, S. Adamović, I. Oroš, J. Krstić, Appl. Clay Sci., 65–66, (2012); [4] J. Ranogajec, P. Kojić, O. Rudić, V. Ducman, M. Radeka, J. Mater. Civ. Eng., 24, pp. 1254–1260, (2012), ISSN: 0899-1561; [5] M. Hadnađev, J. Ranogajec, S. Petrović, S. Markov, V. Ducman, R. Marinković-Nedučin, Phil. Mag. and Phil. Mag. Lett. (2010).

Name: Abhi Ray
Country: Australia
Date and place of birth: 1946, Calcutta, India
Present position and address: Professor Of Materials Science and Chemistry, University of Technology, Sydney
Email: A.Ray@uts.edu.au
Education and scientific degrees: Ph.D. (University of NSW, Australia), M.Sc. (Calcutta Unversity, India)
Workplaces: retired from University of Technology, Sydney and currently a member of the Standards Australia's Working Group for creating a new Australian Standard for Portland Cement manufacture in Australia with reduced carbon dioxide since nominated in 2010
Main fields of interest: materials chemistry, glass science and technology; specialization in X-ray and thermal analyses of inorganic materials, cement chemistry, glass technology, mineralogy and mineral chemistry, ceramic-based building materials
Awards and acknowledgments: 2007: "Crystal Achievement Award" for the most innovative technological article by the US National Glass Association (NGA) for the article "Glass Breaks from Nickel Sulphide Inclusions" published in the October 2006 issue of "Glass Magazine; 2005: "Branauer Award" from the American Ceramic Society for the best research publication in the field of cement chemistry in the Journal of American Ceramic Society
Professional activities: Journal Referee for J. Cement and Concrete Research, J. Materials Science, J. Materials in Civil Engineering, J. Geosciences, J. Environmental Management, J. Environmental Science and Technology, J. Thermal Analysis and Calorimetry, J. American Ceramic Society, Thermochimica Acta, J. Australian Ceramic Society; Grant Referee for Australian Research Council (1994–), UTS Research Grants–Research Excellence and Early Career Researchers (2003–2012), Academy of Finland—Research Council for Natural Sciences and Engineering (2011–); Research Suppervisor (supervised 25 Ph.D. candidates at the University of Technology, Sydney).

Name: Michael Reading
Country: UK
Date and place of birth: 1956, Birmingham, UK
Present position and address: Research Director, Cyversa, 223c Earlham Road, Norwich NR2 3RQ, UK
Email: mike@mike-reading.com
Website: www.mike-reading.com
Education and scientific degrees: Applied Chemistry, Salford University B.Sc. Hons. (1974–1978); Ph.D. (1983); Professor (2004)
Workplaces: ICI (1984–1997); Loughborough University (1997–2004); University of East Anglia (2004–2010); Cyversa (2010–)
Relevant categories in thermal analyses: fields (materials characterisation for polymers and pharmaceuticals); methods (modulated temperature DSC, hot stage microscopy, scanning probe microscopy, local thermal analysis, photothermal spectroscopy and other chemical imaging methods)
Awards and acknowledgments: French Government Scholarship (1979); Royal Society of Chemistry TMG Young Scientist award (1988); Mettler Award (2000); Elected Fellow of the North American Thermal Analysis Society (2007); GlaxoSmithKlien International Achievement Award; Awards given for inventions: Pitcon Gold Award (1998); UK Millenium Product Award (1999); Micrscopy Today Innovation Award (2011) and five R&D 100 awards
Publication record: papers (70), books (2), book chapters (12), patents (25)
Equipments: MTDSC, Nano-TA, Nano-IR
5 most important publications: [1] Thermal Scanning Probe Microscopy in the development of pharmaceuticals, X. Dai, J. G. Moffat, J. Wood and M. Reading, Adavnced Druge Delivery Reviews vol.64, issue 5 April 2012, p 449–460. Review; [2] Modulated Temperature Differential Scanning Calorimetry, Eds. M. Reading and D. J. Hourston (2010), Springer. Chapter 1 Theory and Practice of Modulated Temperature Differential Scanning Calorimetry, A. A. Lacey, D. M. Price and M. Reading; [3] Thermal Analysis of Polymers, Fundamentals and Applications, Eds. J. Menzel and B. Prime (2009) Wiley. Chapter 7, Micro and Nanoscale Local Thermal Analysis, V. V. Gorbunov, D. Grandy, M. Reading and V. V. Tsukruk; [4] Micro-thermal Analysis of Polymers: Current Capabilities and Future Prospects, M. Reading, D. M. Price, D. B. Grandy, R. M. Smith, L. Bozec, M. Conroy, A, Hammiche and M. P. Pollock, Macromol. Symp., 167 (2001), 45; [5] Handbook of Thermal Analysis and Calorimetry, Ed. M. E. Brown (1998) Elsevier. Chapter 8, Controlled Rate Thermal Analysis and Related Techniques, M. Reading.

Name: John Redfern
Country: UK
Date and Place of Birth: 1930, Hemel Hempstead, Hertforshire, UK
Education and scientific degrees: Chemist, University of London B.Sc. (1951–1954), P.G.C.E (1954–1955), Ph.D. (1959), Hon Batchelor's Degree (University of Surrey, 2011
Workplaces: Teacher, Wotton Underwood (1950–1951), Lecturer/Senior Lecturer, Battersea College of Technology (1956–1965), Research Director, Stanton Instruments (1965–1969), Managing Director, Stanton Redcroft (1970–1987), Chairman Stanton Redcroft (1987–1988), Director Thermal Scientific (1983–1986), Consultant to Polymer Laboratories Ltd (1987–1988), Director, Fire Instrumentation and Research Equipment (1988–2001), Director Interscience Communications Ltd (1985–2001)
Main Fields of Interest: ion exchange, research into thermal analysis, manufacture and marketing of a range of thermal analysis equipment including thermobalances, differential thermal analysers, etc., manufacture and marketing of fire testing equipment
Professional activities: Chartered Chemist, Fellow of The Royal Institute of Chemistry, Co-Founder (With Dr Robert MacKenzie) of International Conferences on Thermal Analysis from 1965 also served on sub-committees to do with Publications, Nomenclature, and Standardization. Member of the Organising Committee and Co-Founder of Interflam (International Flammability Conferences) (1991–2001), Chairman of the Thermal Methods Group 1967–1968
Publication record: papers (50), book chapters (2)
5 most important publications: [1] 'Differential Thermal Analysis' Vols. 1&2 published by Academic Press, Ed; [2] 'Proceedings of 1st ICTA 1965 published by Macmillans; [3] 'For better Thermal Analysis' published in 1991; [4] Editorial Board of the Journal of Thermal Analysis published by Akademiai Kiado and Wiley monthly.

Name: Géza Regdon Jr.
Country: Hungary
Date and place of birth: 1959, Szeged, Hungary
Postal address: H-6720 Szeged, Eötvös u. 6., Hungary
Email: geza.regdon@pharm.u-szeged.hu
Website: http://www.pharm.u-szeged.hu/phtech/english/home
Researcher ID: F-2304-2013
Education and scientific degrees: Pharmacist diploma, Med. Univ. of Szeged (1982); Univ. doctorate, Medical Univ. of Szeged (SZOTE) (1984); Specialist pharmacist (pharmaceutical technology) (1986); Ph.D. degree, Semmelweis Medical Univ. (1996); Habilitation, Univ. of Szeged (SZTE) (2004); associate professor, SZTE, Dep. of Pharm. Technol. (2005–)
Workplaces: Univ. of Szeged, Dep. of Pharmaceut. Technol. (1982–1984); Pharmacy of University (1984–1988); Univ. of Szeged, Dep. of Pharmaceutical Technology (1988–)
Main fields of interest: preparation and evaluation of film-coated dosage forms and free films; thermoanalytical examination of active ingredients, auxiliary materials and free films (DSC, MTDSC, TG, TG-MS); theoretical and experimental examination of factors influencing film-forming (temperature, plasticizers, type and quantity of polymers, etc.)
Relevant categories in thermal analyses: fields (inorganic, materials, pharmaceutical, polymer); methods (TG-MS, DSC (normal and modulated program), kinetics, extremely high temperature (above 1,000 °C); specific heat, calorimetry)
Awards and acknowledgments: 'Lajos Dávid' prize (1987), 'Antal Végh' NIVO prize (2013)
Professional activities: Secretary of the Thermoanalytical Section of the Hungarian Chemical Society, (2006–); Vice-chairman of the Section of Pharmaceutical Technology of the Hungarian Pharmaceutical Society, (2009–)
Publication record: papers (116), books (2), citations (366), h-index (12), sum of impact factors (65.6)
Equipments: Mettler DSC 821e with modulated program, Mettler TGA/DSC1 instrument with Pfeiffer mass spectrometer, Leica thermomicroscope
5 most important publications: [1] K. Nikowitz, K. Pintye-Hódi, G. Regdon jr.: Eur. J. Pharm. Sci. 48, 563–571 (2013); [2] G. Regdon Jr., D. Hegyesi, K. Pintye-Hódi: J. Therm. Anal. Calorim. 108, 347–352 (2012); [3] T. Hekmatara, G. Regdon jr., P. Sipos, I. Erős, K. Pintye-Hódi: J. Therm. Anal. Calorim. 86, 287–290 (2006); [4] Károlyházy, L., Regdon, G. jr., Éliás, O., Beke, Gy., Tábi, T., Hódi, K., Erős, I., Mátyus, P.: J. Mol. Struct. (Theochem) 666–667, 667–680 (2003); [5] Á. Gombás, P. Szabó-Révész, M. Kata, G. Regdon jr., I. Erős: J. Therm. Anal. Calorim. 68, 503–510 (2002).

Name: Perla Relkin
Country: France
Date and place of birth: 1946, Morocco
Present position and address: Professor Emeritus at AgroParisTech-Centre de Massy, 1 avenue des Olympiades, 91300 Massy/France
Email: perla.relkin@agroparistech.fr; perlarelkin@gmail.com
Website: www.agroparistech.fr
Education and scientific degrees: Master degree (Physics-1972/Univesity of Paris Diderot); Ph.D. degree (Physics-1975/ University of Paris Diderot); D.Sc (Physical Chemistry-1986/University P&M Curie, Paris/France).
Workplaces: Education Nationale (1977 to 1986); AgroParisTech –Institute for Science and Life Engineering and Environment/Assistant Professor/ Physicochemistry of proteins (1987 to 1994); Full professor/Food Biophysics (1995–2014); Professor Emeritus (1st September, 2014)
Main fields of interest: Food Science and Structure Engineering; concept and methods of thermodynamics and kinetics applied to structural changes in food ingredients, model and complex food systems; combined physicochemical approaches to the study of food structure stability and identification (spectroscopy, static & dynamic scattering; light & electron microscopy; thermal analysis).
Relevant categories in thermal analyses: methods (DSC, DTGA, DMTA; rheology; DRX)
Professional activities: Teaching Food Biophysics to master students, supervising Ph.D. students; Sci. Avidsor; Member of the Editorial Board of J Thermal Analysis and Calorimetry (2003–), Member of the French Association of Calorimetry and Thermal Analysis, AFCAT- (1996–2012), Executive Committee Chairperson of the 33rd Journées de Calorimetrie and Analyse Thermique; Member of the Expert committee for additive and technological ingredients, French Agency for Health Security of Foods (2003 to 2009); Reviews of manuscripts for international journals on thermal analyses; Food Science and Engineering; Chairperson of the 6th International symposium on Delivery of Food Functionality: Physically inspired approaches from the nano to micro scales (in preparation); Co-chairing ICTAC working group on TACAF, Thermal Analysis Combined Approaches to Food (2005–)
Publication record: papers (69), ca. 100 papers in peer reviewed journals, book chapters (10), h-index (22), sum of impact factors (107.3), citations (1200)
5 most important publications: [1] Relkin, Perla (1996) Thermal unfolding of beta-lactoglobulin, alpha-lactalbumin, and bovine serum albumin. A thermodynamic approach; Critical Reviews in Food Science and Nutrition, 36 (6), 565–601; [2] Relkin, Perla (1998) Reversibility of heat-induced conformational changes and surface exposed hydrophobic clusters of beta-lactoglobulin: their role in heat-induced sol-gel state transition; Internaitonal Journal of Biological Macromolecules, 22 (1), 59–66; [3] Relkin, Perla; Yung, Jin-Mi; Kalnin, Daniel; Ollivon, Michel (2008) Structural behaviour of lipid droplets in protein-stabilized nano-emulsions and stability of alpha-tocopherol; Food Biophysics, (2) 163–168; [4] Relkin, Perla; Jung, Jin-Mi; Ollivon, Michel (2009), Factors affecting vitamin degradation in oil-in-water nano-emulsions; Journal of Thermal Analysis and Calorimetry, 98 (1) 13–18; [5] P. Relkin (2009) Heat-induced phase transformations of protein solutions and fat droplets in oil-in-water emulsions: thermodynamic and kinetic approaches. In "Calorimetry in Food Processing: Analysis and Design of Food Systems", G. Kaletunc (Ed), Blackwell Publishing.

Name: Ghomareza Rezaei Behbehani
Country: I.R. Iran
Date and place of birth: 1958, Ramhormoz, Iran
Email: grb402003@yahoo.com
Education and scientific degrees: Physical chemistry, University college Dublin, (1991–1994); Ph.D. (1995); associated prof. (2007); prof. (2013)
Workplaces: Chemistry Department, IKIU, Qazvin, Iran
Main fields of interest: physical and biophysical chemistrty, prediction of competetive, noncompetetive inhibition, non-cooperativity, positive and negative cooperativity using his own treatment of the extended solvation model
Relevant categories in thermal analyses: Methods (isotermal titration calorimetry, UV, fluorescence, phosphorescence and FTIR)
Publication record: number of papers (100), books (3), citations (175), h-index (14)
5 most important publications: [1] G. Rezaei Behbehani, A. A. Saboury, L. Barzegar, O. Zarean, J. Abedini, M. Payehghdr, J. Therm. Anal. Calorim., 101 (2010) 379–384; [2] G. Rezaei Behbehani, A. A. Saboury, A. Taherkhani, L. Barzegar, M. Mollaagazade, J. Therm. Anal. Calorim., 105 (2011) 1081–1086; [3] G. Rezaei Behbehani, M. Mehreshtiagh, A. A. Saboury, L. Barzegar, J. Solution Chem., 41 (2012) 581–588; [4] G. Rezaei Behbehani, A. A. Saboury, J. Thermal Anal. Calorim. 89 (2007) 859–863; [5] G. Rezaei Behbehania, W.E. Waghorne, Thermochim. Acta, 478 (2008) 1–5.

Name: Simone Pereira da Silva Ribeiro
Country: Brazil
Date and place of birth: 1981, Rio de Janeiro, Brazil
Present position and address: Chemist researcher. Instituto de Química—Pólo de Xistoquímica. Rua Hélio de Almeida 40, sl. C205. Cidade Universitária, ilha do Fundão, CEP: 21941-614. Rio de Janeiro, RJ, Brazil
Email: spsilva@iq.ufrj.br
Education and scientific degrees: Chemist Technician, ETFQ-RJ (1996–2000), Chemist, Universidade do Estado do Rio de Janeiro—UERJ (2000–2004), Master Degree, Chemist Intitute Universidade Federal do Rio de Janeiro—UFRJ (2005–2007), Ph.D., UFRJ (2008–2012)
Workplaces: Nitriflex, Indústria e comércio (2000–2001). Pólo de Xistoquímica-Universidade Federal do Rio de Janeiro-UFRJ (2001–2013)
Main fields of interest: development of polymeric nanocomposites with fire retardant properties
Relevant categories in thermal analyses: fields (materials, organic, polymer, nano, polymeric materials); methods (TG, DSC, DTA, TGA/DSC, TGA/FTIR)
Professional activities: Development of polymer nanocomposites with flame retardant properties chair stage of the World Cup in Brazil. (2006–2008). Member of the National Institute for Research and Development and Innovation in Complex Functional Materials (INOMAT) (2008–2014). Journal reviewer: Journal of Thermal Analysis and Calorimetry, Journal of Applied Polymer Science and Industrial and Engineering Chemistry Research
Publication record: papers (5), citations (15), h-index (2)
Equipments: DSC Q1000-Mettler, DSC Serie-7 PerkinElmer and TGA-51 Shimadzu
5 most important publications: [1] Ribeiro, S. P. S., Estevao, L. R. M., Nascimento, R. S. V. Journal of Thermal Analysis and Calorimetry., v.87, pp. 661–665, 2007; [2] Ribeiro, S. P. S., Estevao, L. R. M., Nascimento, R. S. V. Science and Technology of Advanced Materials., v.9, pp. 1–7, 2008; [3] Ribeiro, S. P. S., Estevao, L. R. M., Pereira, C., Rodrigues, J., Nascimento, R. S. V. Polymer Degradation and Stability. v.94, pp. 421–431, 2009; [4] Ribeiro, S. P. S., Estevao, L. R. M, Novák, C., Nascimento, R. S. Veiga. Journal of Thermal Analysis and Calorimetry, v.106, pp. 535–539, 2011; [5] Ribeiro, S. P. S., Estevao, L. R. M., Pereira, C., Nascimento, R. S. V., Journal of Applied Polymer Science v. 130, pp. 1759–1771, 201.

Name: Alan Riga
Country: USA
Date and place of birth: 1937, Cleveland, Ohio, USA
Present position and address: CASE WRU Professor/6325 Aldenham Dr. Cleveland Ohio 44143-3331 USA
Email: alanriga@hotmail.com
Education and scientific degrees: BS MS Ph.D. Case WRU
Workplaces: Sr. Chemist, Lubrizol 1960–1997; Professor Cleveland State Univ. 1998–2012; Professor, Case WRU 2010–2016; Professor Notre Dame College Ohio 2011–2016; Professor University of Toledo, College of Pharmacy 2000–2016; Sr. Material Scientist TechCon Inc. 1998–2016 Cleveland Ohio 44143
Main fields of interest: thermal analysis petrochemical, polymers, pharmaceuticals, forensics, biology malaria, inorganic and organic polar ionic conductivity by DEA and DSC; biomedical engineering, amorphous crystalline content of drugs and polymers by DEA and DSC
Relevant categories in thermal analyses: fields (drugs, polymers, forensics); methods (DSC, TGA, TMA, dielectric analysis)
Awards and acknowledgments: ASTM International Charles Dudley Award 2009–2010; Distinguished Leadership Award/Cleveland Technival Societies Council 2003; "International Scientist Award", North American Thermal Analysis Society and Mettler-Toledo Company, Cleveland Ohio, 1998; "Fellow and Award of Merit" the American Society for Testing and Materials, Philadelphia, PA., 1995; "NATAS Outstanding Achievement (Mettler-Toledo) Award, 1995.; "Fellow of the Society", North American Thermal Analysis Society, Atlanta, GA., 1992; "Fellow of the Society", Society of Plastics Engineers, Inc., Brookfield, CT., 1992; Outstanding Member in the American Society for Testing and Materials, "People", Standardization News, Philadelphia, PA. 1991.; "Honorary Member", Brazillian Thermal Analysis Society, Pocos de Caldas, Brazil, 2002
Publication record: papers (188), books (6), patents (5)
Equipments: DSC, MDSC, TGA, TMA, DMA, DEA
5 most important publication: [1] Alan Riga et al., Solid and Liquid State Studies of a Wide Range of Chemicals by Isothermal and Scanning DEA, accepted 2013 by Journal Thermal Analysis and Calorimetry; [2] Alan Riga et al. "Quantification of Crystalline and Amorphous Properties of Pharmaceutical Solids by Thermal Analytical Methods, 38th North Am. Therm. Anal. Soc., August, 2010. T-67 NATAS; [3] Chris Ivanoff, Timothy Hottel and Alan Riga, Macroesis: A Model for Transporting Drugs intoTeeth Electrokinetically is Tested in an vitro Bleaching Study, Journal of American Dental Association 2011, 24, no.5, 259–253; [4] Dhruthiman R. Mantheni, and Alan T. Riga Solid State Studies of Drugs and Chemicals by Dielectric and Calorimetric Analysis, Journal of Thermal Analysis and Calorimetry, paper II, accepted 2013; [5] Mekki Bayachou, Alan Riga, and Salaam Saleh, Probe Functionaqlization with Rhop-3 Antibody: Towards a Rhop-3 Antigen Immunosensor for the Detection of Malaria, Journal of Analytical and Bioanalytical Chemistry, 2012, 108, no.1, 227–233.

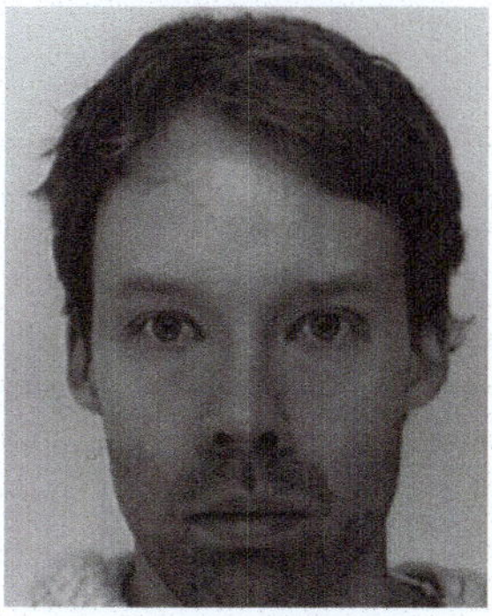

Name: Joakim Riikonen
Country: Finland
Date and place of birth: 1981, Kokkola, Finland
Present position: Post doctoral researcher, University of Eastern Finland
Email: joakim.riikonen@uef.fi
Website: www.uef.fi/en/pharmaceuticalphysics/joakim-riikonen
Researcher ID: A-6217-2010
Education and scientific degrees: M.Sc. (Physics) 2006, Ph.D. (Applied Physics) 2012.
Workplaces: University of Turku (2005–2008), Finland; University of Eastern Finland, Finland (2009–); CNRS, Mulhouse, France (2014)
Main fields of interest: thermoporometry and mesoporous materials, especially porous silicon, in drug delivery applications
Relevant categories in thermal analyses: fields (inorganic, materials, organic, pharmaceutical); methods (TG, DSC, microcalorimetry)
Awards and acknowledgments: Vitali Parkhutik Prize 2012
Publication record: papers (25), books (2), h-index (9)
Equipments: TG, DSC, TXRD
5 most important publications: [1] J. Riikonen, J. Salonen, V.-P. Lehto, Utilising thermoporometry to obtain new insights into nanostructured materials—Review part 1, Journal of Thermal Analysis and Calorimetry, volume 105, issue 3, p 811–821 2011; [2] J. Riikonen, J. Salonen, V.-P. Lehto, Utilising thermoporometry to obtain new insights into nanostructured materials—Review part 2, Journal of Thermal Analysis and Calorimetry, volume 105, issue 3, p 823–830 2011; [3] J. Riikonen, J. Salonen, M. Kemell, N. Kumar, D. Murzin, M. Ritala, V.-P. Lehto, A novel method of quantifying the u-shaped pores in SBA-15, Journal of Physical Chemistry C, Volume 113, Issue 37, p 20349–20354, 2009; [4] J. Riikonen, E. Mäkilä, J. Salonen and V.-P. Lehto, Determination of the Physical State of Drug Molecules in Mesoporous Silicon with Different Surface Chemistries, Langmuir, Volume 25, Issue 11, p 6137–6142, 2009; [5] J. Riikonen, M. Salomäki, J. van Wonderen, M. Kemell, W. Xu, O. Korhonen, M. Ritala, F. MacMillan, J. Salonen and V.-P. Lehto, Surface Chemistry, Reactivity and Pore Structure of Porous Silicon Oxidized by Various Methods, Langmuir, volume 28, issue 28, p 10573–10583, 2012.

Name: Jesús María Rincón
Country: Spain
Date and place of birth: 1947, Madrid, Spain
Present position and address: Head of the Vitreous and Ceramics Lab/Group at the IETcc-CSIC; President of the Spanish Society of Scientists (AEC); Inst°. E.Torroja CC Construcción, CSIC, c/Serrano Galvache 4, Madrid- 28033, Spain
Email: jrincon@ietcc.csic.es; rinconjma@gmail.com
Education and scientific degrees: Lic. in Chemistry, Universidad Complutense de Madrid (1964–1969); Ph.D. (1982); CSIC Professor (2001)
Workplaces: Institute of Glasses and Ceramics- CSIC (1970–1994); Visiting Associate Professor in Mats. Sci. Dep. Berkeley, Univ. California, USA (1984–1985); Institute Construction Sciences E. Torroja.- CSIC (1994–)
Main fields of interest: glass-ceramics, glasses, ceramics (compositional and processing design, full characterization of microstructure and properties, applications); thermal methods for investigate nucleation and crystal growth on glasses and ceramics; all electron microscopy and microanalytical methods expert
Awards and acknowledgments: Honorary Member of the Spanish Glass and Ceramics Society (SECV) (1992)
Professional activities: General Secretary of the Spanish Glass and Ceramic Soc. (SECV) and Editor-.in- Chief of the Bol. Soc. Esp. Ceram. Vidr (BSECV) (1988–1992); Chairman and organiser of Spanish National Congresses in Ceramics of SECV (1987–1992): Mérida, Córdoba, Aveivo, Tenerife, Santiago de Compostela and Almería; director of the Summer Monographic Courses of Univ. La Laguna (Tenerife): Glassceramics and Methods for Glasses and Ceramics Characterization (1991–1992); Chairman in monographic sessions in several national and international congresses; invited lecturer international congresses and courses; member TC05 committee on wastes International Commission of Glass (ICG), founder and leader of the Lab/Group of Vitreous and Ceramic Materials at IETcc-CSIC
Publication record: papers (240), books (1), book chapters (250), patents (3), h-index (23)
5 most important publications: [1] J. Ma. Rincón. Principles of nucleation and controlled crystallization of glasses. Polym. Plast. Technol. Engineering, 3 (1992) 3-4, 309–357, invited review; [2] L. Barbieri, A. Karamanov, A. Corradi, I. Lancelloti, M. Pelino, J. Ma. Rincón, Structure, chemical durability and crystallization behaviour of incinerator- based glassy systems, J. Non-Crystalline Solids 354 (2008) 521–528; [3] C. Rubalcaba, Ma. E. Zayas, S. J. Castillo, R. Lozada, M. Pérez-Tello, C. G. Díaz and J. Ma. Rincón, Optical and thermal analysis of ZnO- CdO- TeO2 glasses doped with Nd3+, Optical Materials 33 (2011) 823–826; [4] D. S. Rodríguez, P. A. Arboleda, D. O. Russo, F. Soldera, C. J. R. González- Oliver, J. Ma. Rincón, Thermal behaviour of iron aluminium phosphate glasses containing UO2.67, Glass Technol.:Eur. J. Sci. Technol. A (2013), 54 (3), 111–118; [5] J. Ma. Rincón, J. Caceres, C. J. R. Gonzalez- Oliver, D. O. Russo, A. Petkova and H. Hristov, Thermal and sintering behaviour of basalt glasses and natural basalt powders, J. Therm. Anal. Calorim. 56 (1999) 931–938.

Name: Erich Robens
Country: Germany
Date and place of birth: 1924, Mainz, Germany.
Present position and address: Retired, Schlesierstraße 5, D-61381 Friedrichsdorf, Germany.
Email: Erich.robens@t-online.de
Workplaces: Battelle-Institut e.V. Frankfurt am Main, Germany; Bausparkasse Mainz AG, Mainz and Mainzer Haus Vertriebs GmbH, Mainz, Germany; Jugendberatung und Jugendhilfe e.V. Frankfurt am Main, Germany; Institut für Anorganische und Analytische Chemie, Johannes Gutenberg-Universität, Mainz; Uniwersytet Maria Curie-Skłodowska, Lublin, Poland
Main fields of interests: development of the first automatic gravimetric sorption measuring apparatus, balances, investigation of Lunar regolith, water adsorption
Professional activities: 1948–195ß Assistant chemist at Mogatwerke, Mainz, Germany; 1950–1955 Préparateur at Lycée Français, Mainz; 1956–1987 Physical engineer, shop steward, member of the economic council at Battelle-Institut e.V., Frankfurt am Mainz, Germany; 1985–2008 Guest scientist and student at the Institute for Inorganic and Analytical Chemistry of the Johannes Gutenberg-University, Mainz; 1974–1983 Member (1987–1995 chairman) of the committees "Surface Measurement", German Standardisation Committee (DIN), member of the expert council "Particulate Materials", European Communities Bureau of Reference (BCR), Brussels, Belgium and "Particulate Materials", International Standardisation Organisation (ISO), Geneva, Switzereland; 1972–1997 Member of the board of directors of the "Bausparkasse Mainz AG" (loan and building association); and 1981–1999 of the "Mainzer Haus Vertriebs GmbH"; 1989–2003 Data protector of the "Jugenberatung und Jugendhilfe e.V." (drug abuse society), Frankfurt am Mainz; 2003–2010 Guest Lecturer of Maria Curie-Skłodowska-Universität, Lublin, Poland; 2003–2010 Preparation of a Mars-experiment of the German Zentrums für Luft- und Raumfahrt, Berlin; 2005 NASA principal investigator of the research program of Lunar samples
Publication record: papers (500), book chapters (30), books (6), patents (56)
5 Most important publications: [1] Mikhail, Robens: Microstructure and Thermal Analysis of Solid Surfaces. Wiley, Chichester 1983; [2] Klobes, Robens: Standardization of the Pore Size Distribution. Particle and Particle Systems Characterization (2012); [3] Robens, Adolphs, Bischoff, Goworek, Kutarov, Schreiber, Skrzypiec: Investigation of surface properties of lunar soils. Journal of Geological Sciences, 40 (2012) 43–55; [4] Robens, Rübner, Klobes, Balköse: Water vapour sorption and humidtity—A survey on measuring methods and standards. In: C. T. Okada: Humidity Sensors: Types, Nanomaterials and Environmental Monitoring. Nova Science, Hauppauge, NY, 2011, 1–92; [5] Robens, Jayaweera, Kiefer: Balances—Instruments, Manufacturers, History. Springer, Heidelberg 2014, 730 pp. 763 illus., 359 illus. in color. hardcover, eBook.

Name: Ruben Jesus Sanchez Rodriguez
Country: Brazil
Date and place of birth: 1951, Cuba
Present position and address: Titular Professor, North Fluminense State University, Ave. Alberto Lamego 2000, Campos dos Goytacazes, Rio de Janeiro, Br. (www.uenf.br).
Email: sanchez@uenf.br
Website: http://www.uenf.br/Uenf/Pages/CCT/Lamav/Polimeros/
Researcher ID: F-5999-2010
Education and scientific degrees: Chemestry, Havana University (1970–1975); Doctor in Science and Technology (National Center of Scientific Research, CNIC, Cuba) 1988.
Workplaces: Research of National Center of Scientific Research (Cuba)1991, Federal University of Rio de Janeiro (Brazil) 1992–1993, North Fluminense State University (Brazil) 1994 up today.
Main fields of interest: polymer, composites, nanomaterials, ceramic (properties and applications)
Relevant categories in thermal analyses: fields (polymer, composites, nanomaterials, ceramic); methods (TG, TG/MS, DSC, DMA and Photothermal methods)
Awards and acknowledgments: Awards; Post-graduate meeting. 2008, 2009, 2010 and 2013, Brazilian Polymers Congress, CBPol 2006. Research Productivity (CNPq) 2003, 2007, 2010, 2014. Award Recovery of Economic Value of Metallic Wastes, 65° ABM Congress 2012. Honored Professor of Materials Engeneering; 1998–2000, 2002, 2003
Professional activities: Organizer; "10 Years of Engineering in UENF", 2003. Member of the organizing committee POLYCHAR 15—World Forum on Advanced Materials, 2007. International Symposium on Biopolymers, Santos Brazil 2014. Member of the Brazilian Polymer Society (ABPol). Leader of the Polymer Research Group and Professor of Undergraduate and Postgraduate Courses in Material Science and Engineering (UENF)
Publication record: papers (84), book chapters (4), patents (3), Complete works published in proceedings of conferences (131), h-index (11).
Equipments: TGA 5000 TA/MS, SDT 2960 TA, QDSC TA, DMA TA, QDMA TA
5 most important publications: [1] Rodríguez, R. J. S., Amaral, C. R., Garcia, F. G., Barros, Jr., L., Impact of aliphatic amine co-monomers on DGEBA epoxy networks properties. Polym. Eng. Sci. v. 12, pp. 17, 2013; [2] Rodríguez, R. J. S, Monteiro, S. N., Calado, V., Margem, F. M., Thermogravimetric behavior of natural fibers reinforced polymer composites An overview. Mater. Sci. Eng. A. v. 557, pp. 17–28, 2012; [3] Rodríguez, R. J. S; Souza, D., Castillo, T. E.;. Effects of hydroxyvalerate contents in thermal degradation kinetic of cellulose acetate propionate/poly(3-hydroxyalkanoates) blends. J. Therm. Anal. Calorim. v. 109, pp. 1353–1364, 2012; [4] Rodríguez, R. J. S; Carvalho, E. A. S.;Tavares, M. I. B.; Lamônica, A. C. Characterization and Properties of Hydrophilic Cellulose Acetate Propionate Derivative. J. Polym. Environ. pp. 610–620, 2010; [5] Rodríguez, R. J. S; Schripsema, J., da Silva L. F., Taciro, M. K., Pradella, J. G. C, Gomez, J. G. C. Medium-chain-length polyhydroxyalkanoic acids (PHAmcl) produced by Pseudomonas putida IPT 046 from renewable sources. Eur. Polym. J. v. 39, pp. 1385–1394, 2003.

Name: Manuel Rodríguez De Rivera
Country: Spain
Date and place of birth: 1957, Madrid, Spain
Present position and address: Departamento de Física, Universidad de Las Palmas de Gran Canaria. Campus Universitario de Tafira. E-35017 Las Palmas de Gran Canaria, Spain
Email: mrodriguez@dfis.ulpgc.es
Education and scientific degrees: Industrial Engineer, Escuela Técnica Superior de Ingenieros Industriales de Las Palmas de Gran Canaria, Spain (1981), Ph.D. (1984), Professor Adjunto (1984), Professor Catedrático (1993)
Workplaces: Departamento de Física. Universidad de Las Palmas de Gran Canaria
Main field of interest: calorimetric Instrumentation (theory and their practical applications)
Publication record: papers (34)
5 most important publications: [1] F. Socorro, M. Rodríguez de Rivera, J. P. Dubès, H. Tachoire and V. Torra. Computer-controlled experimental set-up and signal processing in calorimetric studies of liquid mixtures. Measurement Science and Technology, 1 (1990) 1285–1290; [2] F. Socorro, I de la Nuez, M. Rodríguez de Rivera. Calibration of isothermal heat conduction calorimeters: case of flow calorimeters. Measurement, 33 (2003), 241–250; [3] Manuel Rodríguez de Rivera, Fabiola Socorro and José S. Matos. Heats of Mixing Using an Isothermal Titration Calorimeter: Associated Thermal Effects. Int. J. Mol. Sci. 2009; 10, 2911–2920; [4] Ch. Jesús, F. Socorro, M. Rodríguez de Rivera, New approach to Tian's equation applied to heat conduction and liquid injection calorimeters. J. Therm. Anal. Calorim. 2012; 110: 1523–1532; [5] Ch. Jesús, F. Socorro and M. Rodríguez de Rivera. Development of a calorimetric sensor for medical application. Part III. Operating methods and applications. J. Therm. Anal. Calorim. 2013; 113: 1009–1013.

Name: Vladimir Rogov
Country: Russia
Date and place of birth: 1945, Dnepropetrovsk, USSR
Present position and address: Russia, Novosibirsk 630090, Prospekt Akademika Lavrent'yeva, 5. Boreskov Institute of Catalysis.
Email: rogov45@mail.ru
Education and scientific degrees: Novosibirsk State University (1963–1968), Ph.D. (1976)
Workplaces: Boreskov Institute of Catalysis, Novosibirsk State University
Main fields of interest: heterogeneous catalysis
Relevant categories in thermal analyses: fields (inorganic, nano, ceramics); methods (TG, DTA, DSC, calorimetry, microcalorimetry, instrument development)
Publication record: papers (122), citations (885), h-index (15)
Equipments: SENSYS DSC TG Setaram, STA 409 Netzsch
4 most important publications: [1] V. Sadykov, V. Rogov, E. Ermakova et al. Mechanism of CH_4 Dry Reforming by Pulse Microcalorimetry: metal nanoparticles on perovskite/fluorite supports with high oxygen mobility, Thermochimica Acta, 567 (2013) 27; [2] A. Bobin, V. Sadykov, V. Rogov, et al. Mechanism of CH4 Dry Reforming on Nanocrystalline Doped Ceria-Zirconia with Supported Pt, Ru, Ni, and Ni-Ru. Topics in Catalysis, 56 (2013) 958; [3] S. Pavlova, V. Sadykov, V. Rogov et al. Structural features and transport properties of La(Sr) Fe(1-x)Ni(x)O(3)/Ce(0.9)Gd(0.1)O(2) nanocomposites—advanced materials for IT SOFC cathodes. Heat Transfer Engineering, 34 (2013) 904; [4] V. Sadykov, N. Sazonova, V. Rogov et al. Partial oxidation of methane on Pt-supported lanthanide doped ceria-zirconia oxides: effect of the surface/lattice oxygen mobility on catalytic performance, Catalysis Today, 169 (2011) 125.

Name: Blanca Rojas De Gascue
Country: Venezuela
Date and place of birth: 1963, Caracas, Venezuela
Present position and address: Polymers Laboratory Coordinator. Regular full-time professor in Polymers Laboratory, Orient University (UDO). Institute for Research in Biomedicine and Applied Sciences (IIBCA-UDO) (1997-currently), Cerro del Medio, Av. Universidad, Cumaná, Estado Sucre, Venezuela.
Email: blanca_gascue@yahoo.com
Website: http://www.iibcaudo.com.ve
Education and scientific degrees: Materials Engineer, Simón Bolívar University (1986) Masters in Engineering, Simón Bolívar University (1993) Ph.D. in Materials Chemistry, Complutense University of Madrid (1997)
Workplaces: Polymers Laboratory, Orient University (UDO). Institute for Research in Biomedicine and Applied Sciences (IIBCA-UDO), Cumaná, Estado Sucre, Venezuela.
Main fields of interest: thermal properties of polyolefins, successive self-nucleation and annealing (SSA), crystal morphology and hydrogels
Relevant categories in thermal analyses: fields (polyolefins preferably polyethylene, hydrogels, biopolymers); methods (differential scanning calorimetry (DSC), dynamic conditions, successive self-nucleation and annealing (SSA))
Awards and acknowledgments: Stimulus Program Investigator (2011) and 2013. Scholarship for doctoral studies from AECI (Spanish Agency for Latin American Cooperation) 1994–1997
Professional activities: Project Coordinator titled:Provide support to the Service of the University Hospital of Traumatology Patricio Antonio Alcalá (Huapa) Cumana to assess the quality of Biomaterials and propose employing new developments (2013–). Member of the board of the Venezuelan Association of Polymers (ASOVENP). Coordinator of the XII National Colloquium Polymers, CVP2007
Publication record: papers (54 journal papers, 250 works presented in Congress)
Equipments: Differential scanning calorimetry (DSC), Transmission Electron Microscope (TEM), Scanning electron microscopy (SEM), Infrared Spectrometer
5 most important publications: [1] N. González, J. Contreras, F. López-Carrasquero, A. El-Halah, C. Torres, JL. Prin, J. Benítez, B. Rojas de Gascue. Interciencia, 38 (6): 430 (2013); [2] B. Rojas de Gáscue, J. Prin, G. Hernández, EM. Vallés, AT. Lorenzo y AJ. Müller. J. Therm. Anal. Calorim., 103:669 (2011); [3] B. Rojas de Gáscue, M. Ramírez, JL. Prin, C. Torres, L. Bejarano, H. Villarroel, L. Rojas, M. Murillo, I. Katime. Rev. LatinAm. Metal. Mat., 30 (1), 28 (2010); [4] H. Villarroel, JL. Prin, M. Ramírez, G. Bolívar, L. Rojas de Astudillo, I. Katime, E. Laredo, B. Rojas de Gáscue. Rev. Iberoam. Polim., 11(7), 625 (2010); [5] B. Rojas de Gáscue, B. Méndez, J. L. Manosalva, J. López, V. Ruiz Santa Quiteria, A. Müller. Polymer 43, 2151 (2002).

Name: Andrei Rotaru
Country: Romania
Date and place of birth: 1983, Craiova, Romania
Present position and postal address: Senior Researcher II, INFLPR—Institute for Laser, Plasma and Radiation Physics, Bvd. Atomistilor 409, Magurele (Ilfov), Bucharest, Romania
Email: andrei.rotaru@inflpr.ro
Website: www.inflpr.ro, www.ceec-tac.org
Researcher ID: A-9726-2014
Education and scientific degrees: BSc in Chemistry and in Physics at University of Bucharest (2006), Dr. Eng. in Mechanical Engineering at University "Politehnica" of Bucharest (2011), Ph.D. in Chemistry at University of St Andrews (2013); Assitant Researcher at INFLPR (2006), Scientific Researcher III at INFLPR (2011), Senior Researcher II at INFLPR (2014)
Workplaces: INFLPR—National Institute for Laser, Plasma and Radiation Physics
Main fields of interest: thermophysical properties of materials, physical chemistry of solids, thermal analysis, heterogeneous kinetics, solid state sciences, Electroceramics, Thin films
Relevant categories in thermal analysis: fields (dyes, liquid crystals, fuels and biofuels, ferroelectrics, relaxors, thermoelectrics); methods (TGA, DSC, DTA, impedance and dielectric spectroscopy, XRD, SEM, AFM, FTIR)
Awards and acknowledgments: ICTAC Travel Grant Award for Young Scientists, ICTAC14, Brazil (2008), Fellowship of the Roberto Rocca Educational Program (2008)
Professional activities: President of the Central and Eastern European Committee for Thermal Analysis and Calorimetry (CEEC-TAC) since 2011; Co-chairman of the CEEC-TAC1 (Craiova, 2011), CEEC-TAC2 (Vilnius, 2013), CEEC-TAC3 (Ljubljana, 2015) conferences; Regional EdiChairman of the Lifetime Prediction of Materials and Energetics Committee at ICTAC since 2014; Regional Editor at the Journal of Thermal Analysis and Calorimetry since 2012; Member of ICTAC since 2006
Publication record: papers (35), books (1), citations (240), h-index (11), sum of impact factros (47)
Equipments: Hi-Res TGA, MDSC, High-pressure DSC, Furnaces up to 1,250 and 1,400 °C
5 most important publications: [1] A. Rotaru, A. J. Miller, D. C. Arnold, F. D. Morrison; Towards novel multiferroic and magnetoelectric materials: dipole stability in tetragonal tungsten bronzes; Philosophical Transactions of the Royal Society A, 372, 20120451, 2014; [2] A. Rotaru; Thermal analysis and kinetic study of Petroşani bituminous coal from Romania in comparison with a sample of Ural bituminous coal; Journal of Thermal Analysis and Calorimetry, 110, 3, 1283–1291, 2012; [3] A. Rotaru, D. C. Arnold, A. Daoud-Aladine, F. D. Morrison; Origin and stability of dipolar response in a family of tetragonal tungsten bronze relaxors; Physical Review B, 83, 184302, 2011; [4] A. Rotaru, C. Constantinescu, A. Mândruleanu, P. Rotaru, A. Moldovan, K. Győryová, M. Dinescu, V. Balek Matrix assisted pulsed laser evaporation of zinc benzoate for ZnO thin films and non-isothermal decomposition kinetics; Thermochimica Acta, 498, 1-2, 81–91, 2010; [5] A. Rotaru, M. Gosa; Computational thermal and kinetic analysis. Complete standard procedure to evaluate the kinetic triplet form non-isothermal data; Journal of Thermal Analysis and Calorimetry, 97, 2, 421–426, 2009.

Name: Petre Rotaru
Country: Romania
Date and place of birth: 1950, Craiova, Romania
Present position and address: Professor, University of Craiova, 13 AI Cuza Street, 200585-Craiova, Romania
Email: protaru@central.ucv.ro, petrerotaru@yahoo.com
Website: http://cis01.central.ucv.ro/physics/
Researcher ID: A-9784-2014
Education and scientific degrees: BSc in Physics and in Chemistry at University of Craiova (1969–1973); Ph.D. in Physics at Institute of Atomic Physics Bucharest (1985); Ph.D. in Chemistry at Polytechnic Institute of Bucharest (1996); Professor at University of Craiova (2011)
Workplaces: Research Center for Chemical Fertilisers Craiova (1973–1978); University of Craiova (1978–)
Main fields of interest: thermal analysis and calorimetry; thermophysical properties of solids; heterogeneous kinetics and catalysis; materials characterisation
Relevant categories in thermal analyses: fields (liquid crystals, metal-organic complexes; catalysis, catalysts and porous materials); methods (TGA, DSC, DTA, XRD, FTIR, BET)
Professional activities: Member of the International Organising/Scientific Committee of the Central and Eastern European Conferences for Thermal analysis and Calorimetry; Member of the Commission for Thermal Analysis and Calorimetry of the Romanian Academy
Publication record: papers (93), books (4), patents (5), citations (130), h-index (9), sum of impact factors (59)
Equipments: PerkinElmer Diamond TG/DTA/DSC system coupled with FTIR spectrometer Spectrum 100; High-temperature furnaces
5 most important publications: [1] S. Degeratu, P. Rotaru, S Rizescu, N. G. Bizdoaca, Thermal study of a shape memory alloy (SMA) spring actuator designed to insure the motion of a barrier structure, Journal of Thermal Analysis and Calorimetry, 111 (2013) 1255; [2] A. Moanta, C. Ionescu, P. Rotaru, M. Socaciu, A. Harabor, Structural characterization, thermal investigation, and liquid crystalline behavior of 4-[(4-chlorobenzyl)oxy]-3,4′-dichloroazobenzene, Journal of Thermal Analysis and Calorimetry, 102 (2010) 1079; [3] P. Rotaru, R. Scorei, A. Hărăbor, M. D. Dumitru, Thermal analysis of a calcium fructoborate sample, Thermochimica Acta, 506 (2010) 8; [4] P. Rotaru, S. I. Blejoiu, M. Stanciu, G. Stoenescu, M. Mateescu, V. Voiculescu, The influence of the tabletting pressure upon the textural and diffusional properties of a Fe_2O_3–Cr_2O_3 catalyst, Microporous and Mesoporous Materials, 83 (2005) 159; [5] P. Rotaru, S. I. Blejoiu, R. Constantinescu, N. Pometescu, F. Uliu, O. Bunescu, Perfectly stirred catalytic reactor, Applied Catalysis. A: General, 166 (1998) 363.

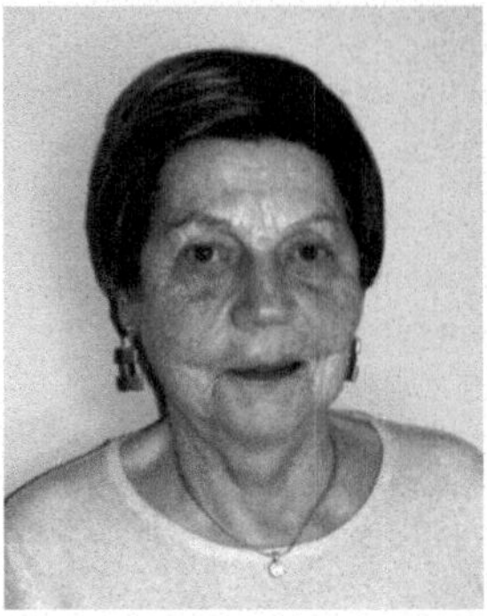

Name: Françoise Rouquerol
Country: France
Date and place of birth: 1937, Versailles, France
Present position and address: Emeritus Professor, Aix-Marseille University MADIREL, Centre de St Jérôme, 13397 Marseille Cedex 20, France
Email: francoise.rouquerol@univ-amu.fr
Education and scientific degrees: Scientific studies (Physics and Chemistry) at the Sorbonne (Paris University). Licence ès-Sciences Physiques (1958); D.Sc (1965)
Workplaces: Physical Chemistry Laboratory of the Sorbonne, Paris University (1959–1966), Provence University, Marseille (1966–2001), MADIREL Joint Laboratory (CNRS-Provence University Joint Laboratory), Marseille (2001–)
Main fields of interest: application of SCTA to heterogeneous kinetics, thermodynamic study of adsorption. nanomaterials (characterisation and preparation)
Relevant categories in thermal analyses and calorimetry: fields (inorganic, materials, nano, minerals, ceramics, cement, adsorbents); methods (TG, EGA, kinetics, microcalorimetry)
Awards and acknowledgments: Bronze Medal of CNRS: 1966
Professional activities: Assistant-Professor in Paris University (1966–1978) and in Provence University, Marseille; Senior Lecturer (1978–79); Professor (1979–2002); Emeritus Professor (2002–). French National Representative at the IUPAC Commission of Thermodynamics (1987–1999) and at the IUPAC Teaching Chemistry Committee (2000–2004). President of the Teaching Division of the Société Française de Chimie (2000–2008)
Publication record: papers (180), books (7), citations (1384), h-index (20)
Equipments: Sample Controlled TG, Sample Controlled Evolved Gas Analysis, Gas adsorption microcalorimetry
5 most important publications: [1] F. Rouquerol, J. Rouquerol, K. S. W. Sing, P. Llewellyn and G. Maurin: Adsorption by powders and porous solids: principles, methodology and applications, 2nd Ed., Academic Press, 2014, 626 p; [2] F. Rouquérol, G. Chambaud, R. Lissillour, Abdou Boucekkine, Renaud Bouchet, Florence Boulc'h, Virginie Hornebecq: Chimie Générale: 7° édition du cours de chimie physique, Dunod, Paris, 2013, 652 p; [3] P. Llewellyn, F. Rouquérol, J. Rouquerol: SCTA and adsorbents, Hot topics in Thermal Analysis and Calorimetry (2003) 3 pp. 135–173; [4] F. Rouquerol, J. Rouquerol, K. S. W. Sing: Applications of Adsorption by Porous Solids (Chapter 6, pp 77–84) in Chemical Thermodynamics. Chemistry for the 21st century, Edited by T. M. Letcher, IUPAC Blackwell Science, 1999; [5] S. Bordère, F. Rouquerol, J. Rouquerol, J. Estienne, A. Floreancig: Kinetical possibilities of controlled transformation rate thermal analysis (CRTA). Application to the thermolysis of hexahydrated uranyl nitrate, J. Therm. Anal. 1990, 36, 1651–68.

Name: Jean Rouquerol
Country: France
Date and place of birth: 1937, Paris, France
Present position and address: Directeur de Recherche Emérite, Aix Marseille University, MADIREL Laboratory, Centre de St Jérôme, 13397 Marseille Cedex 20, France
Email: jean.rouquerol@univ-amu.fr
Education and scientific degrees: Scientific studies (Physics and Chemistry) at the Sorbonne (Paris University). Licence ès-Sciences (1959); D.Sc (1964)
Workplaces: Physical Chemistry Laboratory of the Sorbonne, Paris (1959–1966); CNRS Research Center for Thermodynamics and Microcalorimetry, Marseille (1966–2001); MADIREL Joint Laboratory (Aix Marseille University-CNRS), Marseille (2001–)
Main fields of interest: thermal preparation, modification and characterization of technological adsorbents, with special attention to sample controlled thermal analysis (SCTA), microcalorimetry and adsorption (gas or liquid)
Relevant categories in thermal analyses: fields (inorganic, materials, nano, minerals, ceramics, cement); methods (TG, EGA, DSC, kinetics; microcalorimetry, instrument development)
Awards and acknowledgments: Bronze Medal of CNRS (1971); Silver Medal of the French Society for the Development of National Industry (1971); NATAS Mettler Award (1988); Doctor Honoris Causa of UNED University in Madrid, Spain (2003); ICTAC Honorary Member (2010); Grand Prix de la Section Provence, Société Chimique de France (2013); AICAT-SETARAM Award (2013)
Professional activities: Director of CNRS Research Center for Thermodynamics and Microcalorimetry in Marseille (1990–2001), President of French Association for Calorimetry and Thermal Analysis (AFCAT) (1986–1989), Chairman of IUPAC Commission on Colloids and Surface Chemistry, including Catalysis (Commission I-6) (1992–1995), Chairman of the ESTAC Committee (1991–1994), and, at ICTAC, Chairman of the Scientific Commission (1982–1996), Vice-President (1996–2000) and President (2000–2006)
Publication record: papers (300), books (10), patents (4), citations (10300), h-index (31)
Equipments: Sample Controlled TG, Sample Controlled Evolved Gas Analysis, Gas adsorption microcalorimetry 77, 87, 300 K, Liquid adsorption microcalorimetry
5 most important publications: [1] T. O. Sörensen and J. Rouquerol Eds.: Sample Controlled Thermal Analysis (SCTA) Kluwer Academic Publishers, Dordrecht, 2003, 252 p; [2] F. Rouquerol, J. Rouquerol, K. S. W. Sing, P. Llewellyn and G. Maurin: Adsorption by powders and porous solids: principles, methodology and applications, 2nd Ed., Academic Press, 2014, 626 p; [3] J. Rouquerol, I. Wadsö, T. J. Lever and P. J. Haines: Developments in Nomenclature (Chapter 2) In "Handbook of Thermal Analysis and Calorimetry", Volume 5, "Further Advances, Techniques and Applications", M. Brown and P. Gallagher Eds, Elsevier, Amsterdam, 2007, pp 13–54; [4] F. Rouquerol, J. Rouquerol and P. Llewellyn: Thermal Analysis (Part B, Chapter 2.12) In: Handbook of Clay Science, 2nd Ed. F. Bergaya and G. Lagaly Eds. Developments in Clay Science, Volume 5, Elsevier, 2013; [5] J. Rouquerol and P. Boivinet: Calorimetric measurements (Chapter 27) in "Differential Thermal Analysis", R. Mackenzie ed., London, Academic Press, 1972, Vol.2, pp. 23–46.

Name: Ewa Rudnik
Country: Poland
Date and place of birth: 1962, Warsaw, Poland
Present position and address: professor, Deputy Director for Science, Scientific and Research Centre for Fire Protection—National Research Institute, Nadwiślańska 213, 05-420 Józefów, Poland
Email: erudnik@cnbop.pl
Education and scientific degrees: M.Sc., Chem. Eng., Warsaw University of Technology (1986), Ph.D. (1997), D. Sc. (2009).
Workplaces: Industrial Chemistry Research Institute (1986–2011); Scientific and Research Centre for Fire Protection—National Research Institute (2011–)
Main fields of interest: thermal properties of polymers; thermal characteristics of renewable raw materials; degradation of materials
Relevant categories in thermal analyses: fields (materials, polymer); methods (TG, DTA, DSC, cone calorimetry)
Awards and acknowledgments: Marie Curie Fellow (2007–2009); European Commission expert (participation in the expert evaluators panels for the 6th and 7th EU Framework Programme) (since 2004)
Professional activities: Chairwoman of Editorial Advisory Board of Safety and Fire Technique, member of Polish Society on Calorimetry and Thermal Analysis (PTKAT)
Publication record: papers (72), books (1), book chapters (3), patents (4), citations (456), h-index (10)
Equipments: DSC, cone calorimeter
5 most important publications: [1] Rudnik E., Compostable polymer materials, Elsevier, Oxford, 2008; [2] Rudnik E., Briassoulis D.: Degradation behavior of poly(lactic acid) films and fibers in soil under Mediterranean field conditions and laboratory simulations testing, Industrial Crops and Products 2011, 33, 648–658; [3] Rudnik E.: Thermal properties of biocomposites, Journal of Thermal Analysis and Calorimetry, 2007, 88, 495–498; [4] Rudnik E., Matuschek G., Milanov N., Kettrup A.: Thermal stability and degradation of starch derivatives, Journal of Thermal Analysis and Calorimetry 2006, 85, 267–270; [5] Rudnik E., Matuschek G., Milanov N., Kettrup A.: Thermal properties of starch succinates, Thermochimica Acta 2005, 427/1–2, 163–166.

Name: Zofia Rzączyńska
Country: Poland
Date and place of birth: 1947, Lublin, Poland
Present position and address: Associate Professor, Faculty of Chemistry; Maria Curie Sklodowska University; Head of Department of General and Coordination Chemistry; UMCS 20-031 Lublin, Poland; Maria Sklodowska Sq. 2
Email: z.rzaczynska@poczta.umcs.lublin.pl
Education and scientific degrees: (1971–1974) tutor of chemistry; (1974–1982) assistant, Ph.D. 1982, (1982–1994) senior assistant, D.Sc. 1994; (1994–) associate Professor Maria Curie Sklodowska University.
Workplaces: Faculty of Chemistry Maria Curie Sklodowska University Lublin (1974–)
Main fields of interest: relation—crystal structure and thermal stability of lanthanide complexes; synthesis of microporous coordination framework, spectroscopic investigation of lanthanide compounds
Relevant categories in thermal analyses: fields (inorganic, materials, nano, complex, organic, pharmaceutical); methods (TG, DTA, EGA, DSC, kinetics, specific heat, calorimetry)
Awards and acknowledgments: Medal of Minister of Science and Higher Education (1999); Medal of the Committee of the National Education (Medal KEN)(2002); Medals from International Centre for Diffraction Data (ICDD) (2002–2010)
Professional activities: Chairmen and organiser of International Conference ICVMTT33, Zamość, Poland 2011, Member of Polish Chemical Society (PTChem); v-chair Scientific Society of Lublin (LTN), Member of Polish Society of Calorimetry and Thermal Analysis (PTKAT)
Publication record: papers (101), books (6), citations (230), h-index (11), sum of impact factors (95)
Equipments: thermoanalytical instruments: Setsys 16/18 Setaram; TA Q 5000
5 most important publications: [1] Z. Rzączyńska, A. Ostasz, S. Pikus: J. Thermal. Anal. Calorim.82(2005)347; [2] Z. Rzączyńska, J. Sienkiewicz-Gromiuk, H. Głuchowska: J. Thermal. Anal. Calorim.101(2010)213; [3] R. Łyszczek, Z. Rzączyńska, A. Kula, A. Gładysz –Płaska: J. Anal. Appl. Pyrol. 92(2011)347; [4] A. Bartyzel, Z. Rzączyńska, A. Danczowska-Burdon: J. Coord. Chem. 63(2010) 46; [5] I. Rusinek, J. Sienkiewicz-Gromiuk, L. Mazur, Z. Rzączyńska: J. Inorg. Organomet. Polym. 23 (2013)1068.

Name: Pedro E. Sánchez-Jiménez
Country: Spain
Date and place of birth: 1976, Sevilla, Spain
Present position and address: Researcher at Instituto de Ciencia de Materiales de Sevilla., Instituto de Ciencia de Materiales de Sevilla. Américo Vespucio 49, 41092. Sevilla, Spain
Email: pedro.enrique@icmse.csic.es
Education and scientific degrees: B.Sc Chemistry, University of Seville (1999). B.Sc Biochemistry (2001); Ph.D. Chemistry (2008)
Workplaces: Instituto de Ciencia de Materiales de Sevilla, Spain (2003–08). University of Colorado at Boulder, USA (2008–2009). Institut Europeen des Membranes, Montpellier, France (2010). Instituto de Ciencia de Materiales de Sevilla, Spain (2011-current)
Main fields of interest: reactivity of solids, both thermally and mechanically induced. polymer degradation; solid state kinetics; thermal characterization and improvement of heat transfer fluids.
Relevant categories in thermal analyses: fields (inorganic, materials, nano, polymer, ceramics, cement, other (thermal fluids)); methods (TG, DTA, DSC, kinetics, extremely high temperature (above 1,000 °C); specific heat, calorimetry, instrument development, other (sample controlled thermal analysis))
Publication record: papers (41), book chapters (1), citations (496), h-index (14)
Equipments: DSC (also High Pressure), TGA (From Vacuum to High Pressure and coupled to MS and FTIR), DTA, TMA, Microwave furnaces
5 most important publications: [1] P. E. Sánchez-Jiménez, J. M. Valverde. Applied Energy. 118(1), pp. 92. 2014; [2] P. E. Sánchez-Jiménez, A. Perejón. J. Phys. Chem. C. 116(21), pp. 11797. 2012; [3] A. Perejón, P. E. Sánchez-Jiménez. J. Phys. Chem B. 115(8), pp. 1780. 2011; [4] P. E. Sánchez-Jiménez, L. A. Pérez-Maqueda. Pol. Deg. Stab. 95, pp 733. 2010; [5] J. M Criado, P. E. Sánchez-Jiménez et al. J. Therm. Anal. Cal. 92(1), pp. 199. 2009.

Name: Viorel Zoltan Sasca
Country: Romania
Date and place of birth: 1949, Salard, Romania
Present position and address: Head of Heterogeneous Catalysis Group, Scientific reasearcher 1st degree, Institute of Chemistry Timisoara-Romanian Academy, Bd. M. Viteazu 24, 300223 Timisoara, Romania
Email: viorelsasca@yahoo.co.uk; vsasca@acad-icht.tm.edu.ro
ORCID: 0000-0001-7795-9816
Education and scientific degrees: Chemical engineer, Technical University of Timisoara; Ph.D.
Workplaces: Arad Chemical Fertilizers Plant (1972–1981), Institute of Chemistry Timisoara of Romanian Academy (1981–)
Main fields of interest: design of catalysts for utilization of waste and renewable substances; the study of solid catalysts with regard to composition, thermal behavior, structure and texture; acidity and redox properties; the design and construction of lab installation for catalysts preparation and evaluation of catalytic activity; kinetics of solid state reactions and heterogeneous catalytic reactions
Relevant categories in thermal analyses: fields (kinetics of mixed oxides formation and decomposition of catalysts/catalyst precursor, catalysts lifetime, adsorption-desorption reactions, heterogeneous redox reactions); methods (TGA/DTG-DTA, DSC, Flynn-Wall-Ozawa isoconversional, Friedman isoconversional, standard test method for decomposition kinetics by thermogravimetry, ASTM 1641-04 and standard practice for calculating Thermal endurance of materials from thermogravimetric decomposition data-ASTM E1877-00)
Professional activities: Member of Chemical Society of Romania, member of Romanian Catalysis Society
Publication record: papers (58), books (1), patents (9)
Equipments: TGA/SDTA 851/LF/1100-Mettler+MS Thermo Star- Pfeiffer Vacuum and DSC 823^e
5 most important publications: [1] V. Sasca, O. Verdes, L. Avram, A. Popa, A. Erdohelyi, A. Oszko, Appl Catal A: Gen 2013 451 50–57; [2] V. Sasca, O. Verdes, L. Avram, A. Popa, Rev Roum Chim 2013 58 (4-5) 451–461; [3] V. Sasca, N. Doca, A. Popa and N. Jaeger, React. Kinet. Mech. Catal. 2012 105(1) 207–221; [4] V. Sasca, Orsina Verdeş, Livia Avram, A. Popa, P. Barvinschi and M. Mracec, Rev Roum Chim, 2011 56 (5), 501–516; [5] V. Sasca, Livia Avram, Orsina Verdes and A. Popa, Appl Surf Sci 2010 256 5533–5538.

Name: Igor Savchenko
Country: Russia
Date and place of birth: 1984, Kartali, Russia
Present position and address: Senior Researcher, Russia, Novosibirsk, 630090, Lavrentiev avenue 1.
Email: savchenko@itp.nsc.ru
Website: http://www.itp.nsc.ru/
ORCID: 0000-0003-1934-9846
Education and scientific degrees: Bachelor of Physics, Master of science, Novosibirsk State University, Ph.D.
Workplaces: Kutateladze Institute of Thermophysics, Siberian Brunch of the Russian Academy of Sciences, Senior Researcher
Main fields of interest: experimental investigation of thermophysical properties condensed matter
Relevant categories in thermal analyses: fields (inorganic materials, ceramics, liquid metals, coolants, construction materials); methods (DSC, laser flash method, drop calorimetry, extremely high temperature, instrument development)
Professional activities: Organaiser of the 13th Russian conference on thermophysical properties, 30th Siberian Thermophysical Seminar, 11th Russian School-Conference for Young Scientists "Actual issues of thermal physics and physical fluid dynamics"
Publication record: papers (12), citations (14), h-index (3), sum of impact factors (6)
Equipments: Netzsch DSC 404 F1 Pegasus, Netzsch LFA 427, Massive drop calorimeter
5 most important publications: [1] Stankus, S. V., Savchenko, I. V. Laser flash method for measurement of liquid metals heat transfer coefficients, Thermophysics and Aeromechanics. 2010. vol. 16, № 4, pp. 585–592; [2] Savchenko, I. V., Stankus, S. V., Agazhanov, A. Sh. Heat transfer coefficients of liquid indium in the temperature range 470–1,275 K Thermophysics and Aeromechanics. 2010. vol. 17, № 1, pp. 121–125; [3] Savchenko, I. V., Stankus, S. V., Agadjanov, A. Sh. Measurement of Liquid Tin Heat Transfer Coefficients within the Temperature Range of 506–1,170 K, High Temperature. 2011. vol. 49, № 4, pp. 506–511; [4] Stankus, S. V., Savchenko, I. V., Agadzhanov, A. Sh., et al. Thermophysical Properties of MPG-6 Graphite, High Temperature. 2013. vol. 51, № 2, pp. 179–182; [5] Savchenko I. V., Stankus S. V., Agazhanov A. Sh. Measurement of the Thermal Conductivity and Diffusivity of Molten Lead in the Interval 601–1,000 K, Atomic Energy. 2013. vol. 115, № 2, pp. 83–87.

Name: Christoph Schick
Country: Germany
Date and place of birth:1953, Gross-Luesewitz, Germany
Present position and address: Professor, University of Rostock, Institute of Physics, Wismarsche Str. 43–45, 18051 Rostock, Germany
Email: christoph.schick@uni-rostock.de
Website: http://www.polymerphysik.uni-rostock.de/
Researcher ID/ORCID: C-1154-2009/0000-0001-6736-5491
Education and scientific degrees: Dipl. Physicist, Dr. rer. nat. habil.
Workplaces: (1976–1979) Physics Department, Technical University of Leuna-Merseburg, research fellow; (1979–1992) Physics Department, Pedagogical University Güstrow, senior research fellow, first assistant to professor; (1992–) Physics Department, University of Rostock, professor
Main fields of interest: polymer physics, calorimetry, glass transition, melting and crystallization
Relevant categories in thermal analyses: fields (inorganic, materials, nano, organic, pharmaceutical, polymer, biology, glass, ceramics, other (metals, alloys)); methods (TG, DSC, thermomechanical analysis, kinetics, cryo, specific heat, calorimetry, microcalorimetry, instrument development, other (chip calorimetry))
Awards and acknowledgments: Mettler Toledo Award of the North American Thermal Analysis Society (NATAS) (2006), The James J. Christensen Memorial Award in Recognition of Outstanding Contributions to the Innovative Development and Use of Calorimetric Equipment, CALCON, USA (2008), The 2010 AICAT-SETARAM Award in recognition of outstanding contributions to advance the physical knowledge and knowledge generating tools related to ordering, crystallisation, aggregation or organization of synthetic polymers, AICAT, Italy (2010), The 2011 AFCAT Calvet Price, France, Wissenschaftspreis der GEFTA 2014, Germany
Professional activities: Editor "Thermochimica Acta", German Society for Thermal Analysis (GEFTA); Board member, Organizer of the biannual "Laehnwitzseminar on Calorimetry"
Publication record: papers (273), book chapters (4), patents (3), h-index (37)
Equipments: Different differential scanning calorimeters (DSC), Calvet calorimeter, Chip calormeter (AC and fast scanning), thermographimetric analyzers
5 most important publications: [1] Cebe, P.; Hu, X.; Kaplan, D. L.; Zhuravlev, E.; Wurm, A.; Arbeiter, D.; Schick, C., Beating the Heat—Fast Scanning Melts Silk Beta Sheet Crystals, Scientific Reports (2013) 1130; [2] Androsch, R.; Di Lorenzo, M. L.; Schick, C.; Wunderlich, B., Mesophases in polyethylene, polypropylene, and poly(1-butene), Polymer 51 (2010) 4639–4662; [3] Tress, M.; Erber, M.; Mapesa, E. U.; Huth, H.; Müller, J.; Serghei, A.; Schick, C.; Eichhorn, K.-J.; Voit, B.; Kremer, F., Glassy Dynamics and Glass Transition in Nanometric Thin Layers of Polystyrene, Macromolecules 43 (2010) 23 9937–9944; [4] Sargsyan, A.; Tonoyan, A.; Davtyan, S.; Schick, C., The amount of immobilized polymer in PMMA SiO_2 nanocomposites determined from calorimetric data, Europ. Polymer J. 43 (2007) 3113–3127; [5] Hempel, E.; Hempel, G.; Hensel, A.; Schick, C.; Donth, E., Characteristic Length of Dynamic Glass Transition near Tg for a Wide Assortiment of Glass-Forming Substances, J. Phys. Chem. B 104 (2000) 2460–2466.

Name: Alberto Schiraldi
Country: Italy
Date and place of birth: 1946, Bitonto, Italy
Present position and address: Full Professor of Physical Chemistry, Dept of Food Environmental Nutrition Sciences, University of Milan, Via Celoria 2, 20133 Milano, Italy
Email: alberto.schiraldi@unimi.it
Website: www.unimi.it
Education and scientific degrees: graduation in Chemistry
Workplaces: Dept of Food Environmental Nutrition Sciences, University of Milan,
Main fields of interest: food and bio-science: water activity, metastable states, dispersed systems, microbial growth
Relevant categories in thermal analyses: fields (phase transitions, conformational transitions, shelf life assessment, microbial growth); methods (micro- and nano-DSC, TG, ITC)
Awards and acknowledgments: AICAT-SETARAM award 2009
Professional activities: member of the Editorial Board of J. Polish Food Nutrition Sci.; President of AICAT (It. Ass. Calorimetry Therm. Anal) 1999–2004; organiser of XIII ICTAC conference in 2004
Publication record: papers (130), books (3), average citations per article (9.82), h-index (19)
Equipments: Mettler DSC 20, PerkinElmer DSC, SETARAM C80, SETARAM microDSCIII; SETARAM DAM, SETARAM nano-DSC, SETARAM nano-ITC, SETARAM DSC-TG, TA Modulated DSC 2920, Modulated Adiabatic Calorimetry (assembled in Pisa)
5 most important publications: [1] A. Schiraldi, M. Signorelli and D. Fessas, Knudsen thermogravimetry approach to the thermodynamics of aqueous solutions. J. Chem. Thermodyn. 62 (2013) 79–85; [2] Ausili, A. Pennacchio, M. Staiano J. D. Dattelbaum, D. Fessas, A. Schiraldi, S. D'Auria. Amino acid transport in thermophiles: Characterization of an arginine-binding protein from Thermotoga maritima. 3. Conformational dynamics and stability. J Photochem. Photobiol. B: Biology, 118 (2013) 66–73; [3] L. Introzzi, T. O. J. Blomfeldt, S. Trabattoni, Si. Tavazzi, N. Santo, A. Schiraldi, L. Piergiovanni, S. Farris, Ultrasound-assisted pullulan/Na^+ -montmorillonite oxygen barrier nanocomposite coatings for food packaging applications Langmuir, 28 (2012) 11206–11214; [4] S. Farris, L. Introzzi, P. Biagioni, T. Holz, A. Schiraldi, L. Piergiovanni. Wetting of Biopolymer Coatings: Contact Angle Kinetics and Image Analysis Investigation. Langmuir. 27 (2011 May) 7563–7574; [5] P. Pani, A. Schiraldi, M. Signorelli, D. Fessas. Thermodynamic Approach to Osmo-dehydration. Food biophysics, 5:3(2010) 177–185.

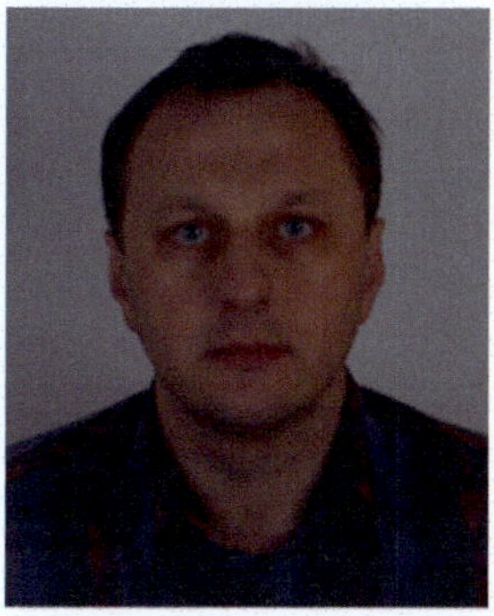

Name: David Sedmidubský
Country: Czech Republic
Date and place of birth: 1966, Prague, Czech Republic
Present position and address: Professor of inorganic chemistry; Head of the at the Deparment of Inorganic Chemistry, Institute of Chemical Technology Prague (ICT Prague); Technická 5, 166 28 Prague, Czech Republic
Email: sedmidub@vscht.cz
Website: www.vscht.cz/ach/ustav-osobni_sedmidub.html
Education and scientific degrees: Chemical Engineer (ICT Prague, 1989), Ph.D. (ICT Prague, 1993), Associate Professor (ICT Prague, 2003), Professor (ICT Prague, 2007)
Workplaces: Institute of Chemical Technology Prague (since 1995), Institute for Transuranium Elements—Joint Research Centre—European Commission (2003–2004, 2008–2009), Institute of Physics—Academy of Sciences of the Czech Republic (1989–2003)
Main fields of interest: measurement and theory of heat capacity, cohesion (chemical bonding) in solids, thermodynamics of materials, phase diagram determination (modeling and experiment), crystal chemistry, magnetism and electric transport, electronic structure calculations of solids
Relevant categories in thermal analyses: fields (high temperature calorimetry, low temperature calorimetry, phase transitions, thermogravimetry, heat transport); methods (DTA, HF DSC—incremental and continuous methods (C_p measurement and enthalpies of phase transitions), TGA (nonstoichiometry under controlled atmosphere), high temperature drop calorimetry (relative enthalpies and heats of solution), relaxation time methods (low temperature Cp measurement), laser flash method for thermal conductivity measurement)
Professional activities: member of the executive board of the Thermal analysis group, Chemical thermodynamics group and Inorganic chemistry group of the Czech chemical society
Publication record: papers (104), book chapters (3), patents (6), citations (662), h-index (15)
Equipments: Multi-HTC 96 (Setaram), DSC 404C Pegasus (Netzsch), STA PT1600 (Linseis), LFA 1000 (Linseis)
5 most important publications: [1] D. Sedmidubský, O. Beneš, R. J. M. Konings, High Temperature Heat Capacity of $Nd_2Zr_2O_7$ and $La_2Zr_2O_7$ Pyrochlores, J. Chem. Thermodyn. 37 [10] (2005) 1098; [2] D. Sedmidubský, R. J. M. Konings, P. Novák, Calculation of Enthalpies of Formation of Actinide Nitrides, J. Nucl. Mater. 344 (2005) 40–44; [3] J. Leitner, P. Chuchvalec, D. Sedmidubský, A. Strejc, P. Abrman, Estimation of Heat Capacities of Solid Mixed Oxides, Thermochimica Acta 395 [1–2] (2003) 25–43; [4] J. Leitner, A. Strejc, D. Sedmidubský, K. Růžička, High temperature enthalpy and heat capacity of GaN, Thermochimica Acta, 401 [2] (2003) 169–173; [5] J. Leitner, P. Voňka, D. Sedmidubský, P. Svoboda, Application of Neumann-Kopp Rule for the Estimation of Heat Capacity of Mixed Oxides, Thermochimica Acta 497 (2010) 7–13.

Name: James Dimitrios Constantine Seferis
Country: USA
Date and place of birth: 1950, Athens, Greece
Email: jcseferis@aol.com
Education and scientific degrees: B.Sc., Chemical Engineering, University of Colorado (1969–1973); Ph.D.: University of Delaware (1973–1977)
Workplaces: University of Washington (1977–2007); Boeing Steiner Professor Chemical Engineering; Management Science; Environmental Health—Polymer Composites; Chair, Advanced Materials, Manufacturing and Management Processes; Distinguished Affiliate Professor, Sung Kyun Kwan University, Seoul, Korea (2006–Present); Visiting: University of Naples, Italy; University of Tokyo; University of Hamburg; Ecole Polytechnic Federal de Lausanne; National Technical University of Athens; University of Delaware; University of Texas; University of Nevada. Founder/Chairman of the Board and CEO, Polymeric Composites Laboratory; Glocal/F.R.E.E.D.OM. (1982-Present)
Main fields of interest: carbon fiber reinforced polymeric composites; dynamic thermal analysis for heterogeneous anisotropic viscoelastic polymeric systems; scaling phenomena from molecular to airplane structures
Awards and acknowledgments: METTLER Award and Fellow NATAS and SAMPE; Distinguished Honorary Member, Mediterranean Thermal Analysis Society; Corresponding Member, Academy of Athens, Greece; Alexander von Humboldt Foundation Award Professor; White House First Presidential Young Investigator Award.
Professional activities: Consultant to over fifty organizations, including technical and strategic management. Organized 100 International symposia workshops on polymers, composites, management practices, and teaming. Currently chairman of the Board and Director of public/private partnership consortium with direct operations in seven countries. Edited Special Journal Issues.
Publication record: papers (455), books (9), citations (4296)
Equipments: Full set of thermal analysis equipment by Netzsch.
5 most important publications: [1] "Crystallization Kinetics Polyetheretherketone (PEEK) Matrices," J. C. Seferis, et al., Polymer Engineering and Science (26 (22)1574–1581 (1986); [2] "The Evolution of An Aerospace Material: Influence of Design, Manufacturing and In-Service Performance," J. C. Seferis, et al., Composite Structures 27 193–206 (1994)., [3] "Model High-Performance Adhesive Systems," J. C. Seferis, et al., J. Applied Polymer Science 66 1953–1963 (1997); [4] "Matrix Hybridization in the Interlayer for Carbon Fiber Reinforced Composites," J. C. Seferis, et al., Polymer Composites 31 (11) 1965–1976 (2010); [5] "Prepreg Manufacturing," J. C. Seferis, et al., Wiley Encyclopedia of Composites –Second Edition, Editors L. Nicolais and A. Bozzacchiello, Vol. 4 2381–2391 (2012).

Name: Jaroslav Šesták
Country: Czech Republic
Date and place of birth: 1938, Držkov, Czech Republic
Present position and address: Emeritus scientist, Institute of Physics, Academy of Sciences, Cukrovarnická 10, CZ-16200 Prague, and Team leader, New Technology Research Centre of West Bohemian Region, University of West Bohemia, Universitni str. 8, CZ-30114 Pilzen.
Email: sestak@fzu.cz
Website: www.fzu.cz/~sestak; www.thermotics.eu (e-books available)
Education and scientific degrees: M.Eng., Ph.D., Doc. hab., Dr.Sc., Prof., Dr.h.c., Emeritus
Main subject of interest: thermodynamics, solid state chemistry and physics, nonequilibrium, material science
Relevant categories in thermal analyses: fields (materials, biocompatible and nano-composites, polymer, glass, ceramics, cement); methods (TG, DTA, DSC, thermophysical measurements, theory, quenching, kinetics); interdisciplinary education
Awards and acknowledgments: NATAS (USA 1974), Kurnakov (USSR 1985), Bodenheimer (Israel 1987), ICTAC (England 1992), Hanus (Czech Chemical Society, 1998) and Heyrovsky (Czech Academy of Sciences, 2000), Patras University (Greece 2007); Honorary chair of ESG/ICG glass international conferences (Slovakia 2004 and 2008), Member of the Czech Engineering Academy (2004), Board of trustees of Technical University in Liberec (2009), Doctor honoris causa of Pardubice University (2010), Emeritus of Czech Academy Science (2011); Medals 2013: Russian Academy and StPetersburg University; GEFTA/Netzsch Award (Berlin 2014); Prague honorary medal (2014)
Professional activities: invited speaker and charter member of the first TA conference where ICTA was shaped (London 1965), early book on TA (1971), cofounder of Thermochimica Acta (1970), of Journal of Mining and Metallurgy (1995), and of Global Journal of Analytical Chemistry (2010), Councilor of ICTA, former Chair of the Czech TA society, program Chair of the 8th ICTA (Bratislava 1985), coined field of kinetic phase diagrams (book 1991), member of Editorial Board of JTAC and of Journal Applied Glass Science (Wiley), backing keystones of School of Energy Sciences, Kyoto University (1996), of Faculty of Humanities, Prague Charles University (1999), of Institute of Interdisciplinary Studies, West Bohemian University in Pilzen (2000) and of Prague branch of New York University (2000)
Publication record: papers (271), books (16), patents (1), citations (2290), average citations (49), h-index (24), invited lectures (>200), standalone art photoexhibitions (24)
5 most important publications: [1] J. Šesták, G. Berggren, Study of the kinetics of the mechanism of solid- state reactions at increasing temperatures Thermochim. Acta 1971; 3: 1-12; [2] JJ. Mareš, J. Stávek, J. Šesták, Quantum aspects of self-organized periodic chemical reaction. Journal of Chemical Physics 2004; 121: 1499–1503; [3] J. Šesták, Rationale and fallacy of thermoanalytical kinetic patterns: how we model subject matter, J Therm Anal Calorim, 2012; 110: 5-16; [4] J. Šesták, Science of Heat and Thermophysical Studies: a generalized approach to thermal analysis. Amsterdam: Elsevier 2005; [5] J. Šesták, P. Hubík, JJ. Mareš (edts). Glassy, amorphous and nano-crystalline materials: thermal physics, analysis and properties and J. Šesták, P. Šimon (edts). Thermal analysis of micro-, nano- and non-crystalline materials. Berlin: Springer 2011 and 2013.

Name: Dipak Kumar Setua
Country: India
Date and place of birth: 1956, Midnapur, West Bengal, India.
Present position and address: Director of Advanced Centre of Research in High Energy Materials (ACRHEM), University of Hyderabad, Gachibowli, Hyderabad 500 046, Andhra Pradesh, India.
Email: dksetua@rediffmail.com
ORCID: 0000-0003-0860-3458
Education and scientific degrees: B.Sc (Chemistry), Calcutta University (1978); M.Sc (Physical Chemistry), IIT Kharagpur (1981); Ph.D (Rubber Technology), IIT Kharagpur (1985); Post-doc (Polymer Engineering), University of Akron, Ohio, USA (1989–1991).
Workplaces: Sc. 'G', Additional Director, DMSRDE (DRDO), Kanpur 208013, U.P., India. (1984–).; Director ACRHEM, Univ. of Hyderabad (UoH), Hyderabad 500 046, A.P. India (2013–)
Main fields of interest: polymer blends, smart elastomer, composite, polymer nanocomposites, thermal analysis of polymeric materials and rubbers
Relevant categories in thermal analyses: fields (nano materials, polymers, rubbers, composites); methods (TGA, DSC, DTA, EGA, thermo-mechanical analysis, dynamic mechanical analysis, dielectric Analysis, TGA-MS)
Awards and acknowledgments: "Rubber Technology Innovation Award" of AIRIA, India, 2013; NETZSCH (NGB, Germany) ITAS (BARC, India) Award, 2012; K. M. Philip Award of IRMRA, India 2000; DRDO Cash Award, 1997, 1994, 1987; Dunlop Award of IRMRA, India 1984
Professional activities: Member of Editorial Board, J. Applied Polymer Science (2002–); J. Plastic and Polymer Technology (PAPT); Member of Society of Plastic Engineers (SPE), USA; Research Advisory Committee of Shriram Institute for Industrial Research, Delhi; Governing Council Member of Indian Rubber Manufacturers Research Association (IRMRA); Polymer Society of India, Indian Thermal Analysis Society (ITAS, BARC)
Publication record: papers (65), book chapters (7), patents (2), h-index (15), sum of impact factors (2.5)
Equipments: TGA, DSC, DTA, EGA, TMA, DMA, DEA, TGA-MS
5 most important publications: [1] Studies on Phase Morphology and Thermo-Physical properties of Nitrile Rubber Blends, K. Agarwal, M. Prasad, A. Chakraborty, C. B. Vishwakarma, R. B. sharma, D. K. Setua, J. Therm. Anal. Calorim., 104, 1125, 2011; [2] Thermal and morphological analysis of thermoplastic polyurethane clay nanocomposites: Comparison of efficacy of dual modified laponite vs. commercial montmorillonite, M. Mondal, P. K. Chattopadhyay, S. Chattopadhyay and D. K. Setua, Thermochim. Acta, 510, 185, 2010; [3] On the Use of Micro Thermal Analysis to Characterize Compatibility of Nitrile Rubber Blends, D. K. Setua and Y. N. Gupta, Thermochim. Acta, 462, 32, 2007; [4] D. K. Setua, Y. N. Gupta, S. Kumar, R Awasthi, A. Mall and K. Shekhar, J. Appl. Polym. Sci., 100, 677, 2006; [5] Y. N. Gupta, A. Chakraborty, G. D. Pandey and D. K. Setua, J. Appl. Polym. Sci., 92, 1737, 2004.

Name: Carmelo Sgarlata
Country: Italy
Date and place of birth: 1972, Catania, Italy
Present position and address: Assistant Professor of Analytical Chemistry, Department of Chemical Sciences, University of Catania, Viale Andrea Doria 6, 95125, Catania, Italy
Email: sgarlata@unict.it
Website: http://www.dipchi.unict.it/cat/docenti/sgarlata-carmelo/
Education and scientific degrees: Master degree ("Laurea") in Chemistry, University of Catania (1999); Doctoral degree in Chemical Sciences, University of Catania (2004)
Workplaces: Postdoctoral Research Assistant, University of Catania (2004–2009); Postdoctoral Fellow, Lawrence Berkeley National Laboratory and University of California, Berkeley, CA, USA (2008–2010); Assistant Professor of Analytical Chemistry, University of Catania (2010–)
Main fields of interest: speciation and determination of thermodynamic parameters of host-guest and metal-ligand systems; energetics of reaction through nano-calorimetry; molecular recognition and self-assembly in aqueous solution; fluorescent chemosensors for metal ions of biological/environmental relevance; sequestering of pollutants through polymeric membranes; patterning of gold surfaces with organic layers
Relevant categories in thermal analyses: fields (analytical, supramolecular, metal complexes, solution equilibria); methods (calorimetry ITC, calibration methods, thermodynamic parameters refinement)
Awards and acknowledgments: Award "for the contribution to the development of nanoscience and nanotechnology", Accademia Gioenia of Catania (2007); Travel grant for research projects abroad, University of Catania (2008)
Professional activities: Secretary of the Board of GICAT (Italian Group of Calorimetry and Thermal Analysis, 2014–2016).
Publication record: papers (23), citations (238), h-index (9), sum of impact factors (81.4)
Equipments: Nano-ITC and DSC (TA Instruments)
5 most important publications: [1] C. Sgarlata, J. S. Mugridge, M. D. Pluth, B. E. F. Tiedemann, V. Zito, G. Arena, K. N. Raymond, J. Am. Chem. Soc. 2010, 132, 1005–1009; [2] C. Sgarlata, V. Zito, G. Arena, Anal. Bioanal. Chem., 2013, 405, 1085–1094; [3] C. Bonaccorso, A. Ciadamidaro, C. Sgarlata, D. Sciotto, G. Arena, Chem. Commun., 2010, 46, 7139–7141; [4] C. Bonaccorso, C. Sgarlata, G. Grasso, V. Zito, D. Sciotto, G. Arena, Chem. Commun., 2011, 47, 6117–6119; [5] C. Sgarlata, C. Bonaccorso, F. G. Gulino, V. Zito, G. Arena, D. Sciotto, New J. Chem., 2009, 33, 991–997.

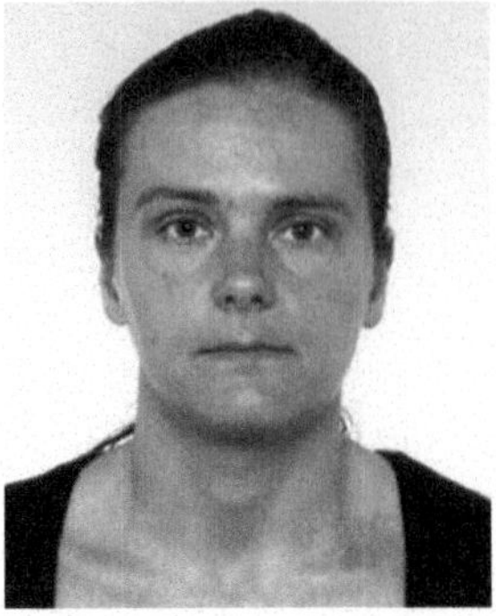

Name: Jana Shánelová
Country: Czech Republic
Date and place of birth: 1973, Ustí nad Orlicí, Czech Republic
Present position and address: Assistant Professor of Physical Chemistry, University of Pardubice; Studentska 573, 53210 Pardubice, Czech Republic
Email: jana.shanelova@upce.cz
Education and scientific degrees: Physical Chemistry, University of Pardubice (1992–1997); Inorganic Chemistry and Technology, University of Pardubice, Ph.D. (2001)
Workplaces: Joint Laboratory of Solid State Chemistry of the Institute of Macromolecular Chemistry of AS CR and University of Pardubice (1997–2001), Department of Physical Chemistry, Faculty of Chemical Technology, University of Pardubice (2001–)
Main fields of interest: application of thermal analysis on chalcogenide glasses and melts
Relevant categories in thermal analyses: fields (inorganic, materials, glass, chalcogenide); methods (DSC, thermomechanical analysis, kinetics, cryo, specific heat, calorimetry, structural relaxation, crystallization, viscosity)
Publication record: papers (15), citations (125), h-index (7), sum of impact factors (27)
Equipments: Pyris 1 DSC (PerkinElmer), DSC 822e (Mettler, Toledo), Q2000 heat flow DSC (TA Instruments), TMA CX04R (R.M.I.)
5 most important publications: [1] Málek J., Shánelová J.: "Viscosity od germanium sulfide melts", J. Non-Cryst. Solids, 243 (1999) 116; [2] Švadlák D., Shánelová J., Málek J., Pérez-Maqueda L., Criado J., Mitsuhashi T.: "Nanocrystallization of anatase in amorphous TiO_2", Termochim. Acta, 414 (2004) 137–142; [3] Shánelová J., Málek J., Alcalá M., Criado J.: "Kinetics of crystal growth of germanium disulfide in $Ge_{0.38}S_{0.62}$ chalcogenide glass", Journal of Non-Crystalline Solids, 351 (2005) 557–567; [4] Shánelová J., Málek J.: "Structural relaxation of As_2Se_3 glass and viscosity of supercooled liquid", J. Non-Cryst. Solids, 351 (2005) 3458–3467; [5] Shánelová J., Koštál P., Málek J.: "Viscosity of $(GeS2)_x(Sb2S3)_{1-x}$ supercooled melts", J. Non Cryst. Solids, 352 (2006) 3952–3955.

Name: Judit Simon
Country: Hungary
Date and place of birth: 1937, Budapest, Hungary
Present position and postal address: Retired from research activity, Editor-in-Chief, J Therm Anal Calorim, until 2014
Email: lexica@chello.hu
Education and scientific degrees: Chemical engineer, Technical University of Budapest (1960); Ph.D. and C.Sc. degrees received in thermal analysis at the Institute for General and Analytical Chemistry of the Technical University
Workplaces: Senior lecturer (1960); Adjunct Professor (1964); Associate Professor (1980) at the Institute for General and Analytical Chemistry of the Technical University
Main fields of interest: applications of simultaneous thermal analysis and coupled methods for different topics including high-temperature reactions, non-isothermal kinetics, polymers, biopolymers, fibres, flammability and retardation processes;her teaching tasks were connected with instrumental analysis including thermal analysis and calorimetry; many students and Ph.D. fellows with thermal analysis and calorimetry were supervised.
Professional activities: The Journal of Thermal Analysis was launched in (1969) based mainly on her idea. She worked one of the editors, later she was appointed Editor-in-Chief; Postdoctoral fellow and Visting Professor at the University of Massachusetts, Department of Polymer Science, USA (1972–1973) and (1989); ICTAC Affiliate Councillor for Hungary (1993–2001); Member of ICTAC Award Committee (1993–1996); Member of ICTAC Nomenclature Committee (1996–2001); President of the Hungarian Thermoanalytical Group (1985–1998), Series Editor of 'The Hot Topics' in Thermal Analysis and Calorimetry, Kluwer Academic Publishers, Dordrecht (2000–)
Awards: ICTA Editor Award (1992); Hungarian Academy of Science Niveau Award (2002) Paulik Award (2007); Honorific member of CEEC-TAC (2011); The ICTAC Distinguished Service Award (2012); Certificate of the Joint Czech-Hungaryan-Polish-Slovak TA Conference (2013)
Publication record: papers (81), books (3), citations (110)
List of the 5 most important publications: [1] L. Erdey, J. Simon, S. Gál, G. Liptay: Thermal properties of analytical grade reagents, Talanta, 13 (1966) 67; [2] J. Simon, B. Androsits, B. Szalai, S. Lantos: The application of Derivatograph in flame retardation research. Determination of ignation temperature, Hung. Sci. Instr., 48 (1980) 1; [3] J. Simon, B. Androsits: Thermoanalytical characterization of flame retardant additive action in solids, Israel J. Chem., 22 (1982) 273; [4] J. Simon, T. Kántor, T. Kozma, E. Pungor: Thermal analysis of Sb_2O_3/organohalidebased flame retardants including atomic absorption detection of the evolved species, J. Thermal Anal., 25 (1982) 57; [5] J. Simon, M. S. Szitányi, T. Kántor: Analysis of various metal oxide/organohalide-based flame retardants, J. Thermal Anal., 32 (1987) 1915.

Name: Peter Šimon
Country: Slovakia
Date and place of birth: 1952, Bratislava, Slovakia
Present position and address: Full professor, Department of Physical Chemistry, Faculty of Chemical and Food Technology, Slovak University of Technology, 812 37 Bratislava, Slovakia
Email: peter.simon@stuba.sk
Education and scientific degrees: Chemical Engineer (M.Sc.), Slovak University of Technology in Bratislava (1971–1976); Ph.D. (1980); Assoc. Prof (1994), D.Sc. (2002); Prof. (2002).
Workplaces: Slovak University of Technology in Bratislava (1980–)
Main fields of interest: physical chemistry of materials, thermodynamics and kinetics of the processes occurring in condensed state, thermoanalytical kinetics, predictions of the material lifetime
Relevant categories in thermal analyses: fields (materials, nano, pharmaceutical, polymer, food); methods (TG, DTA, DSC, kinetics)
Awards and acknowledgments: Best reviewer of JTAC (2012), Medal of STU (2012)
Professional activities: Head of the Slovak Group of Thermal Analysis and Calorimetry; councillor at ICTAC; national representative in the ESTAC committee; regional editor of the Journal of Thermal Analysis and Calorimetry; member of the editorial board of the Journal of Food and Nutrition Research
Publication record: papers (153), books (1), chapters in books (7), patents (6), citations (1220), h-index (22)
Equipments: DSC-7 PerkinElmer, DSC-60 Shimadzu, DSC-25 Mettler-Toledo, TG/DTA Seiko Exstar 6300
5 most important publications: [1] Šimon P., J. Therm. Anal. Calorim. 79 (2005) 703–708: Single-step kinetics approximation employing non-Arrhenius temperature functions; [2] Šimon P., J. Therm. Anal. Calorim. 88 (2007) 709–715: The single-step approximation: Attributes, strong and weak sides; [3] Šimon P., Hynek D., Malíková M. and Cibulková Z., J. Therm. Anal. Calorim. 93 (2008) 817–821: Extrapolation of accelerated thermooxidative tests to lower temperatures applying non-Arrhenius temperature functions; [4] Šimon P., J. Therm. Anal. Calorim. 97 (2009) 391–396: Material stability predictions applying a new non-Arrhenian temperature function; [5] Šimon P., Thomas P., Dubaj T., Cibulková Z., Peller A. and Veverka M., J. Therm. Anal. Calorim. 115 (2014) 853–859: The mathematical incorrectness of the integral isoconversional methods in case of variable activation energy and the consequences.

Name: Sindee L. Simon
Country: USA
Date of birth: 1961
Present position and address: P. W. Horn Professor and Whitacre Department Chair, PO Box 43121, Texas Tech University, Lubbock TX 79409-3121
Email: sindee.simon@ttu.edu
Website: http://www.depts.ttu.edu/che/faculty/faculty.php?name=Sindee%20L.%20Simon
Researcher ID: M-9372-2013
Education and scientific degrees: B.Sc., Chemical Engineering, Yale University, 1983, Ph.D., Chemical Engineering, Princeton University, 1992
Workplaces: Texas Tech University (1999–); University of Pittsburgh (1992–1999)
Main fields of interest: glass physics and behavior at the nanoscale, thermosetting polymers and nanocomposites, calorimetry, dilatometry, and bulk rheology
Awards and acknowledgments: Research/Engineering Technology Award, Society of Plastics Engineers (SPE), 2014.; Scientist of the Year Award, Achievement Rewards for College Scientists (ARCS) Foundation, Lubbock Chapter, coawardee with Gregory McKenna, 2013.; Lubbock Women of Excellence in Science Award, YWCA Lubbock Chapter, 2013.; Service Award, North American Thermal Analysis Society (NATAS), 2012.; Outstanding Professor, Texas Tech AIChE Student Chapter, 2011.; Most Influential Professor, Texas Tech University, Engineering Honors Convocation, 2011.; Fellow, American Physical Society (APS), Elected, December 2010.; Horn Professorship, Texas Tech University, 2010.; Barnie E. Rushing, Jr.,; 2007/08 Faculty Distinguished Research Award, Texas Tech Parents Assoc.; Fellow, Society of Plastics Engineers (SPE), Elected, May 2005.; Fellow, North American Thermal Analysis Society (NATAS), Elected, September 2003
Professional activities: Gordon Research Conference on Polymer Physics, Chair, 2014 meeting. North American Thermal Analysis Society: Conference Chair, 2014; Technical Program Chair, 2012; Conference Chair, 2009; President, 2005; Vice President, 2004
Publication record: papers (91), books (1), citations (2570), h-index (27)
Equipments: DSC, Fast Scanning DSC, TGA
5 most important publications: [1] Q. X. Li and S. L. Simon, "Curing of Bisphenol M Dicyanate Ester under Nanoscale Constraint," Macromolecules, 41 (4), 1310–1317 (2008); [2] P. Badrinarayanan, W. Zheng, Q. X. Li, and S. L. Simon, "The Glass Transition Temperature versus the Fictive Temperature," Journal of Non-Crystalline Solids, 353, 2603–2612 (2007); [3] J. Sun and S. L. Simon, "The Melting Behavior of Aluminum Nanoparticles," Thermochimica Acta, 463 (1-2), 32–40 (2007); [4] S. L. Simon, J. W. Sobieski, and D. J. Plazek, "Volume and Enthalpy Recovery of a Polystyrene," Polymer, 42 (6), 2555–2567 (2001); [5] S. L. Simon, G. B. McKenna, and O. Sindt, "Modeling the Evolution of the Dynamic Mechanical Properties of a Commercial Epoxy during Cure after Gelation," Journal of Applied Polymer Science, 76 (4), 495–508 (2000).

Name: Natalia Nikolaevna Smirnova
Country: Russia
Date and place of birth: 1954, Gorky, Russia
Present position and address: D.Sc., Prof., head of laboratory of chemical thermodynamics Research Institute of Chemistry of Nizhny Novgorod State University of Lobachevskii, Gagarin Prospect 23/5, 603950 Nizhny Novgorod, Russia; office phone (831) 4656450
Email: smirnova@ichem.unn.ru
Education and scientific degrees: Chemist-researcher, Gorky State University of Lobachevskii (1972–1977); D.Phil. (1992); Doctor of Science (2003); Professor (2012)
Workplaces: Research Institute of Chemistry of Nizhny Novgorod State University of Lobachevskii (1977–)
Main fields of interest: the temperature dependence of the heat capacity, temperatures and enthalpies of physical transitions of organic substances, polymer, complexes over the range 5 to 600 K; TG, DSC in the range 100–600 K
Awards and acknowledgments: badge (honorary workers in the science and engineering (2009))
Professional activities: Member of the Organizing Committee of 3 International Conferences: XVIII and XIX International Conference on Chemical Thermodynamics in Russia (RCCT-2011 and RCCT-2013) (October 3–7, 2011, Samara; June 24–28, 2013, Moscow); XIV International Conference on Thermal Analysis and Calorimetry in Russia (RTAC-2013) (September 23–28, 2013, Saint-Petersburg, Russia)
Publication record: papers (183), books (1), book chapters (2), h-index (8)
Equipments: an adiabatic vacuum calorimeter BKT-3 for measurements of the heat capacity, temperatures and enthalpies of physical transitions of substances; a DSC 204 F1 Phoenix differential scanning calorimeter (DSC) (Netzsch, Geraetebau, Germany) for measurements the heat capacities in the ranges 100–600 K; TG F1 209 Iris (Netzsch, Geraetebau, Germany)
5 most important publications: [1] B. V. Lebedev, N. N. Smirnova. 'Chemical Thermodynamics of Polyalkanes and Polyalkenes', Nizhny Novgorod University Publisher, Nizhny Novgorod, Russia 1999. 274 p; [2] B. V. Lebedev, N. N. Smirnova. Chap. 9. The thermodynamics of higher polyolefins. In: Po-lymers and Copolymers of Higher α-Olefins). Hanser Publichers, Munich, Vienna, New York, 1997; [3] B. V. Lebedev, A. V. Markin, N. N. Smirnova. In: investigation of carbon: successes and problems/Compiler T. A. Sladkova, editor-in-chief Bubnov. Nauka, Moscow, 2007; [4] Smirnova N. N. et al., J. Therm. Anal. Calorim. 2013. V. 112. № 3. pp. 1447–1452; [5] N. N. Smirnova et al., J. Chem. Thermodyn. 2012. V. 48. pp. 118–122.

Name: Michio Sorai
Country: Japan
Date and place of birth: 1939, Ryojun-city, China
Present position and address: Emeritus Professor of Osaka University; Fushiodai 1-25-3, Ikeda, Osaka 563-0017, Japan
Email: sorai@chem.sci.osaka-u.ac.jp
Education and scientific degrees: Department of Chemistry, Osaka University (1958–1964); Ph.D. (1968); Research Associate (1964–1981); Professor (1981–2003); Emeritus Professor (2003–).
Workplaces: Department of Chemistry, Graduate School of Science, Osaka University (1964–2003); Technische Hochschule Darmstadt in Germany (1974–1976).
Main fields of interest: molecular thermodynamic studies of phase transitions occurring in magnetic materials, liquid crystals, plastic crystals, organic and organometallic compounds, especially the phase transitions in which electrons are directly involved: spin crossover phenomena, intramolecular electron transfer in mixed-valence compounds, thermochromic phenomena, neutral-ionic transition arising from charge transfer mechanism
Relevant categories in thermal analyses: fields (inorganic, materials, complex, organic); methods (DTA, DSC, cryo temperature; specific heat, calorimetry, instrument development)
Awards and acknowledgments: The Hugh M. Huffman Memorial Award (2001); Honorary Professor of Polish Academy of Sciences (2006); Contribution Award from the Japan Society of Coordination Chemistry (2012); Distinguished Achievement Award from the Japanese Liquid Crystal Society (2013)
Professional activities: Advisory Board of "The Journal of Chemical Thermodynamics" (1990–1995 and 2002–2006); Associate Member of the IUPAC I.2 Commission on Chemical Thermodynamics (1998–2001); President of the Japan Society of Calorimetry and Thermal Analysis (1999–2001); Board of Directors at the International Association of Chemical Thermodynamics (2002–2004); Secretary General of the 14th IUPAC Conference on Chemical Thermodynamics held at Osaka (1996); International Advisory Board of the IUPAC Conferences on Chemical Thermodynamics (2000, 2002, 2004)
Publication record: papers (217), reviews (45), books (4), chapters in books (15)
5 most important publications: [1] M. Sorai, S. Seki: J. Phys. Chem. Solids, 35 (1974) 555–570; [2] M. Sorai: J. Chem. Thermodynamics, 34 (2002) 1207–1253; [3] M. Sorai (Editor-in-Chief), Comprehensive Handbook of Calorimetry and Thermal Analysis, Wiley, Chichester (2004), pp. 534; [4] M. Sorai, Molecular Thermodynamics of Phase Transitions, Asakura Pub. Co., Ltd. (2007), pp. 264; [5] M. Sorai, Y. Nakazawa, M. Nakano, Y, Miyazaki: Chem. Rev., 113 (2013) PR41-PR122. Review.

Name: Antonio Gouveia De Souza
Country: Brazil
Date and place of birth: 1952, Areia-Paraíba. Brazil
Present position and address: Laboratório de Combustíveis e Materiais (LACOM). CCEN—Departamento de Química. Universidade Federal da Paraíba- Campus I, João Pessoa, PB, Brazil, CEP: 58.051-900. Fone/Fax: 55 083 3216 7441.
Email: agouveia@quimica.ufpb.br
Website: www.lacom.quimica.ufpb.br
Education and scientific degrees: Teacher Doctor of Chemistry
Workplaces: Laboratório de Combustíveis e Materiais/ CCEN/Universidade Federal da Paraíba.
Main fields of interest: thermal analysis, kinetics, ceramics, catalysis, biofuels, food and petrochemical
Relevant categories in thermal analyses: fields (kinetics, ceramics, catalysis, biofuels, food and petrochemical); methods (TG, DSC, PDSC, DTA)
Awards and acknowledgments: Thermoanalytical Group of the Hungarian Chemical Society awards Honorary Membership, 2004; Influence of Heating in the thermal, kinetic and reological properties of the rice brain oil, Instituto de Química de São Carlos. 2000.; Estudo térmico de pigmentos cerâmicos CoxZn(7-x)Sb2O12, North American Thermal Analysis Society. 2000
Professional activities: 2000–2002 Vice president of ABRATEC, 2002–2006 President of ABRATEC
Publication record: papers (305), books (1), patents (1), h-index (13)
Equipments: TG, DSC, DTA, PDSC, TMDSC, TG/MS
5 most important publications: [1] Souza, A G. et al.; Caffeic and ferulic acids: An investigation of the effect of antioxidants on the stability of soybean biodiesel during storage. Fuel (Guildford), v. 107, 641, 2013; [2] Souza, A G. et al. Flow properties of biodiesel: correlation between TMDSC and dynamic viscosity. Journal of Thermal Analysis and Calorimetry, v. 7, 10, 2013; [3] Souza, A G. et al. Effect of the composition on the thermal behaviour of the SrSn1–xTixO3. Journal of Thermal Analysis and Calorimetry, v. 1, 2, 2013; [4] Souza, A G. et al. Bioaditivo para biodiesel derivado do gardanol. 2012, Brasil. Patente: Privilégio de Inovação. Número do registro: PI 10 2012, data de depósito: 30/08/2012, título: "Bioaditivo para biodiesel derivado do gardanol", Instituição financiadora: Universidade Federal da Paraiba, FINEP; [5] Souza, A G. et al. Gallic Acid: Thermal and Antioxidant Properties. In: Nova Science Publishers. Handbook on Gallic Acid: Natural Occurrences, Antioxidant Properties and Health Implications. 5ed. New York: Michelle A. Thompson and Parker B. Collins, 2013, v. 5, 1.

Name: Mohammad Reza Sovizi
Country: I.R. Iran
Date and place of birth: 1971, Sabzevar, Iran
Present position and address: No. 6, No. 13, Aban Ave. Ashkestan poure shomali, Andarzgo Street, Tehran, Iran.
Email: mrsovizi@yahoo.com
Education and scientific degrees: SDT University, Isfahan, Iran, B. Sc. in Applied Chemistry, 1993.; University of Ferdosi, Mashhad, Iran, M.Sc. in Analytical Chemistry, 1996.; University of Ferdosi, Mashhad, Iran, Ph.D. in Analytical Chemistry, 2005.; Assistant Professor of Analytical Chemistry, Malek-e- Ashtar University of Technology, 2005.; Associate Professor of Analytical Chemistry, Malek-e- Ashtar University of Technology, 2011.
Workplaces: Department of Chemical and Chemical Engineerig, Malek-e- Ashtar University of Technology, Tehran, Iran.
Main fields of interest: thermal analysis and kinetic investigation of materials
Relevant categories in thermal analyses: fields (inorganic, materials, nano, minerals, complex, organic, pharmaceutical, polymer); methods (TG, DTA, DSC, thermomechanical analysis, kinetics)
Professional activities: the member of ICS
Publication record: papers (20), citations (76), h-index (5)
Equipments: TG and DTA and DSC instruments
5 most important publications: [1] M. R. Sovizi, S. S. Hajimirsadeghi, B. Naderizadeh,; Journal of Hazardous Materials; 168(2009), 1134–1139; [2] M. R, Sovizi, S. G. Hosseini; J Therm Anal Calorim, 2013, Vol. 111, No. 3, pp 2143–2148; [3] M. R. Sovizi; J Therm Anal Calorim; 2010, Vol. 102, No. 1, pp. 285–289; [4] M. R. Sovizi, K. Anbaz; J Therm Anal Calorim, 2010, Vol. 99, No. 2. pp 593–598; [5] Fateme Tajabadi and Yadollah Yamini, Mohammad Reza Sovizi; Microchim Acta; 2013, Vol. 180, 1-2, pp 65–73.

Name: Sergei V. Stankus
Country: Russia
Date and place of birth: 1952, Irkutsk, Russia
Present position and address: Deputy Director, Lavrentyev ave., 1, 630090, Novosibirsk, RUSSIA
Email : stankus@itp.nsc.ru
Website: http://www.itp.nsc.ru
ORCID: 0000-0002-4483-5865
Education and scientific degrees: Physicist, applied mathematician, Novosibirsk State University (1969–1974), Ph.D. (1984), D.Sc. (1992), Professor (2013)
Workplaces: Kutateladze Institute of Thermophysics, Siberian Brunch of the Russian Academy of Sciences
Main fields of interest: experimental investigations of phase trasitions and thermophysical properties of solids, liquids and gases
Relevant categories in thermal analyses: fields (inorganic, materials, organic, ceramics, metals and alloys); methods (DSC, kinetics, extremely high temperature; specific heat, calorimetry, instrument development, gamma-method, dilatometry, laser flash method)
Professional activities: Member of the National Committee of the Russian Academy of Sciences on Thermophysical Properties of Substances, Deputy Editor-in-Chief of the Thermophysics and Aeromechanics, Member of the Editorial Board of the High Temperatures-High Pressures
Publication record: papers (140), citations (300), h-index (10), sum of impact factors (93)
Equipments: Gamma-densitometers, Netzsch DIL 402 C, Netzsch LFA 427
5 most important publications: [1] Stankus S. V., Khairulin R. A. Density and phase diagram of the magnesium–lead system in the region of Mg2Pb intermetallic compound; Thermochimica Acta. 2008. Vol. 474, No. 1-2., pp. 52–56; [2] Khairulin R. A., Stankus S. V., Gruzdev V. A. Liquid-liquid coexistence curve of n-perfluorohexane—n-hexane system, International Journal of Thermophysics. 2007., Vol. 28, No. 4., pp. 1245–1254; [3] Khairulin R. A., Stankus S. V. Application of a γ attenuation technique for the study of phase equilibria in binary liquid systems with a miscibility gap, High Temperatures—High Pressures. 2000., Vol. 32, № 2., pp. 193–198; [4] Khairulin R. A., Stankus S. V. Phase equilibria in the lead—copper liquid system, Journal of Phase Equilibria., 1999., Vol. 20, No. 2., pp. 148–152; [5] Stankus S. V., Tyagel'sky P. V. Crystallization and thermal properties of Al_2O_3–Y_2O_3 melts, J. Crystal Growth., 1996., Vol.167., pp. 165–170.

Name: Mircea Stefanescu
Country: Romania
Date and place of birth: 1946, Icoana-Olt, Romania
Present position and address: Professor—Faculty of Industrial Chemistry and Environmental Engineering, Bulevardul Vasile Parvan no.6, 300223, Timisoara, Romania
Email: mircea.stefanescu@upt.ro
Website: www.chim.upt.ro
Researcher ID: B-9599-2011
Education and scientific degrees: Licensed in Physics and Chemistry, University of Timisoara (1968–1972), Ph.D. (1993), Professor (2007)
Workplaces: Romanian Academy—Chemistry Center, Inorganic Chemistry Department (1973–1979), Politehnica University Timisoara—Faculty of Industrial Chemistry and Environmental Engineering (1979–present)
Main fields of interest: study of carboxylate type complex combinations obtained in the redox reaction between transitional metal nitrates and diols, formation of organic-inorganic hybrid gels based on TEOS and diols and their thermal decomposition, original synthesis method for nanocomposites with controlled magnetic properties
Relevant categories in thermal analyses: fields (inorganic, materials, nano, complex, organic, polymer, composites); methods (TG, DTA, kinetics)
Professional activities: International reviewer, Project leader, Member of organizing committee CEEC-TAC, Romania, 2011, ICTAC membership, Member of the Chemical Society of Romania, Expert evaluator of national and international projects
Publication record: papers (90), books (3), book chapter (1), patent (1), citations (422), h-index (11)
Equipments: Setting up of a thermal analyses laboratory with: Derivatograph 1500-D and Derivatograph C MOM Budapest, Sartorius TG thermobalance, Equipment for DTA
5 most important publications: [1] M. Stefanescu et al., Ultrafine, Perfectly Spherical Ni-Zn Ferrite Nanoparticles, with Ultranarrow Distribution, Isolated in Silica Matrix, Prepared by a Novel Synthesis Method in Liquid Phase, Acta Materialia, 54, (2006), 1248; [2] M. Stefanescu et al., Thermal and FT-IR study of the hybrid ethylene-glycol–silica matrix, J. Sol-Gel Sci. Technol., 2007, 41 (1), 71–78; [3] M. Stefanescu et al., Obtaining of $Ni_{0.65}Zn_{0.35}Fe_2O_4$ nanoparticles at low temperature starting from metallic nitrates and polyols, J. Therm. Anal. Calorim. 2010, 99 (2), 459–464; [4] M. Stefanescu et al., Novel low temperature synthesis method for nanocrystalline zinc and magnesium chromites, Thermochim. Acta, 2011, 526 (1–2), 130–136; [5] M. Stefanescu et al., The Sol-Gel Process: Uniformity, Polymers and Applications, Ed. R. Morris, Nova Publishers, USA, 2011, 289–338.

Name: Oana Elena Stefanescu
Country: Romania
Date and place of birth: 1984, Timisoara, Romania
Present position and address: Ph.D. Eng. SC STADA Hemofarm SRL, Medicines Control Laboratory, Calea Torontalului km.6, Timisoara, Romania
Email: oanaelenastefanescu@yahoo.com
Education and scientific degrees: Diploma engineer Chemistry and Engineering of organic substances, Politehnica University of Timisoara (2002–2007), ERASMUS programme at University Carolo Wilhelmina zu Braunschweig, Germany (2007), Internship at Limbach Institute, Heidelberg, Germany (2008), Training at Charles University of Prague, Czech Republic (2010), Ph.D. Chemical Engineering (2010)
Workplaces: Politehnica University Timisoara—Faculty of Industrial Chemistry and Environmental Engineering (2007–2010), SC Bioclinica SA—Clinical laboratory: HPLC Department (2008–2012), SC STADA Hemofarm SRL, Medicines Control Laboratory (2012-present)
Main fields of interest: organic-inorganic hybrid gels of diol-TEOS type: synthesis and study on the chemical interaction, thermal decomposition of some metal-organic precursors for the obtaining of simple and mixed metal oxides nanoparticles (α-, γ-Fe_2O_3, ferrites, chromites) and nanocomposites (α-, γ-, ε-Fe_2O_3/SiO_2)
Relevant categories in thermal analyses: fields (inorganic, materials, nano, complex, organic, polymer, composites); methods (TG, DTA)
Professional activities: Project leader, Member of organizing committee and book of abstracts co-author at CEEC-TAC, Romania, 2011, ICTAC membership, Winner of one travel grant for young scientists at ICTAC15, Japan, 2012
Publication record: papers (30), book chapter (1), citations (142), h-index (8)
5 most important publications: [1] O. Stefanescu, M. Stefanescu, New Fe(III) malonate type complex combination for development of magnetic nanosized γ-Fe_2O_3, J. Organomet. Chem., 740, 2013, 50–55; [2] O. Stefanescu et al., Preparation of CuFe2O4/SiO2 nanocomposite starting from Cu(II)–Fe(III) carboxylates embedded in hybrid silica gels. J. Therm. Anal. Calorim., 113, 2013 1245–1253; [3] O. Stefanescu et al., Organic-inorganic hybrid gels of diol-TEOS type, Acta Chim. Slov., 59, 2012, 281–288; [4] O. Stefanescu et al., Synthesis and characterization of new hydroxycarboxylate compounds obtained in the redox reaction between $Fe(NO_3)_3$ and diol, Thermochim. Acta, 519, 2011, 22–27; [5] M. Stefanescu, Marcela Stoia, Oana Stefanescu, The Sol-Gel Process: Uniformity, Polymers and Applications, Ed. R. Morris, Nova Publishers, USA, 2011, 289–338.

Name: Leszek Stoch
Country: Poland
Date and place of birth: 1931, Bobowa, Poland
Present position and address: Full professor, Institute od Ceramics and Building Materials,. Postępu str.9, 02-676 Warsaw, Poland, Prof em. AGH Univ.of Science and Technology Kraków
Email: stoch_l@poczta.onet.pl
Education and scientific degrees: AGH University of Science and Technology, Krakow 1949–1955, Msc, Eng., Chemists-Ceramists, Ph.S.(1960), D.Sc.(1966), prof.(1976).
Workplaces: AGH Univ. Science and Technology, Geology Faculty (1951–1980) Faculty of Matertials Engineering and Ceramics (1980–2001), Institute od Ceramics and Building Materials, Warsaw
Main fields of interest: solid state chemistry, materials science, glass science and technology, mineralogy, geochemistry, biomaterials, archeometry
Relevant categories in thermal analyses: fields (sructural thermaochmochistry of solids: glasses, glass-ceramics, minerals, materials, biomaterials, nano-materials); methods (simultaneous TA (derivatography): TG, DSC, DTA. EGA, TD—thermoanalytical applications in geoscience, inorganic chemistry and materials studies)
Awards and acknowledgments: Honorary Member of Polish Cal. TA Soc.and W. Swietoslawski Medal, (1997), Praque Univ. J. Boricki Medal (1990), Prof.of Honour AGH Univ. of Science and Technology
Professional activities: First DTA equipment construction in Poland (1953) and begining of TA researche, Polish Soc.of Cal. Thermal Anal.organizer and v-prezident (1986–1994), Intern. Thermal Anal. Cal. Com.(ICTAC) Council member (1992–2000), Commission of ICTAC in Geoscience member (1993–2000). Internat. Commisison on Glass (ICG) Council member (1993–), European of Glass Sci and Technology Council (ESG) member (1995–)
Publication record: papers (360), books (2), chapters in books (10), patents (14), average citation (7.5), h-index (14)
Equipments: Q-1500, Derivatograph, PerkinElmer DTA-7, STA 449 Netzsch
5 most important publications: [1] L.Stoch, P. Stoch, Significance of crystallochemical factors in chemical reactions into the structure of solids. J. Therm. Anal. Calorim. 109 (2012) 763–766; [2] L.Stoch, P.Stoch, Crystal structure formation in glass from view of HRTEM J. Therm. Anal. Calorim. 88 (2007) 577–582; [3] L. Stoch, Nature and philosophy of thermal processes in minerals and inorganic materials, J. Therm. Anal. 48 (1997) 121–133; [4] L.Stoch, Thermochemistry of solids with flexible structures; [5] L.Stoch, On a model of thermal internal decomposition of solids. Thermochim. Acta, 203 (1992) 259–267.

Name: Muhamed Suceska
Country: Croatia
Date and place of birth: 1954, Han-Pijesak, Bosnia and Herzegovina
Present position and address: Principal Research Scientist, Brodarski institute, Av. V. Holjevca 20, 10000 Zagreb, Croatia and Nanyang Technological University, Energetics Research Institute, 50 Nanyang Avenue, Block N1-B4a-02, Singapore 639798
Email: suceska@hrbi.hr, msuceska@ntu.edu.sg
Researcher ID: 186903
Education and scientific degrees: Chemical Engineer, Technical Military Academy, Zagreb (1977), M.Sc., University of Zagreb (1986), Ph.D. Technical Military Academy, Zagreb (1991). Degrees: Research Associate (1992), Senior Research Associate (1998), Scientific Advisor (2003), Assistant Professor (2005), Associate Professor (2011)
Workplaces: Technical Military Academy, Zagreb (1982–1991), Pires, Zagreb (1991–1993), Brodarski Institute, Zagreb (1993–), Nanyang Technological University, Singapore (2009–)
Main fields of interest: detonation chemistry and physics, kinetics and ageing of energetic materials, numerical modeling combustion, detonation, and self-ignition phenomena
Relevant categories in thermal analysis: fields (energetic materials, polymer); methods (TG, DTA, DSC, thermomechanical analysis, dynamic-mechanical analyses, kinetics)
Awards and acknowledgments: Croatian Annual Award for science (1998)
Professional activities: Organizing Committee: "Central and Eastern European Conference on TA&C", Craiova, Romania (2011), Vilnius, Lithuania (2013); Scientific committee: "New trends in research of energetic materials", Pardubice, Czech Republic, (2000–), "Scientific Conference IPOEX", Ustron Jaszowiec, Poland (2004–), "The International Symposium on Explosion, Shock Wave and Hypervelocity Phenomena", Kumamoto, Japan (2004–); Professional organisations and bodies: Croatian society of chemical engineers (1998–), Croatian Chemical Society (2001–), Associate member of Croatian Academy of Engineering (2002–, International pyrotechnic society (1997–), Croatian representative in NATO expert group for explosives (NATO-PfP AC/326 SG-1) (2003–2009); Member of Editorial board and subject editor: «Central European Journal of Energetic Materials», Poland (2004–); Member of Editorial board: "Global Journal of Analytical Chemistry, Simplex Academic Publisher, India (2009–), "High-Energetic Materials", Poland (2001)
Publication record: papers (83), books (4), citations (252), h-index (7)
Equipment: DSC, TGA + DTA + MS, TMA, DMA
5 most important publications: [1] M. Sućeska, Prop., Explos., Pyrotech., 16 (1991) 197; [2] M. Sućeska, Prop., Explos., Pyrotech., 28 (1999) 280; [3] M. Sućeska, M. Rajić, S. Matečić-Mušanić, S. Zeman, Z. Jalovy, J. Therm. Anal. Cal., Vol. 74 (2003) 853; [4] M. Sućeska, S. Matečić Mušanić, I.Fiamengo Houra, Thermochim. Acta, 510 (2010) 9; [5] M. Sućeska, L. Zhi-Yue, S. Matečić Mušanić, I. Fiamengo, J. Therm. Anal. Cal., 100 (1) (2010) 337.

Name: Parukuttyamma Sujatha Devi
Country: India
Date and place of birth: 1963, Pathiyoor, India
Present position and address: Principal Scientist, Nano-Structured Materials Division, CSIR-Central Glass and Ceramic Research Institute, Kolkata 700 032, India.
Email: psujathadevi@cgcri.res.in, psujathadevi@gmail.com
Website: http://www.cgcri.res.in/page.php?id=136
Researcher ID: G-6970-2011
Education and scientific degrees: BSc (Chemistry, Kerala, 1983), M.Sc. (Chemistry, Kerala 1985), Ph.D. (Chemistry, Indian Institute of Science (IISc), Bangalore, 1991)
Workplaces: Indian Institute of Science, Bangalore, India (1985–1991), City University of New York, USA (1993–1996) State University of New York at Stony Brook (1999–2001) and CSIR –Central Glass and Ceramic Research Instiute, Kolkata, India (1997–present)
Main fields of interest: synthesis and characterization of functional materials for energy and biomedical applications through various thermal and spectroscopic techniques
Relevant categories in thermal analyses: fields (solution phase synthesis of nanomaterials) methods (TGA/DTA/DSC, dilatometric and shrinkage studies)
Awards and acknowledgments: P. K. Kunju Sahib award for Excellence in Chemistry (2010), Materials Research Society of India Medal (2008), Dr. Lakshmi Gold Medal (2005), Rheometric Scientific-ITAS (Indian Thermal Analysis Society) National Award (1998), Dr. R. L. Thakur Memorial National Award for Young Scientists (1997)
Professional activities: Secretary, Materials Research Society of India, Kolkata Chapter (2010–2014), Member, Executive Council, Indian Thermal Analysis Society (2005–2012) Member, National Academy of Sciences, Fellow, West Bengal Academy of Science and Technology
Publication record: papers (80), books (3), patents (3), citations (1494), h-index (21), sum of impact factors (220.5)
Equipments: Simultaneuos Thermogravimetric/Differential Thermal Analyser, Differential Scanning Calorimeter, Impedance Analyser, All Spectroscopic Equipments
5 most important publications: [1] P. Sujatha Devi and M. Subba Rao, (1989). Thermochim. Acta. 153, 181–191.; [2] S. Banerjee, A. Kumar and P. Sujatha Devi, (2011) J. Therm. Anal. Calorim., 104, 859–867; [3] S. Banerjee, P. Sujatha Devi, D. Topwal, S. Mandal, and S. R. Krishnakumar, (2007) Adv. Funct. Mater. 17, 2847–2854; [4] S. Banerjee, A. Bumajdad, P. Sujatha Devi (2011), Nanotechnology, 22, 275506–275513; [5] A. Kumar, P. Sujatha Devi, and H. S. Maiti, (2004) Chem. Mater. 16, 5562–5563.

Name: Petra Šulcová
Country: Czech Republic
Date and place of birth: 1970, Litomyšl, Czech Republic
Present position and address: professor, University of Pardubice, Faculty of Chemical Technology, Department of Inorganic Technology, Studentská 573, 532 10 Pardubice, Czech Republic
Email: petra.sulcova@upce.cz
Education and scientific degrees: Chemical Engineer, Institut of Chemical Technology in Pardubice (1988–1993); Ph.D. (1997) University of Pardubice, Faculty of Chemical Technology; Assoc. Prof. (2002); Prof. (2009) in the area of Chemistry and Technology of Inorganic Materials
Workplaces: University of Pardubice, Faculty of Chemical Technology
Main fields of interest: inorganic technology; synthesis of inorganic materials, especially inorganic pigments, testing of their thermal behaviour and stability; evaluation of colour properties and application for ceramic glazes; study of reactivity of powder materials
Relevant categories in thermal analyses: fields (inorganic, materials, ceramics); methods [TG, DTA (high temperature above 1,000 °C)]
Professional activities: Chairwoman of Czech Group for Thermal Analysis (since 2008), the member of executive board of Chech Chemical Sociaty; the Associate Editor of the J. Therm. Anal. Calorim. (since 2008); the chairwoman and organiser of the 4th Joint Czech-Hungarian-Polish-Slovak Thermoanalytical Conference in Pardubice (2013); since 2010 organization of special session of thermal analysis within Annual Conference of Czech and Slovak Chemical Societies (every 2 years)
Publication record: papers (70), h-index (11)
Equipments: STA (TG-DTA) Jupiter 449/C/6/F (Netzsch, Germany)
5 most important publications: [1] Šulcová P., Trojan M., Šolc Z.: Cerium Dioxide Fluorite Type Pigments, Dyes and Pigments, 37/1 (1998) 65–70; [2] Šulcová P., Trojan M.: Study of $Ce_{1-x}Pr_xO_2$ pigments, Thermochim. Acta 395 (2003) 251–255; [3] Šulcová P.: The synthesis of the $Ce_{0.95-y}Pr_{0.05}Nd_yO_{2-y/2}$ Pigments, Dyes and Pigments, 47/3 (2000) 285–289; [4] Šulcová P.: Thermal stability and colour properties of new pigments based on $BiREO_3$, J. Therm. Anal. Calorim., 109 (2012) 639–642; [5] Llusar M., Vitásková L., Šulcová P., Tena M.A., Badenes J.A., Monrós G.: Red ceramic pigments of terbium-doped ceria prepared through classical and non-conventional coprecipitation routes, J. Eur. Ceram. Soc., 30/1 (2010) 37–52.

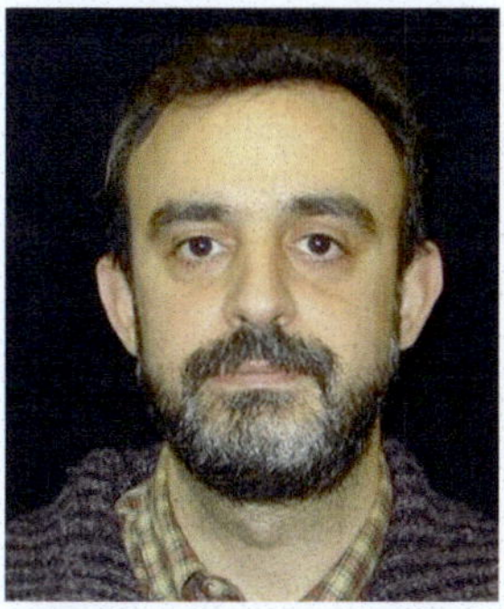

Name: Joan Josep Suñol
Country: Spain
Date and place of birth: 1965, Barcelona, Spain
Present position and postal adress: Professor of Applied Physics (University of Girona). Edifici PII, Campus Montilivi s/n, University of Girona, 17071 Girona, Spain.
Email: joanjosep.sunyol@udg.edu
Education and scientific degrees: B.Sc., the M.Sc. and finally (1996) Ph.D. (always in Physics) at the Autonomous University of Barcelona (UAB), Spain
Workplaces: Autonomous University of Barcelona (1991–1994), INRS-Énergie et Matériaux (1998), University of Girona (1993–)
Main fields of interest: thermal and structural analysis of materials, amorphous and nanocrystalline alloys, shape memory alloys, mechanical alloying
Professional activites: Academic secretary of the GECAT (Calorimetry and Thermal Analysis Society), group of the Royal Spanish Physics Society. Head of the Physics department at the university of Girona (2007–2010). Member of the Spanish Chapter of EPMA (European Powder Metallurgy Association). Coordinator of the Materials Science group of the Girona University, Spain (2007-actually). Associate editor of Journal of Thermal Analysis and Calorimetry
Publication record: papers (140), conferences (180)
5 most important publications: [1] Suñol JJ, Bonastre J; Crystallization kinetics of metallic glasses: Transformation diagrams; JTAC (2010) 102, pp. 447–450; [2] Carrillo F, Colom X, Suñol JJ, Saurina J; Structural FTIR analysis and thermal characterization of lyocell and viscose-type fibres; European Polymer Journal (2004) 40-9, pp. 2229–2234; [3] Sánchez JL, Sánchez T, Santos JD, Pérez MJ, Sánhez ML, Hernando B, Escoda L, Varga R; Martensitic phase transformation in rapidly solidified Mn50Ni40In10 alloy ribbons; Applied Physics Letters (2008) 92-1, 012513; [4] Coll, R. Escoda L, Saurina, J, Sánchez JL, Hernando B, Suñol JJ; Martensitic transformation in Mn-Ni_Sn Heusler alloys; JTAC (2010) 99, pp. 905–909; [5] Suñol JJ, González A, Saurina J; Thermal analysis of two Fe-X-B (X = Nb, ZrNi) alloys prepared by mechanical alloying; JTAC (2003) 72-1, pp. 329–335.

Name: Gopinatha Suresh Kumar
Country: India
Date and place of birth: 1957, Tiruvalla, India
Present position and address: Chief Scientist and Head, Biophysical Chemistrry Laboratory, Head, Business Development Group, CSIR-Indian Institute of Chemical Biology, Kolkata 700 032, India
Email: gskumar@iicb.res.in/gsk.iicb@gmail.com
Website: http://iicb.res.in/divisionwiselistofscientists/chemistry/gsuresh.html
Researcher ID: C-7211-2009
Education and scientific degrees: BSc (Chemistry, Kerala, 1978), M.Sc. (Chemistry, Indore, 1981), Ph.D. (Biophysics, Delhi, 1987)
Workplaces: University of Delhi, India (1981–1987), Hunter College of the City University of New York, USA (1993–1995 and 1999–2001) and CSIR Indian Institute of Chemical Biology, Kolkata, India (1988-present)
Main fields of interest: small molecule-nucleic acid and protein interactions through calorimetry and thermal analysis studies, and spectroscopy
Relevant categories in thermal analyses: fields (DSC, calorimetry, microcalorimetry, biocalorimetry); methods (microcalorimetry)
Professional activities: Secretary, DNA Society of India; Joint Secretary, Chemical Biology Society (India); Member, National Academy of Sciences; Fellow, West Bengal Academy of Science and Technology; Editorial Advisory Board Member of ISRN Spectroscopy J, Int. J. of Phys. Sci., and Open Nat. Product J.
Publication record: papers (155), books (2), citations (3209), h-index (33), sum of impact factors (432.51)
Equipments: Isothermal Titation Calorimeter, Differential Scanning Calorimeter
5 most important publications: [1] Basu, A and Suresh Kumar, G (2014) J. Chem. Thermodyn. 70, 176–181; [2] Hazra, S. Hossain, M and Suresh Kumar, G (2013) Mol. BioSyst. 9, 143–153; [3] A. Kabir, M. Hossain and Suresh Kumar, G (2013) J. Chem. Thermodyn. 57, 445–453; [4] Hossain, M, Khan A. Y, and Suresh Kumar, G (2012) J. Chem. Thermodyn. 47, 90–99; [5] Saha, I., Hossain, M. and Suresh Kumar, G (2010) J. Phys. Chem. B 114, 15278–15287.

Name: Piroska Szabó-Révész
Country: Hungary
Date and place of birth: 1951, Rétközberencs, Hungary
Present position: Head of the Department of Pharmaceutical Technology, University of Szeged
Email: revesz@pharm.u-szeged.hu
Website: www.pharm.u-szeged.hu/pharmtech
ORCID: 0000-0002-5336-6052
Education and scientific degrees: Pharmacist (Medical University of Szeged, 1975), University doctorate (Szeged, 1979), Specialist pharmacist (pharmaceutical technology) (Szeged, 1979), Ph.D. (Budapest, 1992), Habilitation (Szeged, 1996), Doctor of Sciences-D.Sc. (Budapest, 2006), Professor (2004), Head of the Department of Pharmaceutical Technology, University of Szeged (2005)
Workplaces: Department of Pharmaceutical Technology, University of Szeged (1975–)
Main fields of interest: particle engieneering, micronization, nanonization and amorfization of crystalline pharmaceutical agents, solid phase analysis, development of drug delivery systems for nasal and pulmonary administration
Relevant categories in thermal analyses: fields (pharmaceutical); methods (TG, TG-MS, DSC, thermo-microscope, XRPD, RAMAN, FT-IR, NIR)
Awards and acknowledgments: Rector's praise for educational work and for organizing education (2002), Certificates in recognition of project leading in Student Research (1978, 1988, 1999, 2001, 2002, 2003), "Apáczai Csere János" praise from the Educational and Cultural Minister (2010)
Professional activities: supervision of Ph.D. students (15) and diploma work (25), reviewer activity (about 10 journals), national and international project leading, OTKA, DAAD-MÖB projects (Martin-Luther-University, Halle-Wittenberg), HU-SLO Intergovernmental project, 130 lectures at national and international scientific events, more than 60 industrial research-development projects participation, 10 book chapters, 3 university hand-outs
Publication record: papers (184), books (2), patents (3), citations (668), h-index (15), sum of impact factors (188)
Equipments: thermomicroscope, DSC, TG-MS
5 most important publications: [1] Mártha Cs, Jójárt-Laczkovich O, Ulrich J, Szabó-Révész P: Investigation of the crystallinity of sugar alcohols co-ground with polymer excipients. J Therm Anal Calorim 115:(3) pp. 2479–2486. (2014); [2] Ambrus R, Aigner Z, Catenacci L, Bettinetti G, Szabó-Révész P, Sorrenti M: Physico-chemical characterization and dissolution properties of nifluminic acid-cyclodextrin-PVP ternary systems. J Therm Anal Calorim 104:(1) pp. 291–297. (2011); [3] Jójárt-Laczkovich O, Szabó-Révész P: Amorphization of a crystalline active pharmaceutical ingredient and thermoanalytical measurements on this glassy form. J Therm Anal Calorim 102:(1) pp. 243–247. (2010); [4] Gombás A, Szabó-Révész P, Regdon G, Erős I: Study of thermal behaviour of sugar alcohols. J Therm Anal Calorim 73:(2) pp. 615–621. (2003); [5] Gombás A, Szabó-Révész P, Kata M, Regdon G, Erős I: Quantitative determination of crystallinity of alpha-lactose monohydrate by DSC. J Therm Anal Calorim 68:(2) pp. 503–510. (2002).

Name: Irena Szczygieł
Country: Poland
Date and place of birth: 1951, Wągrodno, Poland
Present position and address: Head of Inorganic Chemistry Department, Faculty of Engineering and Economics, Wrocław University of Economics, Komandorska 118/120, 53-345 Wrocław
Email: irena.szczygiel@ue.wroc.pl
Website: http://www.kcn.ue.wroc.pl/
Education and scientific degrees: Chemical Engineer (MSc), Wrocław University of Technology, Faculty of Chemistry (1969–1974); Ph.D. (1978)
Workplaces: Wrocław University of Technology (1974–1978), Ph.D. student; Wrocław University of Economics, Faculty of Engineering and Economics, Department of Inorganic Chemistry (1979–2008), professor assistant and associate professor of Wrocław University of Economics (2008–)
Main fields of interest: synthesis of micro- and nanomaterials for optoelectronics application; binary and ternary phase diagrams of RE oxides; soft magnetic materials; electroless plating; research of practical applications of DTA/TGA/DSC methods
Relevant categories in thermal analyses: fields (inorganic, materials, nano); methods (TGA, DTA, DSC, high temperature (above 1,000 °C); specific heat)
Awards and acknowledgments: Medal of National Education Commission (2013); The first class awards of the Rector of the Wrocław University of Economics for the scientific achievements (2010, 2011, 2012, 2013); Silver Cross of Merit (2008); Individual award of the Rector of the Wrocław University of Economics for habilitation thesis (2007); Individual award of the Rector of Wrocław University of Technology for the doctor thesis (1979) and for scientific activities and cooperation with chemical industry (1976, 1977)
Professional activities: Deputy Member of The Polish Society of Calorimetry and Thermal Analysis Board (PTKAT) 2012–2015; Chairman of the 11th Conference on Calorimetry and Thermal Analysis in Zakopane, Poland (2012) and 10th International Seminar to the Memory of Stanislaw Bretsznajder on Thermal Analysis and Calorimetry in Płock, Poland (2011); Reviewer of the Journal of Thermal Analysis and Calorimetry, Journal of Applied Physics, Journal of Alloys and Compounds, Journal of the American Ceramic Society, Journal of Materials Science
Publication record: papers (62), books (1), patents (3), citations (173), h-index (7), sum of impact factors (83.45)
Equipments: Derivatograph type 3427 with dilatometric attachment (MOM); SETSYS^{TM} (TG-DSC 1500; SETARAM)
5 most important publications: [1] I. Szczygieł, K. Winiarska: "Synthesis and characterization of manganese zinc ferrite obtained by thermal decomposition from organic precursors". Journal of Thermal Analysis and Calorimetry 115 (2014) 471–477; [2] A. Stankiewicz, I. Szczygieł, B. Szczygieł: "Self-healing coatings in anti-corrosion applications". Journal of Materials Science 48/23 (2013) 8041–8051; [3] D. Piotrowska, T. Znamierowska, I. Szczygieł: "Phase equilibria in the $ErPO_4$–K_3PO_4 system". Journal of Thermal Analysis and Calorimetry 113 (2013) 121–126; [4] A. Matraszek, I. Szczygieł, Journal of the American Ceramic Society 95 (2012) 3651–3656; [5] K. Zielińska, A. Stankiewicz, I. Szczygieł, Journal of Colloid and Interface Science 377 (2012) 362–367.

Name: Imre Miklós Szilágyi
Country: Hungary
Date and place of birth: 1979, Szeged, Hungary
Present position and address: Research fellow/assistant professor; Budapest University of Technology and Economics (BME), Department of Inorganic and Analytical Chemistry, MTA-BME Technical Analytical Chemistry Research Group, H-1111 Budapest, Szt. Gellért tér 4., Hungary
Email: imre.szilagyi@mail.bme.hu
Website: www.iaachem.bme.hu/en
ResearcherID/ORCID: E-8325-2010/0000-0002-5938-8543

Education and scientific degrees: Ph.D. in chemistry, BME (2009, summa cum laude); M.Sc. in chemical engineering, BME (2004, honours)
Workplaces: Budapest University of Technology and Economics, Department of Inorganic and Analytical Chemistry, MTA-BME Technical Analytical Chemistry Research Group and MTA-BME Materials Structure Modeling Research Group (2007–); Department of Chemistry, University of Helsinki, Finland (2010–2012)
Main fields of interest: thermal analysis, inorganic chemistry, analytical chemistry, materials science, nanotechnology, phase transitions, atomic layer deposition (ALD), photocatalysis, catalysis, gas sensing, electrochromism, semiconductor oxides, carbon nanostructures, nanocomposites, bionanocomposites, complexes, nanofibers-particles-sheets
Relevant categories in thermal analyses: fields (inorganic, materials, nano); methods (DSC, TG/DTA-MS, TG-FTIR, SEM-EDX, TEM-ED, UV-Vis, Raman, FTIR, XPS, PL, SERS)
Awards and acknowledgments: PerkinElmer-ICTAC Young Scientist Award (2008); E-MRS Graduate Student Award (2009), Career Profile interview in the journal Science (2009, ScienceCareers); Scopus Young Researcher Award Main Prize (2008); Scopus Young Researcher Award in Chemistry (2008); Meisel Tibor Prize (2004); Pro Patria et Scientia I. Prize (2008); Deák Ferenc Scholarship (2008–2009); Innovative Ph.D. Award in Memory of Dr. Tibor Máthé (2009); Young Scientist Awards of the Visegrad Academies (2010) and of the Hungarian Academy of Sciences (2010); János Bolyai Fellowship (2011–2014); CEEC-TAC Grant for Young Researchers and Students (2011); Marie Curie Success Story based on a EU Marie Curie Intra-European Fellowship (2010–2012)
Professional activities: Deputy editor-in-chief of the Journal of Thermal Analysis and Calorimetry; Vice president of the Thermoanalytical Group of the Hungarian Chemical Society; Member of the scientific committees of the 14th ICTAC (2008) and 2nd CEEC-TAC conferences (2013); Member of the Hungarian Chemical Society, ICTAC, ESTAC, E-MRS
Publication record: papers (33 journal papers, 7 proceedings), books (1 book, 1 book chapter), patents (1), citations (>350), h-index (12), sum of impact factors (ca. 70)
Equipments: DSC, TG, TG/DTA, TG-FTIR, TG/DTA-MS (TA Instruments), high temperature XRD (PANanalytical), hot stage microscope, in situ FTIR, in situ Raman
5 most important publications: [1] I. M. Szilágyi et al., Nanotechnology 2013, 24, 245701; [2] I. M. Szilágyi et al., Journal of Catalysis 2012, 294, 119–127; [3] I. M. Szilágyi et al., Journal of Thermal Analysis and Calorimetry 2011, 105, 73–81; [4] I. M. Szilágyi et al. Chemistry of Materials 2008, 20, 4116–4125; [5] I. M. Szilágyi et al., Journal of Thermal Analysis and Calorimetry 2007, 88, 1, 139–144.

Name: Zhi-Cheng Tan
Country: P.R. China
Date and place of birth: 1941, Changsha, Hunan, P.R. China
Present position and address: Professor, Thermochemistry Laboratory, Dalian Institute of Chemical Physics, Chinese Academy of Science, Dalian 116023, China
Email: tzc@dicp.ac.cn
Website: http://www.dicp.ac.cn/
Researcher ID/ORCID: C-8382-2014
Education and scientific degrees: B.Sc. Wuhan Uiversity (1963); Ph.D. Dalian Institute of Chemical Physics, Chinese Academy of Sciences (1968)
Workplaces: Thermochemistry Laboratory, Dalian Institute of Chemical Physics, Chinese Academy of Science
Main fields of interest: heat capacity measurements and adiabatic caloremetry
Relevant categories in thermal analyses: fields (materials, nano, complex, organic, biology, heat storage materials, ionic liquid); methods (TG, DSC, kinetics, specific heat, calorimetry, instrument development)
Awards and acknowledgments: Academic Achievement Award from the Chin. Nat. Congr. Sci. (1978); Technical Invention Prize from the Ministry of Metallurgical Industry (1981); Sci. Tech. Advancement Award from the China State Commission of Sci. and Tech. (1985); Sci. Tech. Advancement Awards from the Chin. Acad. Sci. (1983, 1986, 1990); Special Subsidies from the China State Council (1995); K. C. Wang Education Awards from the K. C. Wang Education Foundation (Hongkong) (1991, 1996); Sci. Tech. Advancement Award from the Chinese Ministry of National Defence (2004); Sci. Tech. Advancement Award from the Liaoning Province, China (2009); Man of The Year 2011 Award from Scientific Chinese Journal (2012); Man of The Year 2012 Award from American Biographic Instiutute (2012); Top 100 Scientists Award—2012 from IBC Cambridge England (2012)
Professional activities: Member of committee: Chinese Society for Measurement (1975–), Chinese Society of Chemical Thermodynamics and Thermal Analysis (1980–2002), Society of Asian Thermophysical Property (1986–); Member of American Association for the Advancement of Science (1993–), Editorial board member of Thermochim. Acta (1993–), Member of New York Academy of Sciences (1996–); Member of ICTAC (1997–); JSPS fellowship (1987–1988); CNRS fellowship (1991–1993);Visiting prof., Thermochemistry Laboratory, Moscow Univ., Russia (1999); Guest prof., Research Center for Molecular Thermodynamics, Osaka Univ., Japan (2000); Guest prof. of Wuhan Univ. (2001–2003); Guest prof. Dalian Jiaotong Univ. (2006–2009); Guest prof. of Liaocheng Univ. (2006)
Publication record: papers (350), chapters in books (3), patents (15), citations (2818), h-index (24)
Equipments: DSC-141, TG-setsys 16/18, Setaram; TA MDSC-Q1000; TGA/SDTA 851e, Mettler-Toledo; adiabatic calorimeter; combustion calorimeter; TAM Air 8-channel Isothermal calorimeter; Isothermal Titration Calorimeter; Isoperibol solution-reaction calorimeter; PPMS Quantum Design (1.9–400 K); High-temp. calorimeter (400–1700 K)
5 most important publications: [1] Z-C. Tan, L-X. Zhou, S-X. Chen: Sci. Sin. B 26 (1983) 1014–1026; [2] Z-C. Tan, F-X. Li: Sci. China B 34 (1991) 560–569; [3] Z-C. Tan, J-B. Zhang:, Sci. China B 42 (1999) 382–390; [4] Z-C. Tan, Y-Y. Di: Prog. Chem. 18 (2006) 1234–1251; [5] Z.-C. Tan, Urs Welz-Biermann: Thermodynamic Properties of Ionic Liquids Chapter 1, pp. 3-36, Ionic Liquids, Theory, Properties, New Approaches, Ed. Alexander Kokorin, Publisher INTECH, 2011.

Name: Yassien Temerk
Country: Egypt
Date and place of birth: 1944, Kous, Egypt
Present position: Prof. of physical and inorganic chemistry, Faculty of science, Assiut university, Assiut, Egypt
Email: Temerk44@yahoo.com
Education and scientific degrees: B.Sc. (1965), M.Sc. (1968), Ph.D. (1971), D.Sc. (2007)
Workplaces: Demonstrator (1965–1971); lecturer of physical chemistry (1971–1975); Associate professor (1975–1979); Professor (1979-until now) Assiut university; Vice dean faculty of science Assiut university (1990–1995); Vice dean faculty of science Assiut university (1995–2001)
Main fields of interest: binding mode and thermodynamics studies on the interaction of the anticancer drugs with DNA, stripping analysis of biological compounds
Awards and acknowledgments: Egyptian award of distinction in chemical research 1981; Order of science and art first rank 1982 (Egypt); Ramachar award of distinction and certificate of merit 1984 Electrochemical soci.of india; Prize of excellence in chemistry academy of scientific research and technology 1988; American biographical record for distinguished activities in chemical research 2003; Prize of Assiut university in basic science 2010; Prize of Masr El-Khier 2011
Professional activities: Member of international seminar in chemistry university of Uppsala, Sweden 1972–1973; Alexander Von Humboldt fellowship Institute for chemie4 kernforshungsanlage Julich Germany 1975–1977; Fulbright fellowship chem. engineering, Case western reserve university Cleveland Ohio USA 1990; membership of bioelectrochemical society Germany; membership of electrochemical society of India; membership of African association of pure and applied chemistry; membership of Egyptian corrosion society; member of the permanent committee for the promotion of professors in chemistry Egypt; member of editorial board of bulletion of faculty of science Assiut university; Contributions to international conferences and scientific visits: 80; Consultant and principle researcher for several projects
Publication record: papers (120), citations (880)
5 most important publications: [1] International Journal of Hydrogen Energy, 35 (2010) 7827–7834; [2] Journal of Solid State Chemistry, 183 (2010) 984–987; [3] Electroanalysis, 23 (2011) 1638–1644; [4] Electroanalysis, 25 (2013) 1381–1387; [5] Analytical and Bioanalytical Chemistry, 405 (2013) 3839–3846.

Name: Maria Rosaria Tiné
Country: Italy
Date and place of birth: 1952, Catania, Italy.
Present position and address: Associate Professor of Biophysical Chemistry—Department of Chemistry and Industrial Chemistry, Via G. Moruzzi 3, 56124, Pisa, Italy
Email: mariarosaria.tine@unipi.it
Website: http://www.dcci.unipi.it/thermolab/
Education and scientific degrees, 1992– Associate Professor of Biophysical Chemistry, teaching Physical Chemistry and Biophysical Chemistry; 1981–1992 Assistant Professor UNIPI; 1977–1981 Fellow Researcher; 1977 Degree in Chemistry summa cum laude (UNIPI)
Main fields of interest: physicochemical properties of chemical and biochemical systems and thermodynamic, kinetic, and calorimetric investigation into: biomaterials, self-assembly systems; materials for biomedical applications; materials used in the cultural heritage; nanoparticles and halloysite nanotube based materials; synthetic carriers and activators of dioxygen and CO; metal complexes; micellar catalysis; thermodynamics of organic mixtures; cheminformatics; improvement of integrated calorimetric techniques
Relevant categories in thermal analyses: fields (materials, nano, polymer, complex, organic, life, food); methods (TG, DTA, EGA, DSC, ITC, kinetics, specific heat, calorimetry, microcalorimetry, instrument development, rheology)
Professional activities: Director of THERMOLAB, Laboratory of Chemical Thermodynamics and Calorimetry (UNIPI); Italian Representative in the ESTAC Committee; Past President and Member of the Executive Board of the Italian Association of Calorimetry and Thermal Analysis (AICAT); Member of ICTAC and ACS; Member of the Italian National University Council; Member of the Steering Committee of the Tuscan Technology District for Cultural Heritage; UNIPI Representative in the Sustainable Housing and Cultural Heritage Tuscan Cluster; Vice-President of the Tuscan Section of SCI; President of the Ph.D. School in Biomaterials (UNIPI); 2002–2007 President of AICAT and member of the Scientific Board of Mediterranean Association of Calorimetry and Thermal Analysis, MEDICTA. Chairman of the XXX AICAT Conference and of the School in Nanocalorimetry for Biological and Biomedical Applications (Pisa, 2008). 2000–2013 Member of the scientific/organizing committees of AICAT and MEDICTA conferences; Guest Editor of 2008 Special Issue of Journal of Thermal Analysis and Calorimetry
Publication record: papers (116), h-index (17)
Equipment: NanoDSC (N-DSC III, CSC); MicroITC (2277 TAM, Thermometrics); DSC (Pyris Diamond, PerkinElmer); TG (Q5000IR, TA Instruments); TGA/EGA/FT-IR; Rheometer (HAAKE RheoStress 6000, Thermo Fisher Scientific)
5 most important publications: [1] Duce C., Bramanti E., Ghezzi L., Bernazzani L., Bonaduce I., Colombini M. P., Spepi A., Biagi S., Tiné M. R., Dalton Transaction 2013, 42, 5975–5984; [2] Duce C., Ghezzi L., Onor M., Bonaduce I., Colombini M. P., Tiné M. R., Bramanti E. Analytical and Bioanalytical Chemistry, 2012, 402, pp. 2183–2193; [3] Tiné, M. R. Coordination Chemistry Reviews, 2012, 256 (1–2), 316–327; [4] Tiné, M. R., Alderighi, M., Duce, C., Ghezzi, L., Solaro, R., Journal of Thermal Analysis and Calorimetry, 2011 103 (1), 75–80; [5] Desii, A., Chiellini, F., Stefano, R. D., Solaro, R. Tiné, M. R., Journal of Polymer Science, Part A: Polymer Chemistry, 2010 48 (4), 986–990.

Name: Mauro Tomassetti
Country: Italy
Date and place of birth: 1943, Rome, Italy.
Present position and address: Graduate in Chemistry (1969) and in pharmacy (1977). Full Professor of Analytical Chemistry (since 2003).
Email: mauro.tomassetti@uniroma1.it
Education and scientific degrees: Full Professor
Workplaces: Department of Chemistry, "Sapienza" University of Rome
Main fields of interest: Analytical Chemistry, thermal Analysis, Biosensors.
Relevant categories in thermal analyses: fields (thermal analysis of drugs, food, cultural heritages (woods, paper, bones), polymers); methods (TG, DTG, DTA, DSC, TMA)
Professional activities: teaching, scientific research
Publication record: papers (235), chapters in books (9), patents (1), citations (2124), h-index (29)
Equipments: TG, DTA, DSC, TMA
5 most important publications: [1] M. Tomassetti, S. Vecchio, L. Campanella, R. Dragone: Food Chemistry, 140 (2013) 700–710; [2] M. Tomassetti, G. Favero, L. Campanella: J. Therm. Anal. Calorim., 112 (2013) 519–527; [3] M. Tomassetti, F. Marini, L. Campanella, A. Coppa: Microchem. Journal, 108 (2013) 7-13; [4] M. Tomassetti, E. Martini, L. Campanella, G. Favero, L. Carlucci, F. Mazzei: J. Pharm. Biomed. Anal., 73 (2013) 90–98; [5] M. Tomassetti, E. Martini, L. Campanella: Electroanalysis, 24 (2012) 842–856.

Name: Vicenç Torra Ferre
Country: Catalonia, Spain
Date and place of birth: 1942, Barcelona, Catalonia, Spain
Present position and address: Prof. of Applied Physics, Polytechnic University of Catalonia, retired Villarroel 162, E-08036 Barcelona, Catalonia
Email: vtorra_1@yahoo.com
Website: http://scholar.google.es/citations?user=IbTzky0AAAAJ
Education and scientific degrees: Ms in Sciences (Physics) 1964, Ph.D. in Sciences (Physics) 1970
Workplaces: University of Barcelona: 1965–1981, University of Balearic Islands in Palma de Mallorca: 1981–1991, Polytechnical University of Catalonia in Barcelona 1991–2012
Main fields of interest: instrumentation and reliability, differential scanning calorimetry, shape memory alloys, thermomechanical analysis
Relevant categories in thermal analyses: fields (smart materials (shape memory alloys) and theoretical study of the behavior of conduction calorimeters); methods (non conventional differential calorimetry)
Awards and acknowledgments: Physics Medal (Spanish Physics Society, 1980)
Professional activities: Honorary Board Member of the JTAC from October 2011; Honoray member, SpanishPhysics Society, 2011; Regional Editor of JTAC 2003–2011, Editorial board of Smart Structures and Systems 2004–2012
Publication record: papers (193), books (4), patents (4), citations (1954), h-index (22)
5 most important publications: [1] Marco, F; Navarro, J; Torra, V; Application of flow calorimetry to study of alloy formation.1. Enthalpies of solution of In, Tl, Cd, Zn, Pb, Ga, Sn, and Bi In Hg At 293.15-K; Journal of Chemical Thermodynamics Volume: 7 Issue: 11 Pages: 1059–1066 (1975); [2] Van Humbeeck, J; Van Hulle, D; Delaey, L; Torra, V; et al.; A 2-Stage martensite-transformation in a Cu-13.99 mass-percent Al-3.5 mass-percent Ni-alloy; Transactions of the Japan Institute of Metals Volume: 28 Issue: 5 Pages: 383–391 (1987); [3] Lovey, Fc; Amengual, A; Torra, V; et al.; On the origin of the intrinsic thermoelasticity associated with a single-interface transformation in Cu–Zn–Al shape-memory alloys; Philosophical Magazine A-Physics of condensed matter structure defects and mechanical properties Volume: 61 Issue: 1 Pages: 159–165 (1990); [4] Lovey FC; Torra V; "Shape memory in Cu-based alloys: phenomenological behavior at the mesoscale level and interaction of martensitic transformation with structural defects in Cu–Zn–Al", Progress in Materials Science, 44 (3) (1999) 189–289; [5] Kirchner, R; de Rivera, MR; Seidel, J; Torra, V; Identification of micro-scale calorimetric devices Part VI. An approach by RC-representative model to improvements in TAM microcalorimeters; Journal of Thermal Analysis and Calorimetry Volume: 82 Issue: 1 Pages: 179–184 (2005).

Name: Kaia Tõnsuaadu
Country: Estonia
Date and place of birth: 1951, Tallinn, Estonia
Present position and address: senior researcher, Ehitajate tee 5, Tallinn, 19086, Estonia
Email: kaia.tonsuaadu@ttu.ee
Website: http://www.ttu.ee/
Education and scientific degrees: 1975 Graduated from Tallinn Technical University as engineer of chemical technology; Doctor's Degree, 1984, Moscow Institute of Chemical Technology; Doctor's Degree, 1995, Tallinn University of Technology
Workplaces: Tallinn University of Technology, Laboratory of Inorganic Materials
Main fields of interest: chemistry and chemical technology (chemistry of inorganic phosphates, thermal reactions, sorption processes on apatites)
Relevant categories in thermal analyses: fields (inorganic, nano, and complex materials, minerals); methods (TG, DTA, EGA-FTIR and MS)
Professional activities: member of the organizing committee of The 1st and 2nd Central and Eastern European Conference on Thermal Analysis and Calorimetry
Publication record: papers (57), patents (1), h-index (9)
Equipments: SetSys-Evo 1600 TG-DTA (Setaram, France) connected to an OmniStar QMS 220 (Pfeiffer, Germany), SetSys TG-DTA-DSC (Setaram) connected to Thermo Nicolet FTIR, LabSys TG-DTA-DSC Setaram.
5 most important publications: [1] Tõnsuaadu K, Gross K. A, Plūduma L, Veiderma M. A review on the thermal stability of calcium apatites. J Therm Anal Calorim, 2012;110:647–659; [2] Tõnsuaadu K, Zalga A, Beganskiene A, Kareiva A. Thermoanalytical study of the YSZ precursors prepared by aqueous sol–gel synthesis route J Therm Anal Calorim 2012;110:77–83; [3] Nemliher J, Tõnsuaadu, Kallaste T. Low-temperature Changes in Crystal Lattis of bioaragonite of Tapes decussatus Linnaeus (Mollusca: Bivalvia), J Therm Anal Calorim 2009;97:27–32; [4] Tõnsuaadu K, Viipsi K, Trikkel A. EDTA impact on Cd2 + migration in apatite—water system. J Hazard Mater, 2008;154:491–497; [5] Peld M, Tõnsuaadu K, Bender V. Sorption and Desorption of Cd^{2+} and Zn^{2+} Ions in Apatite-Aqueous Systems. Environ Sci Technol 2004;38:5626–5631.

Name: Dimitrinka Tsocheva
Country: Bulgaria
Date and place of birth: 1953, Sofia, Bulgaria
Present position and address: Institute of Polymers, Bulgarian Academy of Sciences, Acad. Georgi Bonchev Str., Bl. 103, 1113 Sofia, Bulgaria, Researcher (1986–)
Email: tsocheva@polymer.bas.bg
Education and scientific degrees: Chemical Engineer, University of Chemical Technology and Metallurgy (1971–1976)
Workplaces: Institute of Polymers, Bulgarian Academy of Sciences (1976–)
Main fields of interest: polymers, thermal methods (DSC, TGA)
Relevant categories in thermal analyses: fields (materials, nano, organic, polymer, rubber); methods (TG, DTA, DSC, thermomechanical analysis, kinetics; specific heat, calorimetry, microcalorimetry, instrument development)
Professional activities: Member of Bulgarian Chemical and Polymer Society, Member of the Scintific Commitee of the 1st and 2nd Central and Eastern European Conferences on Thermal Analysis and Calorimetry
Equipments: PerkinElmer DSC 8500, PerkinElmer TGA 4000
5 most important publications: [1] Tsocheva, D., Zlatkov, T., Terlemezyan, L., Journal of Thermal Analysis, 53 (1998) 895; [2] Tsocheva, D., Terlemezyan, L., Journal of Thermal Analysis and Calorimetry, 75 (2004) 739; [3] Tsocheva, D., Terlemezyan, L., Journal of Thermal Analysis and Calorimetry, 81 (2005) 3; [4] Mokreva, P., Tsocheva, D., Ivanova, G., Terlemezyan, L. Journal of Applied Polymer Science, 98 (2005) 1822; [5] Mokreva, P., Tsocheva, D., Ivanova, G., Terlemezyan, L., Journal of Applied Polymer Science, 99 (2006) 75.

Name: Valery L. Ugolkov
Country: Russia
Date and place of birth: 1952, Leningrad, Russia
Present position and address: Grebenshikov Institute of Silicate Chemistry Russian Academy of Sciences, Adm. Makarova emb. 2, 199034 St. Petersburg, Russia
Email: ugolkov@isc.nw.ru
Researcher ID: 6701471637
Education and scientific degrees: Engineer-Chemist-Technologist, St. Petersburg State Technological Institute (1976); Ph.D. (1985)
Workplaces: Institute"Giprocement" (1976–1987); Institute of Problems of Mechanical Engineering Russian Academy of Sciences (1987–2001); Grebenshikov Institute of Silicate Chemistry Russian Academy of Sciences (2001–)
Main fields of interest: studies of processes proceeding in heating or cooling of materials
Relevant categories in thermal analyses: fields (inorganic and organic materials, minerals, metals, fullerens, nanostructures, polymer, glass, ceramics, cement, materials with memory form); methods (TG, DTA, DSC, QMS, thermomechanical analysis, kinetics, cryo, extremely high temperature (above 1,000 °C); specific heat, calorimetry, microcalorimetry)
Publication record: papers (89), books (1), patents (5), citation index (512), h-index (12)
Equipments: STA 429 CD + QMS 403 C; DSC 404 C; DIL 402 C; STA 449 C NETZSCH
5 most important publications: [1] Yu. S. Tver'yanovich, V. L. Ugolkov. Smeared First-Order Phase Transition in Melts. In the book: V. V. Brazhkin et al. (eds), New Kinds of Phase Transitions: Transformations in Disordered Substances. Kluwer Academic Publishers, 2002. Printed in the Netherlands. pp. 209–222; [2] B. V. L'vov, L. K. Polzik, V. L. Ugolkov. Decomposition kinetics of calcite: a new approach to the old problem. Thermochim Acta 390 (2002) 5; [3] B. V. L'vov, V. L. Ugolkov. The self-heating effect in the process of $KMnO_4$ decomposition in vacuum. J Therm Anal Calorim 94 (2008) 453; [4] B. V. L'vov, V. L. Ugolkov. Decomposition of $KMnO_4$ in different gases as a potential kinetics standard in thermal analysis. J Therm Anal Calorim 100 (2010) 145; [5] V. L. Ugolkov. Dilatometry.; Differential Scanning Calorimetry.; Differential Thermal Analysis. In the book: Russian geological encyclopedia. VSEGEI publishing house. vol. 1 (2010) pp. 506; 513; 513–514 (Russia).

Name: Guy Van Assche
Country: Belgium
Date and place of birth: 1969, Willebroek, Belgium
Present position and address: Professor, Vrije Universiteit Brussel, Pleinlaan 2, B-1050 Brussel, Belgium
Email: gvassche@vub.ac.be
Website: http://www.vub.ac.be/MACH/FYSC/
Researcher ID: B-5061-2011
Education and scientific degrees: Chemical Engineer, Vrije Universiteit Brussel (1987–1992); Ph.D. (1998); Professor (2011)
Workplaces: Research group Physical Chemistry and Polymer Science, Vrije Universiteit Brussel (1993–), Kyoto Institute of Technology (1998–1999)
Main fields of interest: development of advanced thermal analysis methods and their application to polymer-based materials
Relevant categories in thermal analyses: fields (materials, nano, polymer); methods (TG, DSC, kinetics, microcalorimetry, instrument development, modulated temperature DSC, rapid-scanning calorimetry, chip calorimetry)
Professional activities: member of the scientific committee of ESTAC11 (Espoo, Finland, 2014); member of the international advisory board of the International Symposium Technologies for Polymer Electronics (Ilmenau, Germany, 2012, 2014); co-organizer session Advanced Instrumentation at ICTAC15 (Higashi-Osaka, Japan, 2012); chairman session New Techniques at ICCT-2010 (Tsukuba, Japan, 2010)
Publication record: papers (88), h-index (18)
Equipments: TG, (MT)DSC, Rapid Heat-Cool DSC, Ultrafast scanning chip calorimetry, AC chip calorimetry, microcalorimetry, isothermal titration calorimetry, DMA, dynamic rheometry, Rheo-DSC, micro- and nano-thermal analysis
5 most important publications: [1] G. Van Assche, A. Van Hemelrijck, H. Rahier, B. Van Mele: Thermochim. Acta, 268 (1995) 121–142; [2] K. Van Durme, G. Van Assche, B. Van Mele: Macromolecules, 37(25) (2004) 9596–9605; [3] J. Zhao, A. Swinnen, G. Van Assche, J. Manca, D. Vanderzande, B. Van Mele: J. Phys. Chem. B, 113(6) (2009) 1587–1591; [4] N. A. Gotzen, G. Van Assche, B. Van Mele: Polymer: 52(19) (2011) 4277–4283; [5] C. Block, A. K. Ghosh, B. Van Mele, G. Van Assche: Thermochim. Acta: 547 (2012) 130–140.

Name: Guy Van den Mooter
Country: Belgium
Date and place of birth: 1964, Willebroek, Belgium
Present position and address: full professor at university of Leuven (KU Leuven), Department of Pharmaceutical and Pharmacological Sciences; Laboratory of Drug Delivery and Disposition
Email: guy.vandenmooter@pharm.kuleuven.be
Website: http://pharm.kuleuven.be/pharbio/index2.htm
Education and scientific degrees: Pharmacist (1987; KU Leuven); Ph.D. in Pharmaceutics (1994; KU Leuven)
Workplaces: Janssen Pharmaceutica Belgium (1995–1996 and 2005–2006); University of Leuven (1996-present)
Main fields of interest: drug delivery systems; exploration of the link between drug formulation, manufacturing process, performance and physical structure.
Relevant categories in thermal analyses: fields (pharmaceutical, polymer, life sciences); methods ((modulated) differential scanning calorimetry; solution calorimetry; hot-stage microscopy)
Awards and acknowledgments: Current status on targeted drug delivery to the gastrointestinal tract ("the Royal College of Physicians", 1993); Association Française pour les Recherches sur les Formes Pharmaceutiques à Libération Modifiées (1996, Paris, France). Arbeitsgemeinschaft fur Pharmazeutische Verfahrenstechnik (1996).
Professional activities: member of the Belgian Society of Pharmaceutical Sciences; the European Federation of Pharmaceutical Sciences (EUFEPS), the Controlled Release Society; the Belgian Pharmacopeia Commission, the American Association of Pharmaceutical Scientists (AAPS); the board of the Thermal Analysis Work Group of the Netherlands (TAWN); member of the Editorial board of International Journal of Pharmaceutics, Journal of Pharmaceutical Sciences, Journal of Pharmacy and Pharmacology
Publication record: papers (more than 200), books (1), patents (7), citations (4800), h-index (37)
Equipments: MDSC Q2000 (TA-Instruments); Diamond DSC (PerkinElmer); Thermometric 2225 Precision Solution Calorimeter/2227 Thermal Activity Monitor (Thermometric AB)
5 most important publications: [1] Van den Mooter, G., Samyn, C., Kinget, R., "The relation between swelling properties and enzymatic degradation of azo polymers designed for colon-specific drug delivery", Pharmaceutical Research, vol. 11, no. 12, 1994, pp. 1737–1741; [2] Van den Mooter, G., Craig, D., Royall, P., "Characterization of amorphous ketoconazole using modulated temperature differential scanning calorimetry", Journal of pharmaceutical sciences, vol. 90, no. 8, 2001, pp. 996–1003; [3] Six, K., Verreck, G., Peeters, J., Brewster, M., Van den Mooter, G., "Increased physical stability and improved dissolution properties of itraconazole, a class II drug, by solid dispersions that combine fast- and slow-dissolving polymers", Journal of pharmaceutical sciences, vol. 93, no. 1, 2004, pp. 124–131; [4] Worku, Z., Paudel, A., Van den Mooter, G., "Can Compression Induce Demixing in Amorphous Solid Dispersions? A case study of naproxen-PVP K25", European Journal of Pharmaceutics and Biopharmaceutics, vol. 81, 2012; [5] Paudel, A., Worku, Z., Meeus, J., Guns, S., Van den Mooter, G., "Manufacturing of solid dispersions of poorly water soluble drugs by spray drying: Formulation and process considerations", International Journal of Pharmaceutics, vol. 453, 2013, pp. 253–284.

Name: Geert Van Den Poel
Country: Belgium
Date and place of birth: 1975, Aarschot, Belgium
Present position and address: Senior Scientist at DSM; Urmonderbaan 22, 6167 RD Geleen, Netherlands
Email: Geert.Poel-vanden@dsm.com
Website: www.dsm-resolve.com
Education and scientific degrees: 1994–1998: Catholic University of Leuven, Diploma: Master Chemistry, option Polymer Chemistry (July 1998); 1998–2003: Ph.D. Research at the Catholic University of Leuven: "Laboratory for Macromolecular Structural Chemistry, Division of Molecular and Nanomaterials"; subject: "Crystallisable thermoplastic/thermosetting polymer blends", Diploma:Ph.D. Sciences, group Chemistry (November 2003); 2003–2005: Postdoctoral Research at DSM Research BV, Performance Materials—R&D, Geleen, the Netherlands and the Catholic University of Leuven—Laboratory for Macromolecular Structural Chemistry, Division of Molecular and Nanomaterials within the framework of a EC-Marie Curie Industry Host Fellowship; 2006–2011: Scientist at DSM Resolve Morphology Group; Thermal Analysis Department—R&D, Geleen, the Netherlands; 2012—Present: Senior Scientist at DSM Resolve Morphology Group; Thermal Analysis Department—R&D, Geleen, the Netherlands
Main fields of interest: polymer chemistry, physics and morphology; thermal analysis and calorimetry
Relevant categories in thermal analyses: fields (materials, nano, pharmaceutical, polymer, food, biology); methods (TG, DSC, kinetics, specific heat, calorimetry, microcalorimetry, instrument development, fast scanning calorimetry)
Professional activities: Member of the TAWN board; Member of ICTAC society
Equipments: 3 heat flux DSCs 823 of Mettler Toledo; 2 TGA-DSCs of Mettler Toledo [with Humidity Chamber]; 1 TGA of Netzsch; 1 Hyper DSC 8500 of PerkinElmer; 1 Hyper DSC Pyris of PerkinElmer; 1 Diamond DSC of PerkinElmer; 1 Flash DSC of Mettler Toledo; 1 TGA-GCMS of PerkinElmer
5 most important publications: [1] Vincent B. F. Mathot, Geert Vanden Poel, Thijs F. J. Pijpers, Benefits and potentials of High Performance Differential Scanning Calorimetry (HPer DSC). In: The handbook of thermal analysis and calorimetry. Volume 5: Further advances, techniques and applications, Brown, Ed., Elsevier (2008) Chapter 8, pp 269–298; [2] DIN Specification 91127: Recommendation for Temperature Calibration of Fast Scanning Calorimeters (FsCs) for Sample Mass and Scan Rate, G. Vanden Poel, A. Sargsyan, V. Mathot, Guy Van Assche, A. Wurm, C. Schick, A. Krumme, D. Zhou, Berlin, Publisher: Beuth Verlag GmbH; [3] Geert Vanden Poel, Vincent B. F. Mathot, High-speed/high performance differential scanning calorimetry (HPer DSC): Temperature calibration in the heating and cooling mode and minimization of thermal lag, Thermochimica Acta, 446 (2006) 41; [4] Vincent Mathot, Marek Pyda, Thijs Pijpers, Geert Vanden Poel, Ernst van de Kerkhof, Sander van Herwaarden, Floor van Herwaarden, Archi Leenaers, The Flash DSC 1, a power compensation twin-type, chip-based fast scanning calorimeter (FSC): First findings on polymers, Special Issue Interplay between Nucleation, Crystallization, and the Glass Transition, Thermochimica Acta, 522(1-2) (2011) 36–45; [5] Microfocus wide-angle X-ray scattering of polymers crystallized in a fast scanning chip calorimeter; Martin van Drongelen, Tamara Meijer-Vissers, Dario Cavallo, Giuseppe Portale, Geert Vanden Poel, René Androsch, Thermochimica Acta 563 (2013) 33–37.

Name: Paul Johan Van Ekeren
Country: The Netherlands
Date and place of birth: 1960, Amersfoort, The Netherlands.
Present position: Independent researcher and lecturer in the field of Thermal Analysis, Calorimetry and Applied Thermodynamics.
Email: P.J.van.Ekeren@gmail.com
Education and scientific degrees: (Physical) Chemistry, Utrecht University, The Netherlands (1978–1984); Ph.D. (1989).
Workplaces: Independent researcher and lecturer (2011–); TNO Defence, Security and Safety (2007–2011); Utrecht University, Chemical Thermodynamics Group (1984–2007).
Main fields of interest: polymorphism, phase diagrams, thermal stability, calibration, instrumentation, (applied) thermodynamics, education
Relevant categories in thermal analyses: fields (inorganic, materials, minerals, complex, organic, pharmaceutical, polymer, food, glass, other (energetic materials)); methods (TG, DTA, EGA, DSC, thermomechanical analysis, kinetics, specific heat, calorimetry, micro-calorimetry, other (HFC, modulated-temperature DSC))
Professional activities: Dutch Society for Thermal Analysis and Calorimetry (TAWN): secretary (1990–1997), president (1997–); PAO-Techniek: lecturer in the course 'Post Academic Thermal Analysis' (1996–); Course Director and lecturer in the 'Thermal Analysis Course with Emphasis on Practical Operations' (1999–); Journal of Thermal Analysis and Calorimetry: member of the Editorial Advisory Board (2000–2003); Regional Editor (2003–); ICTAC: Affiliated Councillor for The Netherlands and Belgium (2004–); Chairman of the Nomenclature Committee (2008–)
Publication record: papers (50)
5 most important publications: [1] P. J. van Ekeren and E. P. Carton, Polyurethanes for potential use in transparent armour investigated using DSC and DMA, J. Therm. Anal. Calorim. 105 (2011) 591–598; [2] E. R. T. Bevers, P. J. van Ekeren, W. G. Haije and H. A. J. Oonk, Thermodynamic properties of lithium chloride ammonia complexes (for application in a high-lift high-temperature chemical heat pump), J. Therm. Anal. Calorim. 86 (2006) 825–832; [3] P. J. van Ekeren, L. D. Ionescu, V. B. F. Mathot and J. C. van Miltenburg, Specific heat capacities and thermal properties of a homogeneous ethylene-1-butene copolymer by adiabatic calorimetry, Thermochim. Acta 391 (2002) 185–196; [4] G. Cornelissen, K. A. Hassell, P. C. M. van Noort, R. Kraaij, P. J. van Ekeren, C. Dijkema, P. A. de Jager and H. A. J. Govers, Slow desorption of PCBs and chlorobenzenes from soils and sediments: relations with sorbent and sorbate characteristics, Environ. Pollut. 108 (2000) 69–80; [5] P. J. van Ekeren, C. M. Hol and A. J. Witteveen, A comparative test of differential scanning calorimeters, J. Therm. Anal. 49 (1997) 1105.

Name: Bruno Van Mele
Country: Belgium
Date and place of birth: 1952, Antwerp, Belgium
Present position and address: Full Professor, Vrije Universiteit Brussel, Brussels Faculty of Engineering (BRUFACE), Pleinlaan 2, B-1050 Brussels, Belgium
Email: bvmele@vub.ac.be
Website: http://www.vub.ac.be/MACH/FYSC/
Education and scientific degrees: Chemical engineer (1975); Ph.D. in Applied Sciences (1982)
Main fields of interest: development of dedicated thermal analysis methodologies to study structure-processing-property relations in advanced (hybrid, nano-structured) polymer systems, with emphasis on fast, sensitive and spatially-resolved analysis; cure mechanism and kinetics (modelling) of thermosets; geopolymers; phase separation in polymer systems (solutions, blends); nanocomposites; thin polymer films and coatings; self-healing polymer systems; polymer systems for energy applications (organic photovoltaics)
Relevant categories in thermal analyses: fields (materials, nano, polymers, geopolymers); methods (TG, DSC, thermomechanical analysis, kinetics; specific heat, microcalorimetry, instrument development (RheoDSC), chip calorimetry, RHC (Rapid Heat Cool DSC))
Professional activities: Head of department Materials and Chemistry (MACH); head of research group Physical Chemistry and Polymer Science (FYSC); co-lecturer in post-academic course on Thermal Analysis in the Netherlands (PAO Techniek)
Publication record: papers (150), citations (2500), h-index (27)
Equipments: (Modulated Temperature) DSC, TGA (coupled with mass spectrometry), TMA, DMA, Micro and Nano-calorimetry, Rheometry, RheoDSC, Micro and Nano-Thermal Analysis, Chip Calorimetry (AC mode and ultrafast scanning), RHC
5 most important publications: [1] "Modulated differential scanning calorimetry: isothermal cure and vitrification of thermosetting systems": Van Assche G., Van Hemelrijck A., Rahier H., Van Mele B., Thermochim. Acta 1995, 268, 121–142; [2] "Low-temperature synthesized aluminosilicate glasses .1. Low-temperature reaction stoichiometry and structure of a model compound": Rahier, H., Van Mele B., Biesemans M., Wastiels J., Wu X., J. of Materials Science 1996, 31, 71–79; [3] "Kinetics of demixing and remixing in poly(N-isopropyl acrylamide)/water studied by modulated temperature DSC": Van Durme K., Van Assche G., Van Mele B., Macromolecules 2004, 37, 9596–9605; [4] "Isotactic polypropylene/carbon nanotube composites prepared by latex technology. Thermal analysis of carbon nanotube-induced nucleation": Miltner H. E., Grossiord N., Lu K., Loos J., Koning C. E., Van Mele B., Macromolecules 2008, 41, 5753–5762; [5] "Phase diagram of P3HT/PCBM blends and its implication for the stability of morphology": Zhao J., Swinnen A., Van Assche G., Manca J., Vanderzande D., Van Mele B., J. Phys. Chem. B 2009, 113, 1587–1591.

Name: Gábor Várhegyi
Country: Hungary
Date and place of birth: 1947, Budapest, Hungary
Present position and address: Independent scientist, Eper u. 2/A, Budapest, Hungary 1112; Advisor to the Institute of Materials and Environmental Chemistry, Research Centre for Natural Sciences; Eper u 2A, Budapest, 1112
Email: varhegyi.gabor@ttk.mta.hu
Website: http://varhegyi-family.net/Gabor/papers
Researcher ID: C-2147-2013
Education and scientific degrees: 1994: Doct. Habil., Budapest University of Technology and Economics; 1992: Doctor of Chemical Science (D.Sc.), Hungarian Academy of Sciences; 1980: Candidate of Chemical Science (C.Sc.), Hungarian Academy of Sciences; 1973: Doctoral degree, Loránd Eötvös University of Sciences; 1970: M.Sc. in Chemistry, Loránd Eötvös University of Sciences, Budapest
Workplaces: Research Centre for Natural Sciences, Hungarian Academy of Sciences: Head of the Group of Thermal Reaction Studies (2008–2013); Head of Department of Environmental Chemistry (1999–2007); Head of Department of Thermal Analysis (1998); Head of Department of Macromolecular Chemistry (1996–1997); Scientific adviser (1992–),
Main fields of interest: thermal decomposition and gasification/combustion properties of biomass materials and other solid fuels; reaction kinetic evaluation of experiments at arbitrary temperature—time functions assuming complex mechanism schemes; developing software for reaction kinetic evaluation, computer control, data acquisition and data processing of measurement system
Relevant categories in thermal analyses: fields (biomass materials, solid fuels, tobacco); methods (TGA, TGA-MS, DSC)
Professional activities: President of the Working Committee for Thermal Analysis of the Hungarian Academy of Sciences
Publication record: papers (100), citations (>3000), h-index (28), sum of impact factors (120)
5 most important publications: [1] G. Várhegyi, Z. Sebestyén, Z. Czégény, F. Lezsovits, S. Könczöl: Combustion kinetics of biomass materials in the kinetic regime. Energy Fuels, 26 (2012) 1323–1335; [2] G. Várhegyi, B. Bobály, E. Jakab, H. Chen: Thermogravimetric study of biomass pyrolysis kinetics. A distributed activation energy model with prediction tests. Energy Fuels, 25 (2011) 24–32; [3] G. Várhegyi, M. J. Antal, Jr., E. Jakab, P. Szabó: Kinetic modeling of biomass pyrolysis. J. Anal. Appl. Pyrolysis, 42 (1997) 73–87; [4] G. Várhegyi, E. Jakab, M. J. Antal, Jr.: Is the Broido—Shafizadeh model for cellulose pyrolysis true? Energy Fuels, 8 (1994) 1345–1352; [5] G. Várhegyi, M. J. Antal, Jr., T. Székely, P. Szabó: Kinetics of the thermal decomposition of cellulose, hemicellulose and sugar cane bagasse. Energy Fuels, 3 (1989) 329–335.

Name: Csaba Béla Várhelyi
Country: Romania
Date and place of birth: 1925, Târgu-Secuiesc (Kézdivásárhely), Romania
Present position and address: retired (1988); Romania, 440–445 Cluj-Napoca, Aleea Scărişoara Nr.1, scara V, parter
Email: varhelyi46@yahoo.com
Education and scientific degrees: diploma chemist University "Bolyai" Cluj (1944–1948); Candidat in chemical science (1962) Ministery of Public Educatoin and Cultural Romania; degrees: assistant 1949–1964, assistant professorand (lecturer) and principal scientific researcher 1964–1988. Ministery of Public Education confers title honorary university professor (1999).
Workplaces: University "Bolyai" Cluj, (1949–1959) on the Inorganic and Analytics Chemistry Department; University Babeş-Bolyai Cluj-Napoca, (1959–1988).
Main fields of interest: organic- and coordination chemistry of transition metals, and their physical-chemical study, especially of their stability constants with thermo-gravimetric measurements
Relevant categories in thermal analyses: fields (complex inorganic and organic, biology); methods (TG, DTA, DTG, DSC, kinetics)
Awards and acknowledgments: Zemplén Géza medal (2000), Gr. Mikó Imre medal, (2006), Arany János medal (2010), common (public) member of Hungarian Academy of Sciences (2000)
Professional activities: member of Hungarian Technical Scientific Society of Transylvania and of Hungarian Chemical Society, founder member of Transylvanian Muzeum Society and honorary member (2003)
Publication record: papers (500), books (6), book chapter (1), patents (3)
Equipments: MOM derivatograph, Du Pont 900 Thermal Analyser (910) type DSC
5 most important publications: [1] J. Zsakó, J. Horák, Cs. Várhelyi, A. Benkő, On the Dioximine Complexes metal s (LXIV) Thermal deamination of some complexes of the type $[Co(Niox.H)_2(amine)_2]X$ Monatshefte f. Chem., (Wien), 112, 945–957 (1981); [2] Cs. Várhelyi, J. Zsakó, G. Liptay, M. Somay, On the Dioximine Complexes … metals (LXVII) New Sulfíto-bis-Dimethylglyoximato-Cobalt (III) Complexes and their thermal Decomposition, Rev. Roumaine Chim., 30, (8), 695–702, (1985); [3] Cs. Várhelyi, J. Zsakó, G. Liptay, Z. Finta, On the Dioximine Comp lexes … metals (LXXVIII), TG and DTA study of the thermal decomposition of some Complexes $M[Co(DH)_2XY]$ and $[Co(DH)_2(H_2O)X]$, J. Thermal Analysis. 32, 785–795 (1987); [4] G. Liptay, Cs. Várhelyi, L. Hiltunen, M. Leskelä, Thermal and structural studies on $[Zn(NH_3)_4S_2O_6]$, Thermochim. Acta, 175, 99–104 (1991); [5] Cs. Várhelyi, J. Zsakó, G. Liptay, Thermal decomposition of some salts of hexathiocyanato-platinic acid, Thermochim. Acta, 203, 297–305 (1992).

Name: Csaba Várhelyi Jr.
Country: Romania
Date and place of birth: 1962, Cluj (Kolozsvár), Romania
Present position and address: lecturer at "Babeş-Bolyai" University, Chemistry and Chemical Engineering Faculty Cluj-N.; 400 594 Cluj-Napoca, Tulcea str. 16/7
Email: vcaba@chem.ubbcluj.ro
Website: www.chem.ubbcluj.ro
Education and scientific degrees: Chemical Engineer (Organic Chemical Technology) at "Babeş-Bolyai" University Cluj-N. (1987), Ph.D. (2000)
Workplaces: 1987–1990 Carbosin Copşa Mică chemical eng., 1990–1995 Carbochim Cluj chemical eng., 1996–1997 Ceroc Cluj researcher, 1997–1999 particular companies Cluj chemical eng., commercial agent, 1999–2008 "Babeş-Bolyai" University, Chemistry and Chemical Engineering Faculty Cluj assistant, 2008– lecturer
Main fields of interest: organic- and coordination chemistry of transition metals, and their physical-chemical study, especially thermo-gravimetric measurements
Relevant categories in thermal analyses: fields (organic, complex, biology); methods (TG, DTA, DTG, kinetics)
Publication record: papers (39), presentations (58), books (2), patents (1)
Equipments: SETARAM LabsysEvo TG-DTA thermal analyzer coupled to a PFEIFFER Vacuo OmniStar quadrupole EGA mass spectrometer (1–300 amu) (sample mass 4–10 mg)
5 most important publications: [1] Cs. Várhelyi jr., Gy. Pokol, Á. Gömöry, A. Ganescu, P. Sohár, Gy. Liptay, Cs. Várhelyi, Journal of Thermal Analysis and Calorimetry, (2006), 83 (3), 701 (On the oximine complexes of transition metals. Part 119. Thermal and spectral studies on $Ni(Diox.H)_2$ type chelate compounds.); [2] Cs. Várhelyi jr., A. Kovács, D. Nemcsok, Z. Németh, E. Kuzmann, A. Vértes, K. Vékey, Cs. Várhelyi, Gy. Pokol, Journal of Coordination Chemistry, (2007), 60 (4), 379. (Spectroscopic and thermal studies of $[Fe(dioximato)_2(amine)_2]$ mixed chelates.); [3] Cs. Várhelyi jr., A. Kovács, Á. Gömöry, Cs. Várhelyi, Gy. Pokol, Gy. Farkas, P. Sohár, Journal of Coordination Chemistry, (2009), 62 (10), 2429 (Comparative spectral and thermal studies of $[Pt(DioxH)_2]$ chelates.); [4] I. M. Szilágyi, A. Deák, Cs. Várhelyi jr., J. Madarász, Gy. Pokol, Á. Gömöry, Cs. Várhelyi, Polyhedron, (2010), 29, 2185 (Structural and thermal study of asymmetric α-dioxime complexes of Co(III) with Cl and methyl-pyridines.); [5] Cs. Várhelyi jr., Gy. Pokol, V. Izvekov, Á. Gömöry, Cs. Várhelyi, L. Kocsis, Studia Universitatis Babeş-Bolyai Chemia, (2011), LVI, 1, 275 (Asymmetric Co(III)-complexes of Ethyl-Methyl-Dioxime.)

Name: Indra K. Varma
Country: India
Date and place of birth: 1939, Ghazipur,UP, India.
Present position and address: Honorary Professor, ITM university, Gurgaon, India.
Email: ikvarma@hotmail.com
Education and scientific degrees: M.Sc. Organic Chemistry, Dphil. Biochemistry (Allahabad University, India), Ph. D. Polymer Chemistry(Glasgow University, UK), D.Sc. Polymer Chemistry (Glasgow university, UK), Honorary Doctorate Degree (KTH, Stockholm, Sweden).
Workplaces: IIT Delhi
Main fields of interest: synthesis and characterization of polymers and thermally stable polymers, sustainable polymers, biodegradable polymers, energetic polymers, advanced fibre reinforced polymer composites
Relevant categories in thermal analyses: fields (polymer (studies on curing behaviour, evaluation of thermal stability of polymers, glass transition temperature, effect of temperature on modulus of polymers, flamability behaviour of polymers)); methods (DSC, TGA, DMA, LOI)
Awards and acknowledgments: Dupont award on thermal characterization of polymers; Certificate of recognition from NASA-Ames Research Center, USA for flame retardant polymers; Honorary Visiting Professor at KTH, Sweden
Professional activities: Fellow of Thermal Analysis Society, India; Fellow of Society of Polymer Science, India; Fellow of National Acadamy of Sciences, India
Publication record: papers (>275), books (1), patents (5)
Equipments: DSC, TGA, DMA, LOI
5 most important publications: [1] Sini. N. K, Jayashree Bijwe, Indra K. Varma, "Renewable benzoxazine monomer from vanillin: Synthesis, characterization, and studies on curing behavior, Journal of Polymer Science Part A: Polymer Chemistry, 52 (2014) 7–11; [2] Sini. N. K, Jayashree Bijwe, Indra K. Varma, "Synthesis of itaconimide/nadimide-functionalized benzoxazine monomers: Structural and thermal characterization", Reactive and Functional Polymers, 73 (2013) 1544–1552; [3] Bimlesh Lochab, Indra K. Varma, Jayashree Bijwe, "Blends of benzoxazine monomers: Effect of structure and composition on polymer properties, Journal of Thermal Analysis and Calorimetry, 111 (2013) 1357–1364; [4] Bimlesh Lochab, Indra K. Varma and Jayashree Bijwe, "Cardanol based Bisbenzoxazines: Effect of Structure on Thermal Behaviour", Journal of Thermal Analysis and Calorimetry, 107 (2012) 661–668; [5]Anne Christine Albertsson, Ulrica Edlund, Indra. K. Varma, "Synthesis, chemistry and properties of hemicelluloses, in Biopolymers—New materials for sustainable films and coatings", D. Plackett, Ed. Wiley, U.K., Chapter 7 (2011) 133–150.

Name: Cornelia Vasile
Country: Romania
Date and place of birth: 1942, Homoraciu, Prahova, Romania
Present position and address: head of research group, P.PONI Institute of Macromolecular Chemistry, 41A Gr. Ghica Voda Alley, Ro 700487, Iasi, Romania
Email: cvasile@icmpp.ro
Researcher ID: C-4194-2012
Education and scientific degrees: M.Sc.: Bucharest University Dept. of Physical Chemistry (1960–1965); Ph.D.: Al. I. Cuza University, Dept. of Physical Chemistry (1968–1971)
Workplaces: Medicine and Pharmacy Institute of Iasi; P.Poni Institute of Macromolecular Chemistry, Iasi; Al. I. Cuza University, Department of Physical Chemistry, Iasi; Laval University, Quebec, Canada, Gh. Asachi Technical University
Main fields of interest: thermal analysis; kinetics of polymer decomposition; destructive and non-destructive recovery of polymer wastes; environmental pollution and protection; cultural heritage preservation; crystalline polymer blends and composites; thermodynamics of polymer solutions and multicomponent polymer systems: thermoresponsive polymers; study of the polymer compatibility and biocompatibility, enzymatic degradation
Relevant categories in thermal analyses: fields (polymer materials thermal and thermoxidative behaviour); methods (DSC, TG/DTG, controlled temperature spectroscopic methods, TG/DTG/DSC/FT-IR/MS, GC-MS, pyrolysis)
Awards and acknowledgments: 1965 Bucharest University award for the scientific research of students; 1969 N. Teclu distinction of the Romanian Academy; 1986, 1987 2004, 2005 Five distinctions awarded at the Inventica Symposia hold at Iasi and Ploiesti-Prahova
Professional activities: 1971–1999 assoc. prof.: Al. I. Cuza Univ. and Polytech. Univ. Iasi; member: 1980–1999 Rom. Assoc. Sci.; 1990–1999 IUPAC representative of Rom. Acad. Chem. Sect. for Affiliate IUPAC Progr. for Young Researchers; 1990- member: Rom. Acad. Comm. for TA&C and Environ. Prot.; 1994–1999 board member of the Mediterranean Network on Sci.and Tech. of Adv. Polym. Based Mater.; 1995- assoc. prof.: LAVAL Univ., Canada; 1995 member: International Organizing Committee of the 4th Mediterranean School for Adv. Polym. Based Mater.; 1996 member: Advisory Committee of Int. J. Polym. Mater. and Scientific Committee of the MED-NET Magazine, Italy; 1998 member: Rom. Assoc. Basic Sciences; 1999 member: International Lignin Inst.; 2000- assoc. prof.: Al. I. Cuza Univ., Dept. Plasma, Optics and Molecular Physics; 1998- assoc. prof. Gh. Asachi Tech. Univ. Iasi, Dept. Macromolecules; 2005- member: member of editorial board of J. Environ. Protection, J. of Polymers, Current Advances in Environmental Science of IRED Conference/ Journals, COST projects E54, FP0904, FP1005, E41, P12, 868, FA0904, M1206
Publication record: papers (372), books chapters (90), patents (45), h-index (21), sum of impact factors (303)
Equipments: DMTA, DSC, TG/DTG/FTIR/MS
5 most important publications: [1] M.-T. Nistor, C. Vasile; J. Therm. Anal. Calorim 111, 1903 (2013); [2] M. Brebu, S. Ucar, C. Vasile, J. Yanik, Fuel 89, 1911–1918 (2010); [3] C. Vasile et al.; Cell. Chem. Technol. 42(4-6), 159–169 (2008); [4] Thermal characterization of lignins. C. Vasile et al. Ch. 7 in New Trends in Natural and Synthetic Polymers Eds. C. Vasile and G. E. Zaikov, Nova Science, New York, 2006, pp. 135–165; [5] Comportarea Termica a Polimerilor, Eds. C. Vasile, et al., Editura Academiei, Bucharest, 1980.

Name: Stefano Vecchio Ciprioti
Country: Italy
Date and place of birth: 1966, Rome, Italy
Present position and address: Assistant Professor at Sapienza University of Rome, I-00161 Rome, via del Castro Laurenziano 7, Italy
Email: stefano.vecchio@uniroma1.it
Website: http://www.sbai.uniroma1.it/~stefano.vecchio/
ORCID: 0000-0002-7864-4266
Education and scientific degrees: Chemist, Sapienza University of Rome (1987–1991); Ph.D. (1992); Assistant Professor (1996–)
Workplaces: Department of Chemistry, Sapienza University of Rome (1994–1995); Department of Chemical Engineering, Sapienza University of Rome (1996–2009); Department of Basic and Applied Science for Engineering, Sapienza University of Rome (2009–).
Relevant categories in thermal analyses: fields (solid state kinetics, thermodynamics of vaporization and sublimation processes, compatibility between active components and excipients along with their application in different fields: pharmaceuticals, life science, coordination chemistry, cultural heritage, food); methods (TG, DTA, EGA, DSC, kinetics)
Professional activities: Regional Editor of the Journal of Thermal Analysis and Calorimetry (2009–), member of the Editorial Board of Thermochimica Acta (2010–), Board member of A.I.C.A.T., Italian Society for Calorimetry and Thermal Analysis (2011–), Board member of G.I.C.A.T., Interdivisional Group of the Italian Chemical Society (2008–2010)
Publication record: papers (78), citations (635), h-index (14)
Equipments: Stanton-Redcroft STA 625 simultaneous TG/DSC, Stanton-Redcroft STA 1500 simultaneous TG/DTA
5 most important publications: [1] M. Tomassetti, A. Catalani, V. Rossi, S. Vecchio: J. Pharm. Biomed. Anal., 37(2005) 949–955; [2] S. Vecchio: J. Therm. Anal. Calorim., 87 (2007) 79, S. Vecchio: Struct. Chem. 24(6) (2013) 1821; [3] S. Vecchio, L. Cerretani, A. Bendini, E. Chiavaro: J. Agric. Food Chem., 57 (2009) 4793; [4] the following three reviews reports the most recent developments dealing with the coupling of thermal analysis techniques with other techniques: S. Materazzi, S. Vecchio: Appl. Spectr. Rev., 45(4) (2010) 241, S. Materazzi, S. Vecchio: Appl. Spectr. Rev., 46(4) (2011) 261, S. Materazzi, S. Vecchio: Appl. Spectr. Rev., 48(8) (2013) 654; [5] S. Materazzi, S. Vecchio, L. W. Wo, S. De Angelis Curtis: J. Therm. Anal. Calorim., 103 (2011) 59.

Name: Venkatarama Venugopal
Country: India
Date and place of birth: 1947, India
Present position and address: Raja Ramanna Fellow, Former Director, Radiochemistry and Isotope Group, Bhabha Atomic Research Centre, Mumbai 400 085, India
Email: vvg1947@yahoo.com
Education and scientific degrees: B.Sc. Univ. of Madras, 1968; M.Sc. Univ. of Madras, 1970; Ph.D. Univ. of Bombay, 1991
Workplaces: Raja Ramanna Fellow, Former Director, Radiochemistry and Isotope Group, Bhabha Atomic Research Centre, Mumbai
Main fields of interest: radiochemistry (thermodynamics of materials), thermal/thermodynamics of plutonium based fuels at high temperature, chemical quality control of fuel, X-ray and solid state chemistry
Awards and acknowledgments: Netzsch-ITAS award for outstanding contributions on the application of thermal analysis for nuclear materials, 2000; Silver medal from the Indian Association of Solid State Chemists and Allied Scientists (ISCAS) in 2001; MRSI Medal, 2004; Indian Nuclear Society award 2006; DAE Award 2008; Dr. Athavle Life time Achievement award for Thermal analysis in 2010; Setaram Life time Achievement award in 2013
Professional activities: Raja Ramanna Fellow and formerly an Outstanding Scientist and the Director of Radio Chemistry and Isotope Group at BARC, Mumbai; Officer-in-charge, Nuclear Material Accounting and Control (NUMAC) cell of DAE. Member of Standing Advisory Group for Safeguard Implementation (SAGSI) to advice Director General IAEA on Safeguards Matters (2008–2011); Chairman of Radiation Technology and Applications Committee (RTAC) of BRNS and a member of BRIT board since 1999 and 2003 till 2011. President of Indian Thermal Analysis Society (ITAS) for two terms and Indian Association of Nuclear Chemists and Allied Scientists (IANCAS) for two terms. He is the vice president of Indian Nuclear Society (INS). Previously regional editor for the Journal of Thermal Analysis and Calorimetry
Publication record: papers (190)
5 most important publications: [1] K. Krishnan, K. D. Singh and V. Venugopal (2000) "Structural and Thermal studies on $PuTe_2O_6$" J Alloy Comp, 307, 114; [2] Y. P. Naik, G. A. Rama Rao and V. Venugopal (2001) "Separation of hydrogen isotopes by gas chromatography using vanadium oxide coated molecular sieve (4A) packed column" J Radioanal Nucl Chem 247, 11; [3] S. C. Parida, S. K. Rakshit, S. Dash, ZileySingh, R. Prasad and V. Venugopal (2003) "Systems Ln-Fe–O (Ln = Eu, Gd): thermodynamic properties of ternary oxides using solid-state electrochemical cells" J Solid State Chem, 172, 370; [4] Abhay Patil, Smruti Dash, S. C. Parida and V. Venugopal (2004) "Thermodynamic studies on $LnCoO_3$ (Ln = Eu, Gd, Tb) by solid-state electrochemical cells" J Alloy Comp, 384, 274; [5] N. L. Misra, K. D. Singh Mudher, V. C. Adya, B. Rajeswari and V. Venugopal (2005) Determination of trace elements in uranium oxide by Total Reflection X-Ray Fluorescence spectrometry" Spectrochim Acta, Part B 60, 834.

Name: Ranjit K. Verma
Country: India
Date and place of birth: 1956, Patna, India
Present position and address: Pro Vice-Chancellor, Patna University, Patna-800 005 (India)
Email: profrkverma@gmail.com; rkvermamagadh@rediffmail.com; pvc@puccmail.ac.in
Website: www.patnauniversity.ac.in
Researcher ID/ORCID: G-5994-2010/0000-0001-9595-7179
Education and scientific degrees: Magadh University; M.Sc. (1973–1975); Ph.D. (1986) Professor (1993)
Workplaces: H.D.Jain College, Ara-802301, India(as Lecturer, 1977–1983); University Department of Chemistry (DST-FIST Sponsored and UGC-BSR Supported), Magadh University, Bodh Gaya (India) (Lecturer, 1983–1986; Reader, 1986–1993; Professor, 1993–2014); Pro Vice-Chancellor, Patna University, Patna (2014–)
Main fields of interest: solid state thermal decomposition, kinetics, calorimetry and nanotisation
Relevant categories in thermal analyses: fields (inorganic, materials, nano, complex, food); methods (TG, DSC, thermomechanical analysis, kinetics, specific heat, calorimetry)
Professional activities: F.I.C. [Elected Fellow, Institution of Chemists (India)]; Secretary, International Confederation for Thermal Analysis and Calorimetry (ICTAC)(2012–); Chairman, Education Committee, ICTAC (2008–2014); Member, Scientific Award Commission, ICTAC (2008); Associate Editor, J Therm Anal Calorim (2013–2015); Regional Editor, J Therm Anal Calorim (2010); Guest Editor, SATAC-2010 and SATAC-2011 Special issues, J Therm Anal Calorim; Hony Editor, J Indian Chem Soc (2007–2010); Vice President, Indian Council of Chemists; Vice President, Indian Thermal Analysis Society; Executive Committee Member, Indian Science Congress Association
Publication record: papers (26), books (3), citations (111), h-index (7), sum of impact factors (25.03)
Equipments: Mettler TG-DSC-1
5 most important publications: [1] Thermal, structural, magnetic and photoluminescence studies on cobalt ferrite nanoparticles obtained by citrate precursor method, J Therm Anal Calorim, 110 (2012) 573–580; [2] Thermal, structural and magnetic studies on chromite spinel, synthesized by citrate precursor method and annealed at temperature 450 and 650 °C, J Therm Anal Calorim, 107 (2012) 197–204; [3] Thermal, XRD and magnetization studies on $ZnAl_2O_4$ and $NiAl_2O_4$ spinels, synthesized by citrate precursor method and annealed at temperature 450 and 650 °C, Therm Anal Calorim, 107 (2012) 205–210; [4] Thermal analysis of 2-oxocyclopentanedithiocarboxylato complexes of iron(III), copper(II) and zinc (II) containing pyridine or morpholine as the second ligand, J Therm Anal Calorim, 94 (2008) 27–31; [5] Non-isothermal dehydration and decomposition of dl-lactates of transition metals and alkaline earth metals- a comparative study, J Therm Anal Calorim, 80 (2005) 351–354.

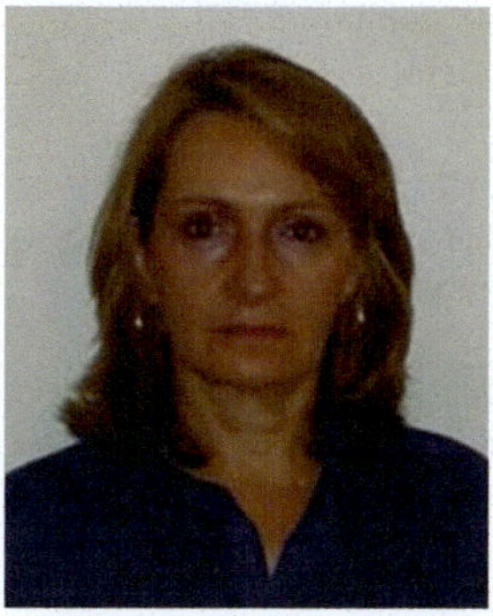

Name: Marilda Mendonça Guazzelli Ramos Vianna
Country: Brazil
Date and place of birth: 1959, São Paulo, Brazil.
Present position and address: Research Associate at Chemical Engineering at Polytechnic School—University of São Paulo; Rua Frederico Guarinon, 125, apto 93C, São Paulo, 05713-460, SP, Brazil.
Email: marilda@pqi.ep.usp.br
Education and scientific degrees: Chemical Engineer, Mackenzie University, Brazil (1978–1982); M.Sc. (2001) and Ph.D. (2005) in Chem. Eng. at University of São Paulo (USP), Brazilian National Council (CNPq) Researcher (2005–).
Workplaces: Chemical Engineering Department, Polytechnic School of São Paulo University, Avenida Professor Luciano Gualberto, trav. 3, 380, São Paulo, 05508-900, SP, Brazil.
Main fields of interest: thermal analysis applied to industrial produtcts and processes and to their effluent and waste thermal characterization and processing
Relevant categories in thermal analyses: fields (inorganic, organic, materials, clays and organoclays, industrial minerals, composite materials, industrial solid wastes and sludges); methods (TG, DTG, DTA, DSC, kinetics, specific heat, calorimetry, conventional TG and DTA instrument)
Publication record: papers (31), citations (95), chapters in books (1), patents (1), h-index (5),
5 most important publications: [1] M. M. G. R. Vianna; J. Dweck; V. F. J. Kozievitch, F. R. Valenzuela-Diaz, P. M. Büchler: Characterization and study of sorptive properties of differently prepared organoclays from a Brazilian natural bentonite. J. Therm. Anal. Calorim., 82 (2005), 595–602; [2] M. M. G. R Vianna, J. Dweck, F. H. Quina, F. M. S. Carvalho, C. A. O. Nascimento: Toluene and naphthalene sorption by iron oxide/clay composites Part I. Materials characterization. J. Therm. Anal. Calorim., 100 (2010), 889–896; [3] M. M. G. R Vianna, J. Dweck, F. H. Quina, F. M. S. Carvalho, C. A. O. Nascimento: Toluene and naphthalene sorption by iron oxide/clay composites. Part II. Sorption experiments. J. Therm. Anal. Calorim., 101 (2010), 887–892.

Name: Lyubomir Vlaev
Country: Bulgaria
Date and place of birth: 1951, Burgas, Bulgaria
Present position and address: Head of the Department of Pysical Chemistry, Assen Zlatarov University, Bul. Prof. Y. Yakimov 1, 8010 Burgas, Bulgaria
Email: vlaev@btu.bg
Education and scientific degrees: Chemical Engeneer, Higher Institute of Chemical Technology, Burgas, Bulgaria (1969–1974); Ph.D.(1984); D.Sc. (2005); Professor (2008).
Workplaces: Department of Pysical Chemistry, Assen Zlatarov University, Burgas, Bulgaria.
Main fields of interest: thermal stability and kinetics of thermal degradation of inorganic compounds and agrowastes
Relevant categories in thermal analyses: fields (inorganic materials and agrowastes); methods (TG, DTA, kinetics)
Publication record: papers (128), chapters in books (3), patents (5), citations (708), h-index (13), sum of impact factors (66.311)
Equipments: Instrument STA 449 F3 Jupiter (Nietzsch, Germany)
5 most important publications: [1] Tanev P., Vlaev L., An attempt at a more precise evaluation of the approach to mesopore size distribution calculations depending on the degree of pore blocking, J. Col. Int. Sci., 160 (1993) 110–116; [2] Vlaev L., Markovska I., Lyubchev L., Non-isothermal kinetics of pyrolysis of rice husk, Thermochim. Acta, 406 (2003) 1–7; [3] Vlaev L., Nikolova M., Gospodinov G., Non-isothermal kinetics of dehydration of some selenite hexahydrates, J. Solid State Chem., 177 (2004) 2663–2669; [4] Vlaev L., Georgieva V. G., Genieva S. D., Products and kinetics of non-isothermal decomposition of vanadium(IV) oxyde compounds, J. Therm. Anal. Calorim., 88 (2007) 805–812; [5] Vlaev L., Nedelchev N., Gyurova K., Zagorcheva M., A comparative study of non-isothermal kinetics of decomposition of calcium oxalate monohydrate, J. Anal. Appl. Pyrol., 81, No.2 (2008) 253–262.

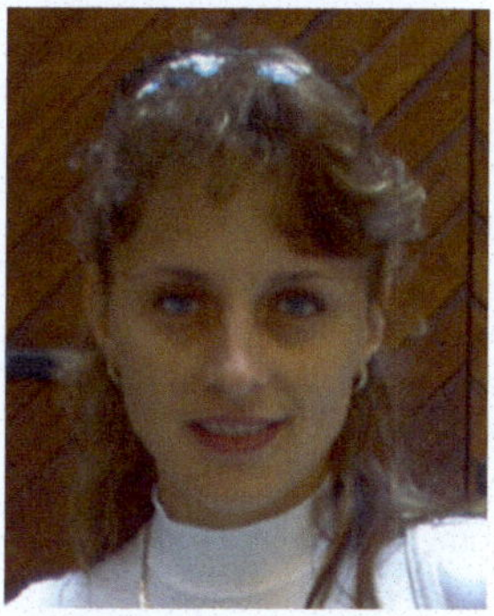

Name: Gabriela Vlase
Country: Romania
Date and place of birth: 1974, Timişoara, Romania
Present position and address: Assoc. Prof., West Univ. Timisoara, Fac. Chemistry, Biology, Geography
Email: gabriela.vlase@cbg.uvt.ro; gabi.vlase@gmail.com
Researcher ID: A-5806-2014
Education and scientific degrees: chem., Ph.D.
Workplaces: West Univ. of Timişoara, Research Center Thermal Analysis in Environmental Problems
Main fields of interest: non-isothermal kinetics, thermal induced interactions
Relevant categories in thermal analyses: fields (pharmaceutical, polymer, complex, materials, food); methods (TG, DTA, EGA, DSC, kinetics)
Professional activities: Member of Research Center Thermal Analysis in Environmental Problems
Publication record: papers (65), books (7), citations (540), h-index (12), sum of impact factors (110)
Equipments: TG/DTG/DTA Diamond PerkinElmer, DSC Diamond PerkinElmer, TMA Diamond PerkinElmer, EGA (FT-IR Spectrum 100 with Multiscope FT-IR Microscope PerkinElmer)
5 most important publications: [1] Vlase, G., Vlase, T., Doca, N., Thermal behavior of some phenitoine pharmaceuticals (2008) Journal of Thermal Analysis and Calorimetry, 92 (1), pp. 259–262; [2] Vlase, G., Vlase, T., Doca, N., Perţa, M., Ilia, G., Plesu, N., Thermal behavior of a sol-gel system containing aniline and organic phosphonates, (2009) Journal of Thermal Analysis and Calorimetry, 97 (2), pp. 473–478; [3] Vlase, G., Vlase, T., Doca, N., Ilia, G., Estimation of flame retardancy effect by thermal analysis using kinetic parameters obtained under non-isothermal conditions in air. (2009) Chinese Journal of Chemistry, 27 (10), pp. 1919–1924; [4] Vlase, G., Vlase, T., Tudose, R., Costişor, O., Doca, N., Kinetic of decomposition of some complexes under non-isothermal conditions, (2007) Journal of Thermal Analysis and Calorimetry, 88 (3), pp. 637–640; [5] Fuliaş, A., Ledeţi, I., Vlase, G., Vlase, T., Physico-chemical solid-state characterization of pharmaceutical pyrazolones: An unexpected thermal behaviour, (2013) Journal of Pharmaceutical and Biomedical Analysis, 81–82, pp. 44–49.

Name: Titus Vlase
Country: Romania
Date and place of birth: 1966, Giarmata, Timiş, Romania
Present position and address: Assoc. Prof., West Univ. Timisoara, Fac. Chemistry, Biology, Geography
Email: tvlase@cbg.uvt.ro; tvlase2004@yahoo.com
Researcher ID: C-4852-2011
Education and scientific degrees: chem. eng., Ph.D.,
Workplaces: West Univ. of Timişoara, Research Center Thermal Analysis in Environmental Problems
Main fields of interest: non-isothermal kinetics, modified NPK method
Relevant categories in thermal analyses: fields (pharmaceutical, polymer, inorganic, materials, food, nano); methods (TG, DTA, EGA, DSC, TMA, kinetics)
Awards and acknowledgments: "Gheorghe Spacu" Award in chemistry given by Romanian Academy (2005)
Professional activities: Scientific/Vice Manager of Research Center Thermal Analysis in Environmental Problems
Publication record: papers (74), books (11), citations (504), h-index (12), sum of impact factors (135)
Equipments: TG/DTG/DTA Diamond PerkinElmer, DSC Diamond PerkinElmer, TMA Diamond PerkinElmer, EGA (FT-IR Spectrum 100 with Multiscope FT-IR Microscope PerkinElmer)
5 most important publications: [1] Segal, E., Budrugeac, P., Carp, O., Doca, N., Popescu, C., Vlase, T., Thermal Analysis-Fundamentals and applications. Kinetic analysis of heterogeneous transformations. (in romanian), Ed. Romanian Academy, Bucarest, 2013; [2] Vlase, T., Vlase, G., Doca, N., Bolcu, C., Processing of non-isothermal TG data: Comparative kinetic analysis with NPK method, (2005) Journal of Thermal Analysis and Calorimetry, 80 (1), pp. 59–64; [3] Vlase, T., Vlase, G., Birta, N., Doca, N., Comparative results of kinetic data obtained with different methods for complex decomposition steps, (2007) Journal of Thermal Analysis and Calorimetry, 88 (3), pp. 631–635; [4] Vlase, T., Doca, N., Vlase, G., Bolcu, C., Borcan, F., Kinetics of non-isothermal decomposition of three IRGANOX-type antioxidants, (2008) Journal of Thermal Analysis and Calorimetry, 92 (1), pp. 15–18; [5] Vlase, T., Vlase, G., Doca, N., Iliescu, S., Ilia, G. Thermo-oxidative degradation of polymers containing phosphorus in the main chain, (2010) High Performance Polymers, 22 (7), pp. 863–875.

Name: Sergey Vyazovkin
Country: USA
Date and place of birth: 1960, Kazan, Russia
Present position and address: Professor, Department of Chemistry, University of Alabama at Birmingham, Birmingham, AL 35294, USA
Email: Vyazovkin@uab.edu
Education and scientific degrees: Ph.D. in Physcal Chemistry, Byelorussian State University (1989); M.S. In Chemistry, Byelorussian State University (1982)
Workplaces: Belorussian State University (1982–1993), Technical University of Vienna (1993–1995), University of Toledo (1995), University of Utah (1995–2001), University of Alabama at Birmingham (2001–)
Main fields of interest: kinetics of physical and chemical processes in the condensed phase
Relevant categories in thermal analyses: fields (inorganic, materials, nano, complex, pharmaceutical, polymer, glass); methods (TG, DSC, kinetics)
Awards and acknowledgments: NATAS Outstanding Achievement (Mettler-Toledo) Award (2004), James J. Christensen Award in Calorimetry (2006), Elected NATAS Fellow (2006)
Professional activities: Editor of Thermochimica Acta; Editorial Board member of Macromolecular Rapid Communication and of Macromolecular Chemistry and Physics; Chair of the ICTAC Kinetics Committee
Publication record: papers (155), citations (9,000), h-index (42)
Equipments: TG, DSC, HP DSC, DMA, TMA, TG-FTIR
5 most important publications: [1] S. Vyazovkin, A. K. Burnham, J. M. Criado, L. A. Pérez-Maqueda, C. Popescu, N. Sbirrazzuoli: ICTAC Kinetics Committee Recommendations for Performing Kinetic Computations on Thermal Analysis Data, Thermochim. Acta 520 (2011) 1; [2] S. Vyazovkin, N. Sbirrazzuoli: Isoconversional Kinetic Analysis of Thermally Stimulated Processes in Polymers, Macromol. Rapid Commun. 27 (2006) 1515; [3] S. Vyazovkin: Modification of the Integral Isoconversional Method to Account for Variation in the Activation Energy, J. Comput. Chem. 22 (2001) 178; [4] S. Vyazovkin, C. A. Wight: Model-free and model-fitting approaches to kinetic analysis of isothermal and nonisothermal data Thermochim. Acta 340/341 (1999) 53; [5] S. Vyazovkin: Evaluation of the activation energy of thermally stimulated solid-state reactions under an arbitrary variation of the temperature J. Comput. Chem. 18 (1997) 393.

Name: Irena Wacławska
Country: Poland
Date and place of birth: 1949, Gorlice, Poland
Present position and address: Professor of AGH, AGH University of Science and Technology, Faculty of Materials Science and Ceramics, PL-30-059 Krakow, 30 Mickiewicza ave., Poland
E-mail: iwac@agh.edu.pl
Education and scientific degrees: Jagiellonian University, Krakow (1972); Ph.D. AGH University of Science and Technology, Krakow (1978); D.Sc. (1998); Professor (2009)
Workplaces: Jagiellonian University, Krakow (1972–1974), AGH University of Science and Technology, Krakow (1974–)
Main fields of interest: chemistry of solids, thermal analysis, thermochemistry of minerals and materials, chemistry of glass
Awards and acknowledgments: Sci. Awards of Rector of AGH University of Science and Technology (1998, 2003, 2008, 2013); Sci. Award of Education Ministry (1999); Sci. Award of Polish Ceramic Society (2001), Gold medal with mention in the Belgian and International Trade Fair for Technological Innovation (2007), Gold medal in the International Competition POLEKO 2007, Award in the 107 International Competition CONCOURS-LEPINE 2008, Distinction in the 8th International Competition ECO 2009 for technological solutions in environmental production
Professional activities: Member of International Confederation for Thermal Analysis and Calorimetry ICTAC (1982–), Member of Polish Society of Thermal Analysis and Calorimetry (1986–), Member of the Board of Polish Society of Thermal Analysis and Calorimetry (1991–1994), Member of the Revision Commission of Polish Society of Thermal Analysis and Calorimetry (2000–2003), Member of the Board of Polish Ceramic Society (1988–2013); Secretary of Series Ceramics of Polish Ceramic Bulletin (1997–2002), Member of Organizing Committee of 5th (1991) and 9th (2003) Polish Conference of Thermal Analysis and Calorimetry
Publication record: papers (188), book (1), patents (3), cititations (391)
List of the 5 most important publications: [1] I. Wacławska, L. Stoch, J. Paulik, F. Paulik: Thermal decomposition of colemanite, Thermochim. Acta, 126 (1988) 307; [2] I. Wacławska: Thermal reactions of solid state amorphized borates and borate glasses, J. Therm. Anal., 49 (1997) 1337; [3] I. Wacławska, M. Szumera: Thermal analysis of glasses for proecological applications, J. Therm. Anal. Calorim., 72 (2003) 1065; [4] I. Wacławska, M. Szumera: Reactivity of silicate glasses in soil environment, J. Alloy. Comp., 468 (2009) 246; [5] I. Wacławska, M. Szumera, P. Stoch: Structural role of Fe in the soil active glasses, Spectrochim. Acta A, 79 (2011) 728.

Name: Tommy Wadsten
Country: Sweden
Date and place of birth: 1934, Stockholm
Present position and address: Research director privat company since 1984 Developmen and Research Wadsten, Rättviksv.50 19271 Sollentuna
Email: wadsten@mbox302,swipnet.se
Education and scientific degrees: 1970 fil. lic. (Ph.D.)
Workplaces: Univ. Stockholm, Brown Univ. USA, Tech. Univ. Budapest, Lenindgad Univ. USSS, Univ. of Dhaka, Bangladesh, Astra-Zenica, Recipharm, Stockholm.
Main fields of interest: Very slow reactions, interaction of water with matter, material sci.
Relevant categories in thermal analyses: methods (DSC, TG)
Professional activities: Exec. Comm. Symp TA Studsvik Sweden Feb 1965; Exec Comm. Symp TA Ursvik Sweden June 1970; Exec Comm. Forth Scand Symp on TA Stockholm 1975; Nordic repres. ICTA Council 4 years; Swedish repres. In NOSTAC many years
Publication record: papers (130), books (2)
Equipments: micro balance, high temp. microscope
5 most important publications: [1] Oxidation of uranium dioxide at room temp. J Nuclear Mat 64 1977; [2] Characterization of a solid residue in an ancient egyptian alabaster jar Mus. Med. Near Eastern Bull 13 1978; [3] Calcium oxalate hydrates on the surface of lichens Lichenologist (3)239–245 1985; [4] Phase diagram and aqueous solub. of the lidocaline-prilocaine binary system, J Pharm. Sciences 73 No 4 1984; [5] Inferences from a corrosion study of a bronze cannon, applied to high level nuclear disposal, Applied Geochemistry Vol 3, 273–280 1988.

Name: Fang Wang
Country: P.R. China
Date and place of birth: 1966, NanJing, P.R. China
Present position and address: Senior engineer, Director of B. Dept., Center for Analysis and Testing, Nanjing Normal University No 1 WenYuan Road, Nanjing, 210023, China
Email: wangfang@njnu.edu.cn
Education and scientific degrees: Nanjing University of Science and Technology, B.E. (1985–1990). Asst. Engineer (1990). Engineer (1996). Senior engineer (2002). Visiting scholar in Rowan University of USA (2013–2014)
Workplaces: Center for Analysis and Testing, Nanjing Normal University (1988-now)
Main fields of interest: investigation and development the thermal analysis methods which are used for various materials, specially polymer and drugs, and study of their theory
Relevant categories in thermal analyses: fields (materials, nano, complex, organic, pharmaceutical, polymer, food, biology, life, glass); methods (TG, DTA, DSC, kinetics, specific heat, calorimetry, microcalorimetry)
Awards and acknowledgments: The large instrument collaboration advanced Award of JiangSu province, China. (2006), Advanced Instrumental Analysis Award of JiangSu province, China (2007, 2008, 2011)
Professional activities: Chairman of Jiangsu Thermal Analysis Society (JTA) of China (2006-now). Membership of Chinese Chemical Society (2014–). Chairman and organiser of the 1st- 5th JTA congress in China (2007, 2008, 2010, 2011, 2013). Ex-chairman and organiser of the 3rd national symposium on Thermal Analysis Kinetics and Thermokinetics of Chinese Chemical Society(TAKT of CCS) and international seminar of advanced TA (2011). National membership present the 15th ICTA in Japan(2012). Member of the committee of 4th national symposium on TAKT of CCS (2013). Chairman and organiser of international seminar of advanced TA in SuZhou (2013), Reviewer of J. Therm. Anal. Calorim. (2013)
Publication record: papers (42), patents (2)
Equipments: DSC, TGA, DTA-TG, DMA (Diamond, Pyris, 7 sreies, PerkinElmer. USA)
5 most important publications: [1] Fang Wang, GePu Guo, et al. J. Therm. Anal. Calorim., 2013, 9(113), 1113–1121; [2] Fang Wang, Lin Zhou, et al., J. Therm. Anal. Calorim., 2010, 102(1): 69–74; [3] Fang Wang, ShenJun Sheng, et al., Acta Phys.-Chim. Sin. 2013, 29 (12), 2505–2512; [4] Fang Wang, Lin Zhou, et al., Chin J. Pharm. Anal. 2009, 29(2): 253–256; [5] Fang Wang, Tu Qiang, et al., Chemical World, 2012, 53(6), 338–341.

Name: Bina N. Wani
Country: India
Date and place of birth: 1958, Mumbai, India
Present position and address: Scientific Officer(G), Bhabha Atomic Research Centre, Trombay Mumbai 400085. India
Email: bnwani@rediffmail.com
Education and scientific degrees: M.Sc.; Ph D.
Workplaces: Chemistry Division, Bhabha Atomic Research Centre, Trombay Mumbai 400085.
Main fields of interest: thermal analysis, material science, catalysis, fuel cell
Relevant categories in thermal analyses: fields (thermal analysis); methods simultaneous TG/DTA, DSC, HT-XRD, dilatometry, electrical conductivity)
Awards and acknowledgments: Netzsch-ITAS 2013, DAE Group Achievment Award 2009
Professional activities: Life member of i) Indian Thermal Analysis, ii) Indian Chemical Society, iii) Indian Society of Electroanalytical Chemistry
Publication record: papers (71), books (2), h-index (9)
Equipments: Simultaneous TG/DTA, HT-XRD, DSC, Impedance Analyzer, FTIR, XRD
5 most important publications: [1] High temperature Thermal Expansion and eledctrical conductivity of $Ln_{0.95}MnO_3$ (Ln = La, Nd, and Gd), R V Wandekar, B N Wani, S R Bharadwaj, J Alloy Comp 433 (2007) 84–90; [2] Preparation Characterisation and the Standard Enthalpy of formation of $La_{0.95}MnO_3$ $Sm_{0.95}MnO_3$, R V Wandekar, B N Wani, S R Bharadwaj, Thermochim Acta 493 (2009) 14–18; [3] Phase Transition in LAMOX type compounds, M Ali(Basu), B N Wani, S R Bharadwaj, J Therm Anal Calorim 96(2) (2009) 463–68; [4] Synthesis, characterisation, TPR-TPOand activity studies on $LaMn_xV_{1-x}O_4$ catalyst, S Varma, B N Wani, N M Gupta, Appl Catal A 205 (2001) 295; [5] Fluorination of Sr_2CuO_3 and high temperature superconducting oxides, B N Wani, L L Lmiller, B Shu, F Borsa, Physica C 272 (1996) 187.

Name: Marek Wesolowski
Country: Poland
Date and place of birth: 1950, Gdansk, Poland
Present position and address: Head of Department of Analytical Chemistry, Medical University of Gdansk, Gen. J. Hallera 107, Gdansk, Poland
Email: marwes@gumed.edu.pl
Education and scientific degrees: M.Sc. in pharmacy, Medical University of Gdansk (1973); Ph.D. in pharmaceutical sciences (1978); D.Sc. in pharmaceutical sciences (1987); Professor (conferred by The President of Poland in 1998)
Workplaces: Department of Analytical Chemistry, Faculty of Pharmacy with Subfaculty of Laboratory Medicine, Medical University of Gdansk (1973 –)
Main fields of interest: physicochemical and analytical studies of the active pharmaceutical ingredients and medicinal products in solid state by thermoanalytical and other techniques (FT-IR, Raman spectroscopy, diffractometry); search for chemical control factors of herbal raw materials by fingerprinting method with application of thermoanalytical, chromatographic and chemometric techniques; application of artificial neural networks and multivariate methods as interpretation tools of the results of chemical analyses
Relevant categories in thermal analyses: fields (inorganic, materials, organic, pharmaceutical, food, lubricating oils); methods (TG, DTA, DSC)
Awards and acknowledgments: Awards of The Ministry of Health (1981, 1987, 1996, 1998, 2006); Awards of The Rector of Medical University of Gdansk (almost every year)
Professional activities: Dean of the Faculty of Pharmacy with Subfaculty of Laboratory Medicine (2002–2008); Chairman of the Scientific and Organizing Committees of two Symposiums on Quality assessment of drug substances and pharmaceutical products with the aid of solid state methods (2008, 2011); President of the Regional Board of the Polish Pharmaceutical Society (1998–2008); Vice-President of the Gdansk Scientific Society (2004–); Member of Polish Society of Calorimetry and Thermal Analysis (1987–)
Publication record: papers (205), books (2), chapters in books (24), conference and other papers (307), citations (453), h-index (11), a sum of impact factors (95.848)
Equipments: heat-flux DSC, model 822^{e}; derivatograph, model OD-103
5 most important publications: [1] M. Wesolowski, B. Suchacz: Anal. Bioanal. Chem., 371 (2001), 323–330; [2] M. Wesolowski, P. Konieczynski: Int. J. Pharm., 262 (2003) 29–37; [3] M. Wesolowski, P. Konieczynski: Thermochim. Acta, 397 (2003) 171–180; [4] M. Wesołowski, J. Erecińska: J. Therm. Anal. Calorim., 82 (2005) 307–313; [5] M. Wesolowski, P. Szynkaruk: J. Therm. Anal. Calorim., 93 (2008) 739–746.

Name: Krystyna Wieczorek-Ciurowa
Country: Poland
Date and place of birth: 1944, Andrychów, Poland
Present position and address: Professor, Cracow University of Technology, 24, Warszawska Str., 31-155 Cracow, Poland
Email: kwc@pk.edu.pl
Website: www.pk.edu.pl
ORCID: 0000-0001-5886-1223
Education and scientific degrees: M.Sc. Jagiellonian University Cracow (1967); Ph.D. Cracow University of Technology (1976); D.Sc. Russian Acad. Sciences (1995); Assoc. Professor (1999); Professor (2011)
Workplaces: Cracow University of Technology, Faculty of Chemical Engineering and Technology (1967–); Postdoc. Fellowships: Technical University of Budapest (Hungary) (1978–1979); University of Oulu (Finland) (1987); Université de Sciences et Techniques de Lille (France) (1989); UN Fellowship: Universität Gesamthochschule, Siegen and Westfalische Wilhelms-Universität, Münster (1990)
Main fields of interest: physicochemistry of inorganic solids using thermoanalytical methods, XRD, IR spectroscopy, SEM-EDS microanalysis; mechanochemical syntheses of functional nano/micro materials (metallic-ceramic composites, catalysts, pigments, biomaterials, dielectrics, tribomaterials, sorbents); combustion processes in technology; Green chemistry
Relevant categories in thermal analyses: fields (Thermal stability of materials: inorganic, ceramic, minerals, nanopowders); methods (TG/DTG, DTA, DSC)
Awards and acknowledgments: Minister of Education Awards (1979, 1981); 17 awards—Rector of Cracow University of Technology (1971–2013); Honorary Award of CUT (1995); Gold Cross of Merit (1988); Kawaler Cross of Merit of Poland Renaissance Order (2000); Medal of Polish Education Committee (2009)
Professional activities: Polish Chem. Soc. Member (1972–); Member of Exec. Council Cracow Div. (1981–1984, 2001–2016); Assoc. Eng. Techn. in Chem. Ind. Member (1980–1988); Member of Exec. Council Cracov Div. (1980–1988); Polish Soc. Calorim. Therm. Anal. Charter Member (1985), Exec. Council Member (1985–2009); ICTAC Member (1994–2012); J. Therm. Anal. Calorim.: Member of Editiorial Advisory Board (1998–1999); Therm. Anal. Cal. Conferences—Exec. Committee Member
Publication record: papers (>120), books (4), patents (11), citations (375), h-index (10), sum of impact factors (>130)
Equipments: TA Thermoanalyser
5 most important publications: [1] K. Wieczorek-Ciurowa: Versatility of the application of thermal analysis to the simulation of gas desulphurisation, Thermochim. Acta, 272 (1996) 233; [2] K. Wieczorek-Ciurowa, D. Oleszak, K. Gamrat: Mechanosynthesis and process characterization of some nanostructured intermetallics-ceramics composites, J. Alloys Comp., 434–435 (2007) 501; [3] K. Wieczorek-Ciurowa: Mechanochemical synthesis of metallic-ceramic composite powders [in] High-energy ball milling: Mechanochemical processing of nanopowders, Ed. M. Sopicka-Lizer, Woodhead Publ. Ltd., UK, 2010; [4] K. Wieczorek-Ciurowa, P. Dulian et al. J. Therm. Anal. Calorim. 101 (2010) 471; [5] P. Baláž, M. Achimovicova, K. Wieczorek-Ciurowa, Chemical Society Reviews, 42 (2013) 7571.

Name: Emmerich Wilhelm
Country: Austria
Date and place of birth: 1941, Wien (Vienna), Austria
Present position and address: Prof. Dr. Dr. h.c. E. WILHELM, former Chairman of the Institute of Physical Chemistry (IPC), retired; IPC, University of Wien (UW), Währinger Straße 42, A-1090 Wien (Vienna), Austria
Email: emmerich.wilhelm@univie.ac.at
Education and scientific degrees: Chemistry, Physics, Philosophy: degrees equivalent to B.Sc., M.Sc. (1959–1965), dissertation (1965–1968), Dr. phil., UW (January 1969); Senior Fulbright Scholar, Wright State University (WSU), OH, USA (1969–1971); Ass. Prof. (1971–1979), Habilitation, UW (1979); Assoc. Prof., WSU, OH, USA (1979–1980); Tenured Assoc. Prof., Vice-Chair, Chair, IPC (1980–2007); Dr. h.c., University B. Pascal, Clermont-Fd., F (2006)
Workplaces: IPC, University of Wien (A); Department of Materials Research and Physics, University of Salzburg (A); University Blaise Pascal, Clermont-Ferrand (F); CRMT/CNRS, Marseille (F); Chemistry Department, Wright State University, Dayton, OH (USA)
Main fields of interest: chemical thermodynamics; calorimetry (enthalpies of mixing, heat capacities); volumetric properties of liquids; gas solubilities; ultrasound and hypersound
Relevant categories in thermal analyses: fields (organic, solution chemistry (including aqueous solutions)); methods (adiabatic calorimetry, flow calorimetry, heat capacity, enthalpy of mixing)
Awards and acknowledgments: Felix-Kuschenitz Award in Thermophysics (Austrian Academy of Sciences, 1991); Hugh M. Huffman Award in Thermodynamics and Calorimetry (Calorimetry Conference, USA, 2002); Dr. h.c. (University Blaise Pascal, Clermont-Ferrand, F, 2006)
Professional activities: Solubility Data Series/IUPAC (Ed. Board, 1976–1981); IUPAC Commis-sion V.8, Solubility Data (1977–2001); IUPAC Commission I.2, Thermodynamics (1985–1993); IUPAC Committee on Legendre Transforms (1994–2001); IUPAC-CODATA Commission (1999–2003); Chemisch-Physikalische Gesellschaft, A, Secretary (1971–1975); Chemical Thermodynamics Advisory Group, DFG Germany (1979–1984); Calorimetry Conference, Board of Directors, USA (1997–1999, 2006–2014); International Council on Materials Education, Board (2001–); Fluid Phase Equilibria (Founding Ed. Board, 1977–1995); Materials Chemistry and Physics (Ed. Board, 1984–1988); International Data Series A (Ed. Board, 1990–1994); Thermochimica Acta (Ed. Board, 1993–2003); co-organizer of 24 international conferences
Publication record: papers (>175), books (2), citations (4116), h-index (32)
5 most important publications: [1] E. Wilhelm, R. Battino: J. Chem. Phys., 55 (1971) 4012–17; [2] E. Wilhelm, M. Zettler, H. Sackmann:, Ber. Bunsenges. Phys. Chem., 78 (1974) 795–804; [3] E. Wilhelm, R. Battino, R. J. Wilcock:, Chem. Rev., 77 (1977) 219–262; [4] E. Wilhelm in: Experimental Thermodynamics, Vol. VII, R. D. Weir and Th. W. de Loos, eds., Elsevier/IUPAC, Amsterdam, 2005, pp. 137–176; [5] E, Wilhelm: J. Solution Chem., 39 (2010) 1777–1818.

Name: Iwona Wilińska
Country: Poland
Date and place of birth: 1973, Łask, Poland
Present position and address: Assistant Professor, Warsaw University of Technology, Faculty of Civil Engineering, Mechanics and Petrochemistry, Institute of Chemistry, Łukasiewicza 17 St., PL-09-400 Płock, Poland
Email: iwona@pw.plock.pl
Education and scientific degrees: M.Sc. (1997) Warsaw University of Technology, Ph.D. (2002) University of Lodz
Workplaces: Warsaw University of Technology, Faculty of Civil Engineering, Mechanics and Petrochemistry, Institute of Chemistry, Płock, Poland (1998–)
Main fields of interest: inorganic chemistry, thermal analysis, cement materials, environmental protection
Relevant categories in thermal analyses: fields (inorganic, materials, minerals, cement); methods (TG, DTA, DSC, calorimetry, complementary methods e.g. XRD, FTIR, SEM)
Awards and acknowledgments: Five Scientific Awards of Rector of Warsaw University of Technology
Professional activities: Member (Secretary) of the Board of Polish Society of Calorimetry and Thermal Analysis (2006–2012), Member of Scientific or Organising Committees of conferences on thermal analysis and calorimetry (Seminars to the memory of Prof. St. Bretsznajder (2007, 2011), Conferences on Calorimetry and Thermal Analysis—CCTA (2009, 2012)), Guest Co-editor of special issue of Journal of Thermal Analysis and Calorimetry vol. 101, no 2, 2010, pp. 413–719
Publication record: papers (31), citations (202), h-index (9)
Equipments: SDT 2960 thermoanalyser (TA Instruments); BMR calorimeter constructed at the Institute of Physical Chemistry, Polish Academy of Sciences
5 most important publications: [1] Pacewska B., Bukowska M., Wilińska I., Swat M., Modification of the properties of concrete by a new pozzolan—a waste catalyst from the catalytic process in a fluidized bed. Cement and Concrete Research, 2002;32:145–152; [2] Pacewska B., Blonkowski G., Wilińska I., Investigations of the influence of different fly ashes on cement hydration. Journal of Thermal Analysis and Calorimetry, 2006;86:179–186; [3] Pacewska B., Wilińska I., Blonkowski G., Investigations of cement early hydration in the presence of chemically activated fly ash. Use of calorimetry and infrared absorption methods. Journal of Thermal Analysis and Calorimetry, 2008;93:769–776; [4] Pacewska B., Wilińska I., Bukowska M., Calorimetric investigations of the influence of waste aluminosilicate on the hydration of different cements. Journal of Thermal Analysis and Calorimetry, 2009;97:61–66; [5] Pacewska B., Nowacka M., Wilińska I., Kubissa W., Antonovich V., Studies on the influence of spent FCC catalyst on hydration of calcium aluminate cements at ambient temperature. Journal of Thermal Analysis and Calorimetry, 2011;105:129–140.

Name: Alina Wojakowska
Country: Poland
Date and place of birth: 1939, Krasnystaw, Poland
Present position and address: Professor at Wrocław Medical University, Alina Wojakowska, 54049 Wrocław, Piechowicka 45, Poland
Email: alina.wojakowska@umed.wroc.pl
Scopus ID: 6701361565
Education and scientific degrees: Wrocław University of Technology (1956–1962), Chemical Egineer and M.Sc.; Wrocław Medical University, Ph.D. (1972); Wrocław University of Technology, D.Sc. (1993)
Workplaces: Institute of Physical Chemistry, Polish Academy of Sciences, (1962–1964); Department of Inorganic Chemistry, Faculty of Pharmacy, Wrocław Medical University (1965–2007)
Main fields of interest:, thermodynamics, phase transitions, phase equilibrium diagrams, molten salts, new solid compounds, solid electrolytes, polymorphism in pharmaceuticals
Relevant categories in thermal analyses: fields (inorganic, materials, minerals, complex, pharmaceutical, glass, ceramics); methods (TG, DTA, DSC, cooling curves, high temperature cryometry, conductivity measurements vs. temperature, instrument development (for own purposes))
Awards and acknowledgments: Minister's Awards for Excellence in research achievements (1975, 1994); Medal of the Committee of National Education (2006); University Awards for outstanding research and teaching; Main achievements: Construction of a number of phase equilibrium diagrams for salt systems involving Cu(I), Ag, Tl(I), Sn(II), Co(II), Zn, Cd and alkali metal halides as well as the Ge-Sb-Sn system; Discovery of the high-temperature modification of $CoBr_2$ and about twenty new ternary compounds as well as a few Ag^+ and Cu^+ superionic conductors formed in the investigated systems; Determination of the activation energy for conduction of the α-CuBr displaying an unusual decrease in conductivity on melting; Description of interactions and structure of ionic liquids based on thermodynamic activity and electrical conductivity determinations)
Professional activities: Member of the Polish Society of Chemistry (PTCh) since 1974; Member of the Polish Society of Calorimetry and Thermal Analysis (PTKAT) since 1994, member of the Executive Board (1997–2009); Reviewer for a few journals
Publication record: papers (125), books (2), patents (1), citations (106), h-index (6-Scopus), sum of impact factors (35)
5 most important publications: [1] A. Wojakowska: J. Chim. Phys. Phys.-Chim. Biol., 90 (1993) 561, Propriétés thermodynamiques des mélanges binaires fondus d'halogénures d'étain(II) avec les halogénures de thallium(I), d'argent(I) et de cuivre(I); [2] A. Wojakowska, A. Górniak, A. Wojakowski: J. Chem. Eng. Data, 49 (2004) 1231, Thermodynamics and equilibrium phase diagrams of the zinc halide silver halide systems; [3] A. Wojakowska, E. Krzyżak: J. Therm. Anal. Calorim., 83 (2006), 597, Factors affecting the general shape of the phase diagram and compound formation in the binary copper(I) halide–alkali metal halide systems; [4] A. Wojakowska, S. Plińska, E. Krzyżak: J. Therm. Anal. Calorim., 90 (2007) 355, Conductometric and DSC evidence for a high temperature phase transition in cobalt dibromide; [5] E. Krzyżak, A. Wojakowska, A. Wojakowski, M. Wołcyrz: J. Therm. Anal. Calorim., 101 (2010) 541, Investigation of the CuBr–LiBr phase diagram.

Name: Shmuel Yariv
Country: Israel
Date and place of birth: 1934, Israel
Present position and address: Professor Emeritus at the Institute of Chemistry, The Hebrew University of Jerusalem, Jerusalem 91904, Israel
Email: shmuel.yariv@gmail.com
Website: http://www.huji.ac.il/dataj/controller/ihoker/MOP-STAFF_LINK?sno=728621&Save_t=
Education and scientific degrees: Chemistry and Geochemistry at the Hebrew University of Jerusalem, M.Sc. (1959); Ph.D. (1964)
Workplaces: Department of Inorganic and Analytical Chemistry, Institute of Chemistry at the Hebrew University of Jerusalem (since 1956)
Main fields of interest: thermo-visible-infrared-XRD and DTA of organo-clay complexes
Relevant categories in thermal analyses: fields (nano, minerals and clay minerals, organo-clay complexes, organic); methods (TG, DTA, EGA, DSC, Thermo-XRD, Thermo-IR and visible-spectroscopy)
Awards and acknowledgments: The Kurnakow Mesal of the Moscow Institute of Chemistry of the Soviet Academy of Science (1991); A special ICTAC Certificate in recognitaion of long and outstanding service as Secretaty of ICTAC (1996); The ICTAC Distinguish Service Award (2000); Special issue of the Journal of Thermal Analysis and Calorimetry, honoring Prof. Yariv (J Therm. Anal. (2003) Volume 71, Number 3. Guest Editor, Prof. Shlomo Shoval, pp. 685–1038). The Robert Mackenzie Memorial Lecture Award (2004)
Professional activities: Assistant Editor of the Israel Journal of Chemistry (1967–1971); Founder and Secretary of the Israel Group for Thermal Analysis (affiliated to the Israel Chemical Society, 1980); Chairman of the Israel Group for Thermal Analysis (1981–1984); Councilor of Israel in the Council of ICTAC (1980–1982); President of the Israel Clay Society (1984–1985); Secretary of the Council of the International Confederation for Thermal Analysis and Calorimetry, ICTAC (1982–1996); Founder and President of the Israel Group for Mechanochemistry, affiliated to the Israel Chemical Society (1992–1997); Councilor of Israel in the Council of the International Confederation for Mechanochemistry, INCOM (1993–2007); Councilor of Israel in the Council of the Organization of Mediterranean Countries for Calorimetry and Thermal Analysis (1998–2006); Councilor of Israel in the Council of the International Confederation for Thermal Analysis and Calorimetry, ICTAC (2000–2008). Member of the Editorial Boards of the following Journals: Journal of Tropical Agriculture (1984–1990); Journal of Thermal Analysis and Calorimetry (1987–2010); Journal of Colloid and Polymer Science (1990–2005)
Publication record: papers (250), books (2), patents (1)
Equipments: Philips Automatic Diffractometer (PW 1710) with a Cu tube anode
5 most important publications: [1] Yariv, S., (2002) IR Spectroscopy and Thermo-IR Spectroscopy in the Study of the Fine Structure of Organo-Clay Complexes. In: "Organo-clay complexes and interactions" (Yariv, S and Cross, H., eds.) Marcel Dekker Inc. New York. pp. 345–462; [2] Yariv, S., (2004) Applied Clay Science, 24:225–236; [3] Yariv, S. and Lapides, I. "The J. Therm. Anal. Calorim., 80 (2005) 11–26; [4] Yermiyahu, Z., Kogan, A., Lapides, I., Pelly, I., and Yariv, S., (2008) J. Therm. Anal. Calorim., 91:125–135; [5] Yariv, S., Lapides, I. and Borisover, M., (2012) J. Therm. Anal. Calorim., 110:385–394.

Name: Etsuo Yonemochi
Country: Japan
Date and place of birth: 1961, Niigata, Japan
Present position and address: Professor, Fine Drug Targeting Research Laboratory, Institute of Medicinal Chemistry, Hoshi University. 2-4-41 Ebara, Shinagawa-ku, Tokyo 142-8501, Japan
Email: e-yonemochi@hoshi.ac.jp
Website: http://polaris.hoshi.ac.jp/kyoshitsu/souzai/indexe.html
Researcher ID/ORCID: F-2364-2011/0000-0001-5255-5129
Education and scientific degrees: Faculty of Pharmaceutical Sciences, Chiba University (1981–1987); Pharmacist (1986); Ph.D. (1991); Professor (2013)
Workplaces: Faculty of Pharmaceutical Sciences, Chiba University (1987–1997), Faculty of Pharmaceutical Sciences, Toho University (1997–2013); Institute of Medicinal Chemistry, Hoshi University (2013–)
Main fields of interest: characterization of pharmaceutical products; Application of thermal method to pharmaceutical formulation
Relevant categories in thermal analyses: fields (pharmaceutical); methods (TG, DTA, DSC, microcalorimetry, ITC, and their simultaneous measurement)
Awards and acknowledgments: FIP Congress, Young Scientist Award (1993); Acad. of Pharm. Sci. Tech. Japan(APSTJ) Asahikasei Research Award (2001)
Professional activities: Advisor of Soc. of Cal. Therm. Anal., Japan; Member of public relations committee of Pharm. Soc. of Japan; Councilor of APSTJ and Japan Soc. Pharm. Machin. Eng.
Publication record: papers (187), books (12), patents (3)
Equipments: DSC, XRD-DSC, ITC
5 most important publications: [1] E. Yonemochi et al., J. Therm. Anal. Calorim, 113 (2013) 1505; [2] Y. Yoshihashi, E. Yonemochi, et al., J. Therm. Anal. Calorim., 99 (2010) 15; [3] E. Yonemochi et al., J. Thermal Anal. Calorim., 85 (2006) 693; [4] E. Yonemochi et al. Thermochim. Acta, 432 (2005) 70; [5] E. Yonemochi, 5.5 Medicine 5.5.2, 5.5.3.: Comprehensive Handbook of Calorimetry and Thermal Analysis, JSCTA Ed., Wiley (2004).

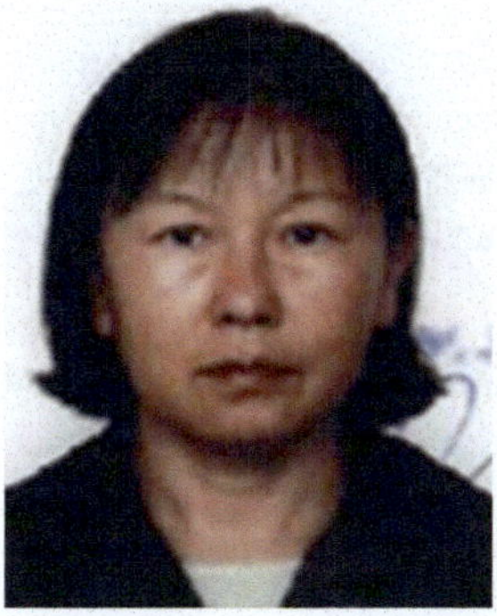

Name: Maria Irene Yoshida
Country: Brazil
Date and place of birth: 1952, Guaíra/SP, Brazil
Present position and address: Chemistry Department of Federal University of Minas Gerais (UFMG), CP.702, CEP:31.270-901, Belo Horizonte, Minas Gerais/Brazil
Email: mirene@ufmg.br
Education and scientific degrees: Chemical Engineer UFMG (1976); Ph.D. in Chemistry UFMG (1983); Associate Professor in Chemistry UFMG (since 1983).
Workplaces: Chemistry Department—Federal University of Minas Gerais (UFMG)—Belo Horizonte, Brazil
Main fields of interest: materials, industrial waste and thermal analysis
Relevant categories in thermal analyses: fields (inorganic, materials, nano, minerals, complex, organic, pharmaceutical, polymer, biology, ceramics, other (industrial waste)); methods (TG, DTA, DSC, thermomechanical analysis, kinetics, cryo, extremely high temperature (above 1,000 °C); calorimetry)
Awards and acknowledgments: Membership of ABRATEC (Associação Brasileira de An. Térmica e Calorimetria) since 1996; Willey prize (Gift Certificate—Wiley-VCH) in 11th CBPOL, 2011; Prize with Honorable Dimension in the 10th. International Congress of Odontology, 2008; Honorable dimension in 23nd. Annual SBPqO Meeting, 2006. Comission Adviser of COPAM (Conselho Estadual de Políticas Ambientais), 2006–2010
Professional activities: Coordinator of Thermal Analysis Laboratory of Chemistry Department UFMG since 1984. Cientific Comission of: CBRATEC (Congresso Brasileiro de Análise Térmica e Calorimetria) since 2000; Organizer Comission of XLIII CBQ Congresso Brasileiro de Química (2003), and of III Simpósio Mineiro de Química (2008); Referee of J. Thermal An. Calorimetry, Thermochimica Acta, Energy and Fuels, Analitica Chimica Acta, etc.
Publication record: papers (102), books (2), patents (3 deposited), citations (284)
Equipments: TG, DSC, TMA and DTA
5 most important publications: [1] Yoshida, M. I.; Silva, V. R.; Silva, M. C.; Sant´Anna S. S.; Carvalho, C. F. J. Therm. Anal. Calorim., 109 (2012) 1429; [2] Yoshida, M. I.; Oliveira M. A.; Gomes, E. C. L.; Mussel, W. N.; Soares, C. D.; Castro, V. D. J. Therm. Anal. Calorim., 106 (2011) 657; [3] Silva, V. R.; Moziewicki, M. A.; Silva, M. C.; Stefani, P. M.; Markovich, N. E., Yoshida, M. I. Polymer Testing, 32 (2013), 438; [4] Pereira, B. G.; Yoshida, M. I.; Soares, C. D.; Cryst. Growth Des., 7 (2007) 2017; [5] Gomes, E.C.L.; Mussel, W. N.; Resende, J. M.; Fialho, S. L.; Barbosa, J.; Yoshida, M. I. J. Braz. Chem. Soc., 4 (2013) 573.

Name: Zhiwu Yu
Country: China
Date and place of birth: 1962, Shanxi, China
Present position and address: Professor of Chemistry, Department of Chemistry, Tsinghua University, Beijing 100084, China
Email: yuzhw@tsinghua.edu.cn
Education and scientific degrees: BSc (1984) and M.Sc. (1987) at Tsinghua University; Ph.D. (1995) at London University; Professor (2002)
Workplaces: University of Illinois at Urbana-Champaign (1996–1998); Tsinghua University (1987–1992, 1998–present)
Main fields of interest: Development and employment of thermal and spectroscopic methods to address issues including phase behavior of model biomembrane systems, unfolding mechanism of proteins, and molecular interactions.
Relevant categories in thermal analyses: fields (complex, biology, life); methods (DSC, kinetics, microcalorimetry)
Awards and acknowledgments: Award from "the Teaching and Research Award Program for Outstanding Young Teachers in Higher Education Institutions of MOE, P.R.C" (TRAPOYT)
Professional activities: ICTAC Councilor (2001–2004); Chairman of the Committee of Chemical Thermodynamics and Thermal Analysis, the Chinese Chemical Society (2002–2006); Member of the Board of Directories, the Chinese Chemical Society (2002–present); Member of the Executive Committee, Acta Physico-Chimica Sinica (2009–present); Regional Editor for Asia, Spectroscopy, Biomedical Applications (2009–2012); Guest Editor, Journal of Thermal Analysis and Calorimetry (Vol.85, No.3, 2006). Secretory of the IUPAC-sponsored International Conference on Chemical Thermodynamics and Calorimetry, Beijing (1989); Vice Chair of the 18th International Conference on Chemical Thermodynamics, Beijing (2004); Chair of the 4th International and 2nd Asian Symposium on Two Dimensional Correlation Spectroscopy, Beijing (2007)
Publication record: papers (99), citations (1348), h-index (19)
Equipments: DSC, ITC
5 most important publications: [1] Z. W. Yu, N. M. Tsvetkova, L. I. Tsonev, P. J. Quinn: Biochim. Biophys. Acta, 1237 (1995) 135; [2] W. Y. Gao, Y. W. Wang, L. M. Dong, Z. W. Yu: J. Therm. Anal. Calorim., 85 (2006) 785; [3] R. G. Wu, L. Chen, Z. W. Yu, P. J. Quinn: Biochim. Biophys. Acta, 1758 (2006) 764; [4] F. G. Wu, N. N. Wang, Z. W. Yu: Langmuir, 25 (2009) 13394; [5] J. J. Luo, F. G. Wu, J. S. Yu, R. Wang, Z. W. Yu: J. Phys. Chem. B, 115 (2011) 8901.

Name: Maria Magdalena Zaharescu
Country: Romania
Date and place of birth: 1938, Cluj-Napoca, Romania
Present position and address: Head of Department, "Ilie Murgulescu" Institute of Physical Chemistry of the Romanian Academy, 202, Splaiul Indepedndetei, 60021 Bucharest, Romania
Email: mzaharescu@icf.ro
Website: www.icf.ro
Education and scientific degrees: Faculty of Chemistry, "Babes-Bolyai" University of Cluj-Napoca, Romania, bachelor degree, 1959; Institute of Chemistry Cluj-Napoca, Romania, Ph.Degree, 1971, Laboriet of Silikatetecknik, University of Lyngby, Demnark, post doctoral scholarship, 1974
Workplaces: "Carbochim" factory (1959–1964), Center of Inorganic Chemistry of the Romanian Academy (1964–1969), Center of Physical Chemistry of the Romanian Academy (1969–1990), "Ilie Murgulescu" Institute of Physical Chemistry of the Romanian Academy, Bucharest, Romania, 1990-present
Main fields of interest: physical chemistry of the oxide systems (reaction mechanisms, phase equilibria studies, structure-properties correlations, glass crystallization studies); sol-gel science, research field that she initiated in Romania
Relevant categories in thermal analyses: fields (oxide and inorganic-organic hybrid materials, nanomaterials, glass, ceramics); methods (TG, DTA, EGA, DSC, extremely high temperature (above 1,000 °C), dilatometrry, high themperature XRD)
Awards and acknowledgements: "Gheorghe Spacu" award of the Romanian Academy for "Phase equilibria studies in oxide systems"—1971; "Scientific Medal"—1982 granted by the Romanian Government; "Scientific Order" 3-rd degree—1983, granted by the Romanian Government; Corresponding Member of the Romanian Academy, since 2001; Honorary Medal "Gheorghe Spacu" of the Romanian Chemical Society—2007
Professional activities: Vice-president of the Romanian Ceramic Society, Member of the Romanian Chemical Society, of the American Ceramic Society, of the International Sol-Gel Society and of the Editorial Board of the Journal of Sol-Gel Science and Technology
Publication record: papers (238), books (8 book chapters), patents (3), citations(1770), h-index (20)
Equipments: TG/DTA apparatus Mettler Toledo GA/SDTA851e coupled with FT-IR Nicolete 6700, Differential Scanning Calorimeter DSC823e Mettler Toledo, X-ray diffraction apparatus Rigaku last IV, with HT 1500 high temperature attachment
5 most important publications: [1] K. A. Simonsen, M. Zaharescu, J. Thermal Anal., 15, 25–35 (1979); [2] A. Brăileanu, M. Zaharescu, D. Crisan, E. Segal, Thermochimica Acta, 2565, 1–13 (1995); [3] M. Zaharescu, A. Jitianu, A. Brăileanu, J. Madarász, Cs. Novák, G. Pokol, J. Therm. Anal. Calorim., 71, 421–428 (2003); [4] O. C. Mocioiu, M. Zaharescu, G. Jitianu, P. Budrugeac, J. Therm. Anal. Calorim., 86, 429–436 (2006); [5] L. Predoană, B. Malic, M. Zaharescu, J. Therm. Anal. Calorim., 98, 361–366 (2009).

Name: Mohamed I. Zaki
Country: Egypt
Date and place of birth: 1950, El-Minia, Egypt
Present position address: Professor Emeritus of Physical Chemistry, Chemistry Department, Faculty of Science, Minia University, El-Minia 61519, Egypt
Email: mizaki@link.net, mizaki@mu.edu.e.g.
Website: www.sccmu.org, www.researchgate.net/profile/Mohamed_Zaki5/
Education and scientific degrees: B.Sc. (1971); M.Sc. (1975); Ph.D. (1979); Professor (1987); Professor Emeritus (2010)
Workplaces: 1971–1975, Assiut University/Egypt; 1975–1993, Minia University/Egypt; 1993–2001, Kuwait University/Kuwait; 2001-to date, Minia University/Egypt
Main fields of interest: synthesis, characterization and development of catalytic solids; surface chemistry and catalysis
Relevant categories in thermal analyses: fields (inorganic; metal oxides; catalysts); methods (TGA/TGR/DTG, DTA, DSC and TPR/TPO/TPD)
Awards and acknowledgments: The 1987 State Encouragious Prize in chemical sciences, Academy of Science and Technology, Cairo/Egypt; First-class Ripon of Excellence in Sciences and Arts, 1995, The Presidency of Egypt; The 2013/2014 Minia University Meritiorious Prize, El-Minia/Egypt
Professional activities: Fellow, The Alexander von Humboldt-Stiftung, Bonn/Germany; Fellow, The Royal Society of Chemistry (CChem, FRSC), London/UK; Alumni, The Fulbright Foundation, Washington/USA; Member, The Egyptian Chemical Society, Cairo/Egypt; Academic Expert, Tempus II/III, EU, Bruxel/Belgium, 2002–2006; Research Associate, CNRS, Universite du Poitiers/France, 1991/92 and 2003; Research Associate, TWAS/CSIR, NCL at Pune/India, 2007
Publication record: papers (172), books (4), patents (2), citations (3813), h-index (32)
Equipments: A Stand-alone Shimadzu TG/DTA/DSC Analyzer (Japan); Quantchrome ChemBET 3000 TPR/TPO/TPD (USA)
5 most important publications: [1] Acid properties of silica and alumina surfaces as probed by thermoravimetry and differential scanning calorimetry of temperature programmed desorption of pyridine, H. M. Ismail, S. A. A. Mansour, M. I. Zaki, Thermochimica Acta, 202, 269 (1992); [2] Thermochemistry of manganese oxides in reactive gas atmospheres: probing catalytic MnOx compositions in the atmosphere of $CO+O_2$, M. I. Zaki, M. A. Hasan, L. Pasupulety, K. Kumari, Thermochimica Acta, 311, 97 (1998); [3] Soot deep oxidation by molybdena and molybdates: A thermogravimetric investigation, M. A. Hasan, M. I. Zaki, K. Kumari, L. Pasupulety, Thermochimica Acta, 320, 23 (1998); [4] Kinetics of formation of barium tungstate in equimolar powder mixture of $BaCO_3$ and WO_3: Thermogravimetric and spectroscopic studies, L. A. Al-Hajji, M. A. Hasan, M. I. Zaki, Journal of Thermal Analysis and Calorimetry, 43, 100 (2010); [5] Temperature-programmed and X-ray diffractometry studies of hydrogen reduction course and products of WO_3 powder: Influence of reduction parameters, M. I. Zaki, N. E. Fouad, S. A. A. Mansour and A. I. Muftah, Thermochimica Acta, 523, 90 (2011).

Name: Dragana Živković
Country: Serbia
Date and place of birth: 1965, Zaječar, Serbia
Present position and address: Full professor at Technical Faculty in Bor, University of Belgrade, VJ 12, 19210 Bor, Serbia
Email: dzivkovic@tf.bor.ac.rs
Website: www.tf.bor.ac.rs
Researcher ID/ORCID: A-4616-2014/0000-0002-2745-5676
Education and scientific degrees: Metallurgical Engineer, University of Belgrade, Technical Faculty in Bor (1989); Mr Sc. (1993); Dr Sc. (1995); Full Prof. (2005); Corresponding Member of the Academy of Engineering Sciences of Serbia (2007)
Workplaces: University of Belgrade, Technical Faculty in Bor, since 1989
Main fields of interest: thermodynamics of alloys; phase equilibria of the multicomponent metal systems; kinetics of the processes in the multiphase systems in extractive metallurgy
Relevant categories in thermal analyses: fields (alloys, materials, inorganic, minerals, complex); methods (calorimetry, DTA, TG, DSC, kinetics)
Awards and acknowledgments: University of Belgrade Award: 1989; Serbian Chemical Society Award: 2001
Professional activities: Chief of the Group of the Metallurgical Engineering, Technical Faculty in Bor (since 2004); vice-dean for scientific work and international collaboration at Technical Faculty in Bor (since 2009); Journal of Mining and Metallurgy, Section B—Metallurgy: Co-Editor (1997–2012) and Editor-in-Chief (since 2012); national representative in Associated Committee for Phase Diagrams and Thermodynamics (since 1999); secretary of the national Committee for Thermodynamics and Phase Diagrams (since 2000); partner of the Research Center for Innovative Materials Design and Application, Powder Metallurgy Research Institute, Central South University, Chagsha, Hunan, China (since 2008)
Publication record: papers (171), books (7), h-index (10), citations (523)
Equipment: Oelsen calorimeter, Q600 SDT TA Instruments
5 most important publications: [1] Živković, D., Manasijević, D., Živković, Ž., Thermodynamic study of Ga-Sn and Ga-Zn systems using quantitative differential thermal analysis (2003) Journal of Thermal Analysis and Calorimetry, 74 (1), pp. 85–96; [2] Živković, D., Katayama, I., Kostov, A., Živković, Ž, Comparative thermodynamic study of GaSb-Sn system (2003) Journal of Thermal Analysis and Calorimetry, 71 (2), pp. 567–582; [3] Živković, D., Manasijević, D., Živković, Ž., Comparative thermodynamic investigation of binary Ga-Bi system: Experimental determination of enthalpies of mixing and activity estimation for liquid Ga-Bi alloys (2005) Journal of Thermal Analysis and Calorimetry, 79 (1), pp. 71–77; [4] Živković, D., Katayama, I., Gomidželović, L., Manasijević, D., Novaković, R., Comparative thermodynamic study and phase equilibria of the Bi–Ga–Sn ternary system (2007) International Journal of Materials Research, 98 (10), pp. 1025–1030; [5] Živković, D., Gomidželovic, L., Manasijevič, D., Talijan, N., Ćosović, V., Calorimetric study and phase diagram investigation of the Au-Ga system (2013) International Journal of Materials Research, 104 (6), pp. 554–560.

Name: Živan D. Živković
Country: Serbia
Date and place of birth: 1949, Leskovac, Serbia
Present position and address: Professor and head of the industrial management department; Technical Faculty in Bor, 19210 Bor, Serbia
Email: zzivkovic@tf.bor.ac.rs
Website: www.tf.bor.ac.rs
Education and scientific degrees: Metallurgical engineer, University of Belgrade (1973); M.Sc. (1975), D.Sc. (1979), both at the University of Ljubljana
Workplaces: University of Belgrade, Technical Faculty in Bor, Serbia (1973–)
Main fields of interest: Thermoanalytical and structural study of complex compounds; thermodynamics of alloys; kinetics of thermal decomposition of complex compounds
Relevant categories in thermal analysis: fields (minerals, inorganic materials and metals); methods (TG, DTG, DTA, DSC; calorimetry and instruments development)
Awards and acknowledgments: Silver medal from Presidency of Yugoslavia, October Prize from City of Bor, Sistek prize from Holding Company RTB Bor, Plaque from University of Belgrade and Technical Faculty in Bor
Professional activities: President of the Serbian Committee for Thermodynamics and Phase diagrams (2000–); Co-president of the Associated Committee for Phase Diagrams and Thermodynamics: Poland, Bulgaria, Czech Republic, Hungary, Slovakia and Serbia (2000–2010); Editor-in-Chief of the Journal of Mining and Metallurgy Section B Metallurgy (1997–2012); Dean of Faculty in Bor (1985–1995); Head of the metallurgical Department (1995–2002); Head of the industrial management department (2002–). Membership conference committee: 12th Molten salt, Beijing, China 2012; 1nd Central and Eastern European Conference on Thermal Analysis and Calorimetry, Crajova, Romania (2012); 2nd Central and Eastern European Conference on Thermal Analysis and Calorimetry Vilinius, Lithuania (2013), etc.
Publication record: papers (140); books (21); citations (314) h-index (10)
Equipments: TG, DTA, DSC, Oelsen calorimeter
5 most important publications: [1] Ž. D. Živković. Determinations of the calibration constang gKs in the quantitative differential thermal analysis, Thermochimica Acta 34 (1979) 91; [2] Ž. D. Živković, Thermal decomposition of hexamine cobalt(II) chloride, J. Thermal Anal., 41 (1994)99; [3] Ž. D. Živković, J. Šestak, Kinetics and mechanism of the MoS oxidation process, J. Them. Anal. Calorim., 53 (1998) 263; [4] Ž. D. Živković, Thermal Analysis, Bor (1984); [5] Ž. D. Živković, B. Dobovišek, Differential thermal analysis, Bor (1984).

Instrument and Parts Manufacturing Companies

Imre Miklós Szilágyi and György Liptay

I.M. Szilágyi (✉) · G. Liptay
Department of Inorganic and Analytical Chemistry, Budapest University of Technology and Economics, Szent Gellért tér 4., H-1111, Budapest, Hungary
e-mail: imre.szilagyi@mail.bme.hu

G. Liptay
e-mail: liptay.g@mail.bme.hu

I.M. Szilágyi and G. Liptay (eds.), *Who is Who in Thermal Analysis and Calorimetry*, Hot Topics in Thermal Analysis and Calorimetry 10,
DOI 10.1007/978-3-319-09486-1_2

METTLER TOLEDO

Company's name: METTLER TOLEDO
Year of foundation: 1945
Postal address: Mettler-Toledo AG, Im Langacher, CH-8606 Greifensee
Email address: info@mt.com
Telephone number: ++41 1 944 77 11
Home page: www.mt.com

Short History: In 1945, a Swiss engineer named Dr. Erhard Mettler started his own precision mechanics company in Küsnacht, a town on the Lake of Zurich in Switzerland. He invented the substitution principle, and developed a single-pan balance capable of being produced in series. Analytical balances with a single weighing pan gradually replaced conventional two-pan balances in the laboratory. Today METTLER TOLEDO specializes in the area of precision instruments for professional use, and is the world's largest supplier of weighing systems, which cover a weighing range from 0.0000001 g up to a thousand tons. The company is also a market leader in a number of related measurement technologies. It offers analytical instruments, systems for automated drug discovery, process analytics instruments, and end-of-line inspection systems for the packaging industry. Besides a large product offering, METTLER TOLEDO serves global customers on a worldwide basis with the largest sales and service network in the industry.

Thermal analysis has played an important role since the early 1960s. Since 1964, METTLER TOLEDO have offered customers innovative thermal analysis solutions, products and services.

Dr. Mettler, the founder of Mettler, strongly promoted thermal analysis. In the early 1960s, Dr. Hans-Georg Wiedemann, an East German scientist, convinced him to combine the advanced weighing technology with a furnace in order to measure weight changes as a function of temperature. This led to the introduction, in 1964, of the first commercially available TGA/DTA system, the TA1.

Main Products: Balances, Thermo Analyzers, Moisture Analyzers, Titrators, pH-Meters, Density-Meters, Refractometers and Pipettes.

Field of Applications

In the Laboratory: Laboratories are a key market for the company. METTLER TOLEDO instruments are used in research, scientific, drug discovery, and quality control labs, for the pharmaceutical, chemical, food, and cosmetics industries, as well as many others. In fact, the company is a global market leader with the three instrument groups most frequently used in the laboratory balances, pipettes, and pH meters—and is a pioneer in the field of—automated chemistry.

In Industry: METTLER TOLEDO supports its customers with comprehensive industrial solutions that cover the various steps in a host of manufacturing processes. Solutions range from receiving raw materials, through various manufacturing processes, in-line process control, and end-of-line packaging control, to logistics and shipping.

Increasingly, these solutions are fully integrated into the customer's IT environment, helping automate their workflows.

In Retail: In food retailing, the company's offering for the management of fresh food ranges from receiving and prepackaging, to in-store solutions for self-service departments, deli counters, and checkout terminals.

METTLER TOLEDO products can all be networked, and collect data at the item level. METTLER TOLEDO software solutions assist customers with pricing and inventory management, and also provide valuable information for merchandising decisions.

List of Important Scientific Publications

Many applications published in the bi-annual UserCom: www.mt.com/ta-usercoms

H. Staub, W. Perron, Isothermal step method for purity determination, Analytical Chemistry, Vol. 46 (1974), p. 128

G. Widmann, R. Riesen, Thermal Analysis, Terms, Methods, Applications, Dr. A. Hütig Verlag, Heidelberg, 1987

H. G. Wiedemann, The investigation of ancient oriental materials and artifacts by thermal analysis, Thermochimica Acta, 148 (1989), p. 95

H. G. Wiedemann, S. Felder-Casagrande, Thermomicroscopy, Chap. 10 in Handbook of Thermal Analysis and Calorimetry, Vol. 1: Principles and Practice, M. E. Brown (Ed.), Elsevier Science B. V., 1998

M. Schubnell, J. E. K. Schawe, Quantitative determination of the specific heat and the glass transition of moist samples by temperature modulated differential scanning calorimetry, Int. J. of pharmaceutics, 217 (2001), p. 173

J. E. K. Schawe, T. Hütter, C. Heitz, I. Alig, D. Lellinger, Stochastic temperature modulation: a new technique in temperature-modulated DSC, Thermochimica Acta, 446 (2006), 147.

Number of Representations All over the World: The company serves its customer base with the largest sales and service network in the industry. Its own sales and service organizations around the globe ensure that the company has factory-trained, experienced, and dedicated specialists serving its customers. In those countries where METTLER TOLEDO has no service organizations of its own, it works in close cooperation with long-standing partners. METTLER TOLEDO products and services are available in more than a hundred countries.

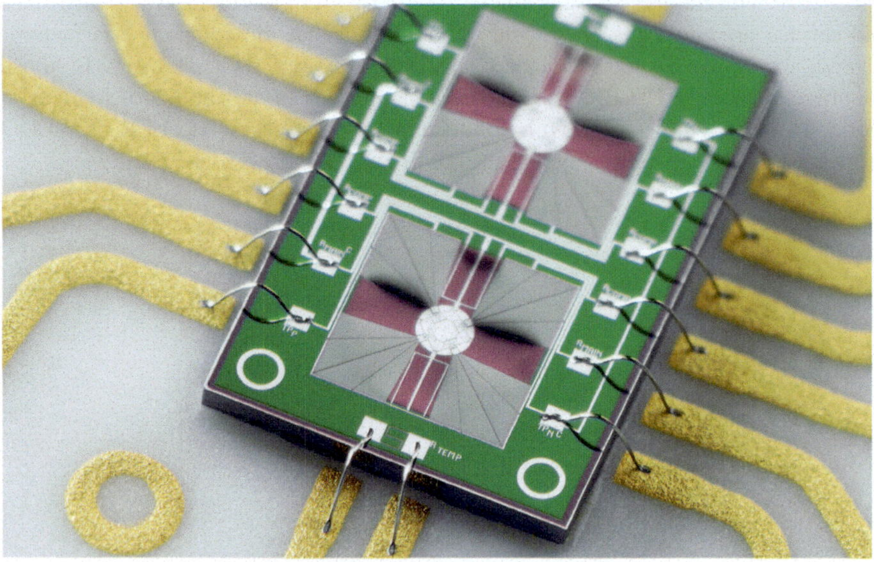

Name: Jürgen E. K. Schawe
Country: Germany
Date and place of birth: 1959, Lübz, Germany
Present position and address: Senior application scientist, Business Unit Materials Characterization at Mettler-Toledo AG. Mettler-Toledo AG, Sonnenbergstrasse 74, CH-8603 Schwerzenbach, Switzerland
Email: juergen.schawe@mt.com
Education and scientific degrees: Ph.D. in solid state physics (1984)
Workplaces: Pedagogical University of Güstrow, Department of mathematics and physics (1986–1990); University of Rostock, Polymer physics group (1990–1992); University of Ulm, Section of Calorimety (1992–1998), Mettler-Toledo AG, Schwerzenbach (1999–)
Professional activities: Member of DPG, GEFTA, NATAS
Awards and acknowledgements: Award for "Applied Chemical Thermodynamics" of the Swiss Society of Thermal Analysis and Calorimetry (STK) (2010)
Main fields of interest: Thermal analysis instrumentation, temperature modulated techniques, fast scanning DSC, structure formation in metastable materials, interplay between physical behavior and chemical reaction kinetics
Publication record: Papers (54), patents (4), citations (1453)
5 most important publications: [1] J. E. K. Schawe: A comparison of different evaluation methods in modulated temperature DSC, Thermochim. Acta, 260 (1995) 1; [2] J. E. K. Schawe, G. R. Strobl: Superheating effects during the melting of crystallites of syndiotactic polypropylene analyzed by temperature-modulated differential scanning calorimetry, Polymer, 39 (1998) 3745; [3] J. E. K. Schawe, A description of chemical and diffusion control in isothermal cure kinetics, Thermochim. Acta, 388 (2002) 299; [4] J. E. K. Schawe, T. Hütter, C. Heitz, I. Alig, D. Lellinger, Stochastic temperature modulation: a new technique in temperature modulated DSC, Thermochim. Acta, 446 (2006) 147; [5] J. E. K. Schawe, A quantitative DSC analysis of the metastable phase behavior of the sucrose-water system, Thermochim. Acta, 451 (2006) 116.

NETZSCH

Company's name: NETZSCH-Gerätebau GmbH
Year of foundation: 1962
Postal address: Wittelsbacherstr. 42 D-95100, Selb/Bavaria, Germany
Phone: +49 9287 881-0, **Fax**: +49 9287 88 11 44
Contact person: Dr. Thomas Denner, Dr. Jürgen Blumm
Email: at@netzsch.com
Website: www.netzsch-thermal-analysis.com
Brief history: NETZSCH-Gerätebau GmbH is part of the NETZSCH Group. The business dates back as far as 1873, when brothers Thomas and Christian Netzsch founded a machine-building company in Selb, Germany. Today NETZSCH employs 3,100 people at more than 20 production and 140 sales and service sites in 35 countries around the world.

The company's involvement in Thermal Analysis began in 1952 with the development and production of DTA equipment, dilatometers and some other testing instruments for ceramic raw materials and products. In 1962, NETZSCH-Gerätebau GmbH became an independent company, now referred to as the 'Analyzing & Testing' business unit within the NETZSCH Group. About 400 people are employed worldwide for the development, production, sales and application of a very broad range of methods and instruments for thermal analysis, thermophysical properties and refractory material testing.

For years, NETZSCH-Gerätebau GmbH has ranked among the worldwide market leaders in thermal analysis as a high-tech enterprise. We provide the groundwork for progress in material properties and production technologies with our applications and testing know-how for a variety of fields, including high-temperature superconductors, polymer- and ceramic-based fuel cells, hydrogen storage, space research, and nano-materials and -technology. Besides the headquarters in Selb, Germany, we also have development, production and assembly sites in Yokohama, Japan, in Boston, USA and in Krakow, Poland. NETZSCH-Gerätebau GmbH has several well equipped application laboratories for demonstration or contract testing and for customer training courses.

Main Products

- Dilatometers
- Thermomechanical analyzers: −260 to 2,800 °C
- DTA: −150 to 2,400 °C
- DSC: −180 to 700 °C; −120 to 1,650 °C
- DMA: −170 to 600 °C
- TGA: −150 to 2,400 °C
- STA (TG-DSC or TG-DTA): −150 to 2,400 °C
- TCT (thermal conductivity testers): −150 to 1,500 °C

- LFA (flash diffusivity methods): −120 to 2,800 °C
- RUL (refractory under load tester): 25 to 1,700 °C
- CIC (creep in compression tester): 25 to 1,700 °C
- HMOR (hot modulus of rupture): 25 to 1,500 °C
- ARC/APTAC/MMC: 25 to 500 °C

Application fields: Polymers, composites, rubbers, pharmaceuticals, food, biomaterials, ceramics, metals, fuels, friction materials, paints, automotive, electronics.
5 most important publications: [1] W.-D. Emmerich, J. Hayhurst, E. Kaisersberger: High temperature dilatometer study of special ceramics and their sintering kinetics; Thermochim. Acta 106 (1986), 71; [2] W.-D. Emmerich, E. Kaisersberger: Simultaneous TG-DTA mass-spectrometry to 1,550 °C; J. Themal. Anal., 17 (1979), 197; [3] E. Kaisersberger, W.-D. Emmerich: New TA-MS coupling system with increased sensitivity for low volatile materials, Thermochim. Acta, 85 (1985), 275; [4] J. Blumm, E. Kaisersberger: Accurate measurement of transformation energetic and specific heat by DSC in the high-temperature region, J. Therm. Anal. Calorim., 64 (2001), 385; [5] J. Opfermann, J. Blumm, W.-D. Emmerich: Simulation of the sintering behavior of a ceramic green body using advanced thermokinetic analysis, Thermochim. Acta, 318 (1998), 213.
Number of representations all over the world: Over 32 NETZSCH sales and service offices in 18 countries in the Americas, Asia and Europe.

Materials have a lot to say
Learn their language with solutions from NETZSCH
NETZSCH is recognized world-wide as a technological leader in the thermal characterization of materials. Thanks to our experienced team, NETZSCH Analyzing & Testing offers the most complete product line, the widest temperature range and the most diverse options, to name a few. The series of patents and international R&D awards which we have received attest to our products' leadership in terms of technique and quality.
Besides the products themselves, NETZSCH also provides localized software, operation manuals and application references. Seminars and users' meetings are also organized by our specialists, and we offer a series of advanced training programs on demand.
NETZSCH
NETZSCH-Gerätebau GmbH
Wittelsbacherstraße 42
95100 Selb
Germany
Tel.: +49 9287 881-0
Fax: +49 9287 881 505
at@netzsch.com
www.netzsch.com

Name: Juergen Blumm
Country: Germany
Date and place of birth: 1968, Koenigshofen i. Grabfeld
Email: juergen.blumm@netzsch.com
Website: www.netzsch.com
Education and scientific degrees: 1995 Dipl.-Phys at the University of Würzburg; 2003 Ph.D. at the University of Würzburg
Workplaces: 1994–1995 Center of Applied Energy Research; 1995—present Netzsch Geraetebau GmbH
Main fields of interest: Thermal conductivity testing, material characterization at high temperatures, technological improvements on thermal instrumentation
Relevant categories in thermal analysis: Fields (ceramics, refractories, metals, building materials, polymers, composites, insulating materials, thermoelectrics, standardization); methods (Transient and stationary methods for thermal conductivity measurements, dilatometry, simultaneous thermal analysis, calorimetry)
Professional activities: VDMA Member of the management board in Ceramic and Building Materials Machines; Member of the advisory boards of the Ceramitec and Analytica exhibitions of the Munich Exhibition Center
Publication record: Papers (41), books (5), patents (4), citations (450)
Equipments: Netzsch LFA 447/427, DSC 204/404, DIL 402, TMA 402, GHP 456, HFM 436, EGA
5 most important publications: [1] Effect of substitutions on the thermoelectric figure of merit of half-Heusler phases at 800 C, S. J. Poon, N. Hickman, T. M. Tritt, J. Blumm—J. Applied Physics, 2006; [2] Heat capacity of α-GaN: Isotope effects, R. K. Kremer, M. Cardona, E. Schmitt, J. Blumm—Physical Review B, 2005; [3] Improvement of the mathematical modeling of flash measurements, J. Blumm, J. Opfermann—High Temperatures. High Pressures, 2002; [4] Thermal conductivity of a volcanic rock material (olivine-melilitite) in the temperature range between 288 and 1,470 K, R. Büttner, B. Zimanowski, J. Blumm—Journal of Volcanology, 1998; [5] Accurate measurement of transformation energetics and specific heat by DSC in the high-temperature region, J. Blumm, E. Kaisersberger—Journal of Thermal Analysis and Calorimetry, 2001.

Name: Ekkehard Füglein
Country: Germany
Date and place of birth: 1967, Neustadt/Weinstraße, Germany
Present position and postal address: Senior Scientist, NETZSCH-Gerätebau GmbH, Applications Laboratory, Wittelsbacherstraße 42, 95100 Selb, Germany
Email: ekkehard.fueglein@netzsch.com/ekkehard.fueglein@th-nuernberg.de
Website: www.netzsch-thermal-analysis.com
Education and scientific degrees: Chemist, Julius Maximilians University Würzburg, Lehrstuhl für Silicatchemie und Materialwissenschaft (1994), Dr. rer. nat., Inorganic Chemistry/Solid State Chemistry, University Würzburg (1998)
Workplaces: Lehrstuhl für Silicatchemie und Materialwissenschaft at Julius Maximilians University Würzburg (1994–1998), Department of Inorganic Chemistry at Technical University Vienna (1998), Linseis Messgeräte GmbH, Selb, Germany (1998–2002), NETZSCH-Gerätebau GmbH, Selb, Germany (2002–)
Main fields of interest: Thermoanalytical instrumentation, coupling techniques, solid state chemistry, solid state kinetics
Relevant categories in thermal analysis: Fields (inorganic, materials, ceramics, polymer, food); methods (TG, DSC, EGA, kinetics, extremely high temperature, specific heat, instrument development)
Professional activities: Member of DKG (German ceramic society), Member of GDCh (German chemical society), Member of GEFTA (society for thermal analysis of German speaking countries), Member of STK (Swiss society of thermal analysis and calorimetry), Member of ICTAC (international confederation on thermal analysis and calorimetry), lecturer at faculty of applied chemistry at Technische Hochschule Nürnberg Georg-Simon-Ohm, Germany, lecturer at Instituto de Materiais Cerâmicos (IMC), Universidade de Caxias do Sul (UCS), Brazil
Publication record: Papers (29), books (7 book chapters), citations (238)
Equipment: Full access to NETZSCH applications laboratory, Selb, Germany
5 most important publications: [1] M. Lerch, E. Füglein, J. Wrba: "Synthesis, Crystal Structure and High Temperature Behaviour of Zr_3N_4", Z. allg. anorg. Chem. 1996, 622, 367; [2] E. Füglein, U. Schubert, "Formation of Mg_2Si from Solid Silicon Monoxide, and Solid-State Comproportionation between Mg_2Si and SiO", Chem. Mater. 1999, 11, 865; [3] E. Marti, E. Kaisersberger, E. Füglein, "Multiscan Differential Scanning Calorimetry (MTSDSC): Thermophysical Procedures for Research, Development, and Quality Control of Substances and Materials", J. Therm. Anal. Calorim. 2010, 101, 1189; [4] E. Füglein, D. Walter,

"Thermal Analysis of Lanthanum Hydroxide", J. Therm. Anal. Calorim., 2012, 110; [5] A. Schindler, G. Neumann, A. Rager, E. Füglein, J. Blumm, T. Denner, "A Novel Direct Coupling of Simultaneous Thermal Analysis (STA) and Fourier Transform-Infrared (FT-IR) Spectroscopy", J. Therm. Anal. Calorim., 2013, DOI 10.1007/s10973-013-3072-9.

Name: Jan Hanss
Country: Germany
Date and place of birth: 1965, Aabenraa, Denmark
Present position and address: Head of Applications Laboratory, NETZSCH-Gerätebau GmbH, Applications Laboratory, Wittelsbacherstraße 42, 95100 Selb, Germany
Email: jan.hanss@netzsch.com
Website: www.netzsch-thermal-analysis.com
Education and scientific degrees: Chemist, University Hamburg (1992), Dr. rer. nat., University Hamburg (1997)
Workplaces: Research assistant at the University of Hamburg (1992–1996, Prof. Krüger), scientist at the University of Hamburg (1997–1998, Prof. Krüger), scientist at the Gerhard-Mercator-University of Duisburg (1998–1999, Prof. Henkel), permanent position at the solid chemistry chair University of Augsburg (1999–2012, final position: Akad. Oberrat), NETZSCH-Gerätebau GmbH, Selb, Germany (2012–)
Main fields of interest: Thermoanalytical instrumentation, coupling techniques
Relevant categories in thermal analysis: Fields (inorganic materials, ceramics, polymer); methods (TG, DSC, STA, EGA, instrument development)
Awards and acknowledgments: 2002 Netzsch-GEFTA Förderpreis
Professional activities: Member of GEFTA (society for thermal analysis of German speaking countries)
Publication record: Papers (21)
Equipment: Full access to NETZSCH applications laboratory, Selb, Germany
5 most important publications: [1] L. Patron, O. Carp, I. Mindru, G. Marinescu, J. Hanss, A. Reller: "Thermal Analysis of some Polynuclear Coordination Compounds as Precursors of Iron Granates ($M_3Fe_5O_{12}$, M = Y^{3+} or Er^{3+})", J. Therm. Anal. Calorim. 2008, 92, 307–312; [2] K. Sommer, S. Andreß, A. Kakschke, R. Wieczorek, S. Hanisch, J. Hanss: „Vanillezucker oder Vanillinzucker“, Chemkon 2009 16 (1), 19–29; [3] J. Barth, G. H. Fecher, M. Schwind, A. Belenau, C. Felser, A. Shkabko, A. Weidenkaff, J. Hanss, A. Reller, M. Köhne: "Investigation of Thermoelectric Properties of LiAlSi and LiAlGe", J of Electronic Materials, 2010, 39, 1856–1860; [4] D. Denysenko, M. Grzywa, M. Tonigold, B. Streppel, I. Krkljus, M. Hirscher, E. Mugnaioli, U. Kolb, J. Hanss, D. Volkmer: "Elucidating Gating Effects for Hydrogen Sorption in MFU-4-Type Triazolate-Based Metal–Organic Frameworks Featuring Different Pore Sizes", Chem. Eur. J. 2011, 17, 1837–1848; [5] M. Grzywa, D. Denysenko, J. Hanss, E.-W. Scheidt, W. Scherer, M. Weil, D. Volkmer: "CuN_6 Jahn-Teller centers in coordination frameworks comprising fully condensed Kuratowski-type secondary building units: phase transitions and magneto-structural correlations" Dalton Trans. 2012, 41, 4239–4248.

PerkinElmer

Company's name: PerkinElmer
Website: www.perkinelmer.com
Company profile: At PerkinElmer, we're taking action to ensure the quality of our environment. Our Environmental Health business develops analytical instrumentation, and support services, all to protect the quality and sustainability of our environment and security of people within their surroundings. PerkinElmer transforms information into actions that improve our safety today and our world tomorrow.

PerkinElmer has been a manufacturer of thermal analysis instruments for over 50 years and developed the power compensated Differential Scanning Calorimeter. We currently offer a complete line of thermal analyzers as well as a wide range of other products both the analytical and life sciences. This wide range of products allows us to also over an extensive suite of hyphenated systems like TG-IR, TG-MS, TG-GCMS, etc from one company.

Name: Kevin P. Menard
Country: USA
Date and place of birth: 1955, Bridgeport CT
Present position and address: Thermal Business Manager, PerkinElmer, 710 Bridgeport Ave, Shelton, Texas 76205
Email: kevin.menard@perkinelmer.com
Website: www.perkinelmer.com
Education and scientific degrees: B.S. Chemistry, B. A. Biology (U. Bridgeport, 1978), Ph.D. Chemistry (Wesleyan U, 1985), MBA (Texas Woman's University, 2008)
Workplaces: Rensselaer Polytechnic Institute, Fina Oil and Chemical, General Dynamics Aerospace, PerkinElmer LAS
Main fields of interest: DMA, hyphenated techniques, environmental effects polymers
Relevant categories in thermal analyses: Fields (polymeric materials, foods, nano, organic, pharmaceutical, metals; methods (DMA, TMA, DSC, TGA, Hyphenated techniques including EGA, dual measurement techniques, and sample modification)
Awards and acknowledgments: Fellowship, Royal Society of Chemistry; Fellowship, American Institute of Chemists; Maro New Technology Award 2001. Best Paper—PolyChar 2001; Best Lecture PolyChar Short Course 1998, 2002, 2006
Professional activities: Society of Rheology, Society of Plastic Engineers (PAD Chair 2006, TPC 2005, Communication Chair—Present), American Chemical Society, American Association of Pharmaceutical Scientists, North American Thermal Analysis Society (Treasurer, 1994–1996), Royal Society of Chemistry, Co-Founder, PolyChar Conferences; ASTM DMA Committee
Publication record: Papers (165), books (1), patents (19)
Equipments: DMA, DSC, STA, TGA, TG-IR, TMA
5 most important publications: [1] Dynamic Mechanical Analysis: A Practical Introduction, 2nd Ed; Kevin P. Menard, Taylor and Lord, NYC: 2008; [2] "Characterization of Amorphous Solids with Weak Glass Transitions by High Ramp Rate Differential Calorimetry" Derrick S. Katayama, John F. Carpenter, Mark C. Manning, Theodore W. Randolph, Peter Setlow, Kevin P. Menard, J. Pharma. Sci., 97, 1011, 2008; [3] "Characterization of Epoxies by High Ramp Rate DSC" Bryan Bilyeu, Witold Brostow, Kevin P. Menard, J. ASTM Int., 2(10), 83, 2005; [4] "Toluene Disproportionation Process", J. Butler, Kevin P. Menard. US 4,665,258: May 12, 1987; [5] "Permeable-wall sample holder and related methods" Kevin P. Menard, US 8,281,673, October 9, 2012.

Company's name: SETARAM Instrumentation (KEP Technologies High Tech Products)
Year of foundation: 1965
Mailing address: SETARAM Instrumentation, 7, rue de l'Oratoire, 69300 Caluire, France
Phone: +33 (0)4 72 10 25 25
Fax: +33 (0)4 78 28 63 55
Contacts: sales@setaram.com; application@setaram.com; support@setaram.com
Website: www.setaram.com
Short history: SETARAM has been active in the area of thermal analysis and calorimetry for roughly 50 years. The know-how resulting from its long experience is being constantly enriched thanks to its policy of permanent innovation. Thus SETARAM has built itself a solid reputation as an expert, more specifically in high temperature thermal analysis, Calvet calorimetry based applications and high pressure or manometry-calorimetry coupled systems.
SETARAM has been part of the group KEP Technologies since 1997 which has always been oriented to the development of high end industrial technologies. Its range of competence and innovating solutions have been recognized by the largest aeronautics and space groups, by the nuclear, renewable energies, chemistry, food, pharmaceutical industries.
Based in France at Caluire (near Lyon) where its completely modernized production facilities, research center and application laboratory are located, SETARAM exports more than 80 % of its instruments all over the world, in most reputed public and private industrial research centers and laboratories. Its international activities are conducted by its subsidiaries in USA (New Jersey) and China (Shanghai and Beijing) but also by its sales offices in Germany, Russia, etc. and its roughly 60 local commercial partners.
In 2008 SETARAM has acquired HY-ENERGY LLC (Newark, CA), a manufacturer of high performance gas sorption scientific instruments for the ever-expanding field of renewable energy research.
SETARAM Instrumentation has developed scientific partnerships with major industrial and academic research centers such as Total, Sanofi-Aventis, French Atomic Center, Institut Néel, Institut Français du Pétrole, etc., to propose devices and accessories at the forefront of technology. Sales and technology partnerships with companies involved in the field of thermal analysis like C-Them (Canada) or

AKTS (Switzerland) allows extending its products range to thermal conductivity measurements and thermokinetics and thermoanalytical software.
Main products: SETARAM supplies the widest range of devices:

- Thermal analyzers:
- SETSYS Evolution, DSC, TGA, TMA, TGA, Simultaneous TGA-DTA/DSC, TGA-EGA, from −150 to 2,400 °C
- Thermal analysis systems for large sample sizes (up to 18 cc): 96 Line
- Calorimeters:
- C80 (amb/300 °C) and other Calvet Calorimeters from −200 to 1,500 °C
- Sensys Evo DSC, TG-DSC (−120/830 °C)
- High sensitivity DSC: μDSC or μSC (−45 to 200 °C)
- Differential Reaction Calorimetry DRC Evolution
- Gas sorption
- Sievert's method with PCTPro E&E (−196 to 500 °C)

Application fields: Ceramics, glasses, chemicals, metallurgy, aerospace, fossil and renewable energy, pharmaceuticals, polymers, food, process safety, etc.
Some most important publications: [1] P. Le Parlouër, C. Dalmazzone, B. Herzhaft, L. Rousseau C. Mathonat, Characterisation of gas hydrates formation using a new high pressure micro-DSC, Journal of Thermal Analysis and Calorimetry, Vol. 78 (2004), 165–172; [2] A. Auroux: Acidic characterization by microcalorimetry and relationship with reactivity, Topics in Catalysis, 4 (1997), 71; [3] Z. Achour-Boudjema, M. Bouroukba, M. Dirand: Binary phase diagram of molecular alloys of the consecutive even-numbered n-alkanes (n-C24H50) and n-hexacosane (n-C26H54), Thermochim. Acta, 276 (1996), 243; [4] G. Barone, P. Del Vecchio, D. Fessas, C. Giancola, G. Graziano: Thermodynamic characterization of RNAase A in the presence of urea and GuHCl, J. Thermal Anal., 41 (1994), 1357; [5] A. Sabbar, A. Zrineh, J.P. Dubès, M. Gambino, J. P. Bros, G. Borzone: The Ag-Bi-In system: enthalpy of formation, Thermochim. Acta, 395 (2003), 47; [6] C. Mathonat, V. Majer, A.E. Mather, J.-P. E. Grolier, Use of Flow Calorimetry for Determining Enthalpies of Absorption and the Solubility of CO_2 in Aqueous Monoethanolamine Solutions, Ind. Eng. Chem. Res. 1998, 37, 4136–4141

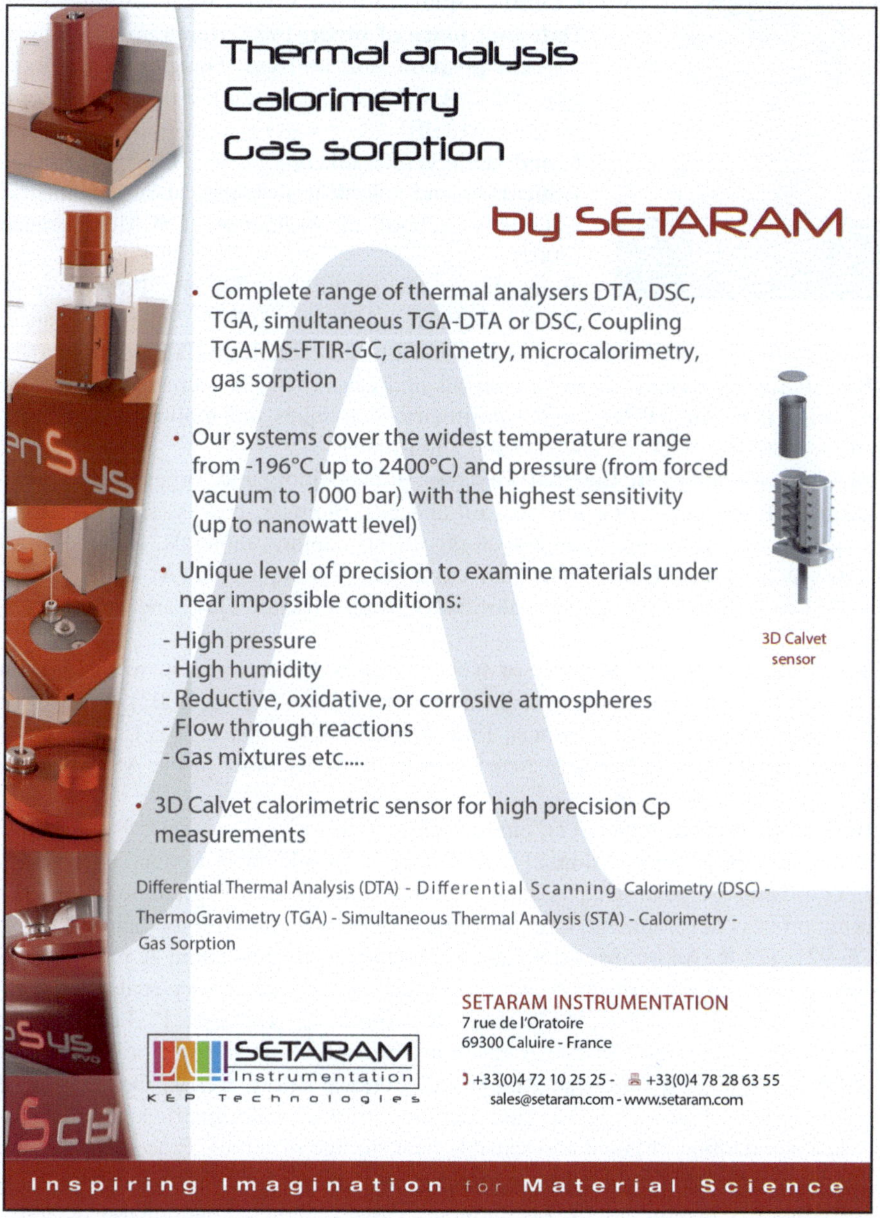
Thermal analysis
Calorimetry
Gas sorption
by SETARAM
• Complete range of thermal analysers DTA, DSC, TGA, simultaneous TGA-DTA or DSC, Coupling TGA-MS-FTIR-GC, calorimetry, microcalorimetry, gas sorption
• Our systems cover the widest temperature range from -196°C up to 2400°C) and pressure (from forced vacuum to 1000 bar) with the highest sensitivity (up to nanowatt level)
• Unique level of precision to examine materials under near impossible conditions:
- High pressure
- High humidity
- Reductive, oxidative, or corrosive atmospheres
- Flow through reactions
- Gas mixtures etc....
• 3D Calvet calorimetric sensor for high precision Cp measurements
3D Calvet sensor
Differential Thermal Analysis (DTA) - Differential Scanning Calorimetry (DSC) - ThermoGravimetry (TGA) - Simultaneous Thermal Analysis (STA) - Calorimetry - Gas Sorption
SETARAM Instrumentation
KEP Technologies
SETARAM INSTRUMENTATION
7 rue de l'Oratoire
69300 Caluire - France
+33(0)4 72 10 25 25 - +33(0)4 78 28 63 55
sales@setaram.com - www.setaram.com
Inspiring Imagination for Material Science

Name: Rémi Andre
Country: France
Date and place of birth: 1977, Remiremont, France
Present position and address: Director of Technology (SETARAM Instrumentation), 7 rue de l'Oratoire 69300 CALUIRE
Email: andre@setaram.com
Education and scientific degrees: M.Sc. Chemistry, Chemistry School of Mulhouse (ESCMu), France (2001)
Workplaces: Swiss Institute for the Promotion of Safety, Switzerland, Intern (2001); Setaram, France, Sales Engineer, Application Engineer, Director of Technology (2001–)
Main fields of interest: Thermal analysis and calorimetry instrumentation; development of innovative thermal and calorimetric techniques and applications; training and promotion of thermal analysis and calorimetry
Relevant categories in thermal analyses: Fields (inorganic, organic, polymers, pharmaceuticals, food, bio products, oil and gas, biomass, process safety, thermal energy storage materials, hydrogen storage, CO_2 capture, minerals, ceramics, catalysts…); methods (DTA, TGA and combined techniques (DTA, DSC, EGA), TMA, DSC, microDSC, Calvet calorimetry, high pressure DSC and calorimetry, drop calorimetry)
Professional activities: Member of ICTAC (since 2004), Member of the French association of Thermal Analysis and Calorimetry AFCAT (since 2002), Member of the American Institute of Chemical Engineering AIChE (since 2012), Member of the American Ceramics Society AcerS (since 2014), Member of the ASTM (since 2013)
Publication record: Papers (7), citation index (18)
5 most important publications: [1] R. André, L. Bou-Diab, P. Lerena, F. Stoessel, M. Giordano, C. Mathonat, A new reaction calorimeter for screening purposes during process development, Organic Process Research and Development 6 (2002), 915–921; [2] R. André, M. Giordano, C. Mathonat, R. Naumann, A reaction calorimeter and calorimetric tools for safety testing on a laboratory scale, Thermochimica Acta 405 (2003), 43–50; [3] R. André, G. Défossé, P. Le Parlouër, Monitoring of biodiesel synthesis reactions by isothermal microcalorimetry, Proceedings of the 19th European Biomass conference (2011, Berlin, Germany); [4] R. André, P. Le Parlouër, S. Moreau, The input of microcalorimetric techniques for the characterization of latent and sensible heat storage materials, Proceedings of the 12th International Conference on Energy Storage (2012, Lleida, Spain); [5] R. André, J. Guérault, Pierre Le Parlouër, Peter Laggner, Phénomènes de polymorphisme étudiés simultanément par microcalorimétrie et diffractométrie des rayons X, Spectra Analyse n°279 (2011).

Name: Christophe Mathonat
Country: France
Date and place of birth: 1968, Cusset, France
Present position and address: Projects Director (SETARAM Instrumentation), 7 rue de l'Oratoire 69300 CALUIRE et CUIRE
Email: christophe.mathonat@kep-technologies.com
Education and scientific degrees: Ph.D. in Physical-Chemistry, Blaise Pascal University—Clermont-Ferrand—France (1995)
Workplaces: Laboratory of Thermodynamics and Chemical Engineering of the Blaise Pascal University—Clermont-Ferrand, France (1992–1996); Chemical Engineering Department of the university of Alberta—Edmonton, Canada (1996); SETARAM—Caluire et Cuire, France R&D Engineer, R&D Manager, Projects Director (1997–Today)
Main fields of interest: Thermal analysis and calorimetry; development of Innovative thermal analyzer and calorimeter; calorimetry applied to nuclear applications; training and promotion of thermal analysis and calorimetry; innovation, project management; team management; risk management
Relevant categories in Thermal Analyses: Fields (inorganic, organic, pharmaceuticals, food, oil and gas, biomass, process safety, hydrogen storage, CO_2 capture, minerals, ceramics); methods (DTA, TGA and combined techniques (DTA, DSC, EGA), TMA, DSC, microDSC, Calvet calorimetry, high pressure DSC and calorimetry, drop calorimetry, large volume calorimetry)
Professional activities: Member of the French association of Thermal Analysis and Calorimetry AFCAT
Publication record: Papers (15), patents (3)
5 most important publications: [1] C. Mathonat, V. Majer, A.-E. Mather, J.-P. Grolier, Fluid Phase Equilibria (1997), 140(1–2), 171; [2] C. Mathonat, C.-J. Wormald, Journal of Chemical Thermodynamics (1998), 30(8), 959; [3] C. Mathonat, C. Chan, Y. Maham, A. E. Mather, Fluid Phase Equilibria (2002), 198 (2), 239; [4] R. André, L. Bou-Diab, P. Lerena, F. Stoessel, M. Giordano, C. Mathonat, 6 (2002) 915–921; [5] G. Jossens, C. Mathonat, F. Bachelet, (2013) Tritium.

TA Instruments

Name: Robert L. Danley
Country: USA
Date and place of birth: 1953 Camden NJ, USA
Present position and address: Fellow Scientist, TA Instruments-Waters LLC 159 Lukens Drive New Castle DE 19720, USA
Email: rdanley@tainstruments.com
Website: www.tainstruments.com
Education and scientific degrees: B.S. Mechanical Engineering (1980), Drexel University; M.S. Mechanical Engineering (1986), Drexel University
Workplaces: Consarc Corporation (1976–1991), TA Instruments (1991–)
Main fields of interest: Invention and development of Tzero® DSC and thermal analysis instrumentation and accessories including TG, DTA, EGA, thermomechanical analysis, cryo and high temperature
Relevant categories in thermal analyses: Methods (TG, DTA, EGA, DSC, thermomechanical analysis, cryo, high temperature, instrument development)
Awards and acknowledgments: Waters Corporate Fellow (2013)
Publication record: Papers (12), patents (22 US patents and numerous related patents in Germany, France, United Kingdom, Japan, etc. The patents cover various thermal instruments and methods but focus mainly on DSC. Many of these patents apply to instruments that are currently in production with more than 1,000 instruments per year installed in laboratories throughout the world.), citations (350)
Equipments: DSC, Modulated DSC, TG, DT/TG, DMA, DTA, thermomechanical analysis
5 most important publications: [1] R .L. Danley: Thermochimica Acta 395 (1), 201–208; [2] J. W. Schaefer, R. L. Danley: US Patent 5,288,147; [3] R. L. Danley: Thermochimica Acta 402 (1), 91–98; [4] R. L. Danley, P. A. Caulfield, S. R. Aubuchon: American Laboratory 40 (1), 9–11; [5] R.L. Danley: EP Patent 1,342,997.

Index of Researchers

I.M. Szilágyi and G. Liptay (eds.), *Who is Who in Thermal Analysis and Calorimetry*, Hot Topics in Thermal Analysis and Calorimetry 10,
DOI 10.1007/978-3-319-09486-1

Index of Instrument and Parts Manufacturing Companies

I.M. Szilágyi and G. Liptay (eds.), *Who is Who in Thermal Analysis and Calorimetry*, Hot Topics in Thermal Analysis and Calorimetry 10,
DOI 10.1007/978-3-319-09486-1

The manufacturer's authorised representative in the EU is Springer Nature Customer Service Centre GmbH, Europaplatz 3, 69115 Heidelberg, Germany. If you have any concerns regarding our products, please contact ProductSafety@springernature.com

Printed and bound by CPI Group (UK) Ltd, Croydon, CR0 4YY
05/10/2026
02228132-0003